Integral Transforms

Integral Transforms

Prof. (Dr.) Sanjeev Kumar
Department of Mathematics
Dr. Bhimrao Ambedkar University
Agra, U.P.

Prof. (Dr.) Narendra Kumar
NIET, NIMS University
Jaipur, Rajasthan

Prof. (Dr.) Anirudh Pradhan
Director, Centre for Cosmology, Astrophysics
& Space Science (CCASS)
GLA University, Mathura, U.P.

Dr. Rakesh M. Patel
Department of Mathematics
Gujarat Arts and Science College
Ahmedabad, Gujarat

I.K. International Pvt. Ltd.

New Delhi

Published by
I.K. International Pvt. Ltd.
4435-36/7, Ansari Road, Daryaganj
New Delhi-110 002 (India)
E-mail: info@ikinternational.com
Website: www.ikbooks.com

ISBN: 978-93-90620-57-9

Published by Krishan Makhijani for I.K. International Pvt. Ltd., 4435-36/7, Ansari Road, Daryaganj, New Delhi-110 002.

Preface

The idea of integral transform was first introduced by the well-known Fourier integral formula. Initially, integral transforms are useful because they offer effective operational approaches for addressing initial value problems and initial-boundary value problems for linear differential and integral equations. In point of fact, the investigation of differential and integral equations that arises in mathematical research was one of the primary impetuses for the development of the operational calculus of integral transforms. The usefulness and interest of integral transforms appear to be more thoroughly established than it ever has been before, in spite of the fact that the demand for mathematical methods to supply both theory and applications for science and engineering is growing at an ever-increasing rate. In this book, our primary objective is to provide a methodical presentation of the fundamental characteristics shared by a variety of integral transforms as well as their applications in the fields of science and technology. In addition, the operational calculus of integral transformations is utilized in the solution of issues pertaining to probability and statistics, as well as integral equations and difference equations, fractional integrals and fractional derivatives, summation of infinite series, and assessment of definite integrals. Students who are interested in studying integral transforms with applications may have access to what appears to be a large number of books. This book, which is an attempt to fulfil that requirement and was written as a consequence of many years of experience teaching advanced undergraduates and first-year graduate students in mathematics, physics, and engineering, was developed as a result of that experience. It is mostly based on a collection of lecture notes that were created by the writers for use in the classes that they taught. This book is intended to serve as an introduction to the theory behind integral transforms as well as their applications to solving issues involving linear differential equations, as well as boundary and initial value problems involving partial differential equations.

The course requires two fundamental requirements in order to participate: a basic understanding of calculus and ordinary differential equations. The book requires only a basic familiarity with complex variables, contour integration, and partial differential equations as its prerequisite for reading it. There are a lot of fresh examples of applications covered in this book that deals with issues in applied mathematics, physics, chemistry, biology, and engineering. Although having a comprehensive understanding of all of these subjects is not required of the reader, having some familiarity with at least some of them would be beneficial. Aside from that, the purpose of the book is to act as a reference for students who are truly interested in advanced study and research in the subject, whether for the subject matter itself or for its applicability to other subjects such as applied mathematics, mathematical physics, and engineering.

Authors

Contents

Chapter 1

Laplace Transforms

1.1 INTRODUCTION

We have seen various operators like D, E, Δ, D^{-1}, ... etc. Operater transforms the function in another form or output. In this chapter, we shall learn about Laplace transform which is important to find solutions/output in various scientific calculations. Several applications in science and technology can be solved by using Laplace transform and give better solution than many other methods like differential equations, and numerical methods.

1.2 INTEGRAL TRANSFORMS

Let $K(p, t)$ be a function of two variables, t and p. Here p is a parameter (may be a real or complex number) which is independent of t. Then the function $f(p)$, which is assumed to be convergent, is defined by the following integral—known as integral transform of function $F(t)$

$$T\{F(t)\} = f(p) = \int_{-\infty}^{\infty} k(p, t)\, F(t)dt \qquad \text{...(i)}$$

$K(p, t)$ is known as kernel of the transform, we may use several symbols in place of p like s, m, ... etc.

Now, integral,

$$F(t) = \int_{a}^{b} q(x, p)\, T\{f(t), p\}dp \qquad \text{...(ii)}$$

is defined as inversion formula to the equation (i).

1.3 LAPLACE TRANSFORMS

If the kernel $k(p, t)$ is defined as given below,

$$k(p, t) = \begin{cases} 0, & \text{if } t < 0 \\ e^{-pt}, & \text{if } t \geq 0 \end{cases}$$

Then, $$L\{F(t)\} = f(p) = \int_0^\infty e^{-bt} F(t)\, dt \qquad \text{...(iii)}$$

The function $f(p)$, defined by equation (iii), is called Laplace transform of the function $F(t)$ and can be denoted by $f(p)$ or $L\{F(t)\}$. It is a function of parameter (or variable) p. Here, parameter p may be a real or complex number.

1.4 SECTIONALLY OR PIECEWISE CONTINUOUS FUNCTION

Let $f(t)$ be a function defined on a closed interval $[a, b]$ and the interval can be subdivided into finite number of subintervals such that $f(t)$ is continuous in every subinterval, in each of which $F(t)$ is continuous and has finite left and right hand limits.

Example 1

$$f(t) = \begin{cases} t^2 + 1, & \text{if} \quad 1 \le t \le \pi/2 \\ \sin t, & \text{if} \quad \pi/2 \le t < \pi \\ |t|, & \text{if} \quad \pi \le t \le 4 \end{cases}$$

Here, function $f(t)$ is a sectionally continuous function in the close interval [1, 4]; interval [1, 4] divided into subintervals $(1, \pi/2)$, $(\pi/2, \pi)$, $(\pi, 4)$ have finite left and right hand limits at the end points: $t = 1, \pi/2, \pi$ and 4 of these subintervals.

1.5 FUNCTIONS OF EXPONENTIAL ORDER

A function $F(t)$ is said to be of exponential order, K as $t \to \infty$, if there is a positive real constant M, and finite number (t_0) such that

$$|F(t)| < Me^{kt}$$

or $$|e^{-kt} F(t)| < M \;\forall\; t \ge t_0$$

If a function $F(t)$ is of exponential order k, it is also β, $\beta > K$.

Function of Class A

A function which is sectionally or piecewise continuous on every finite interval in the range $t \ge 0$ and is of exponential order as $t \to \infty$ is known as a function of class A.

Note: Bounded functions such as sin at, cos at are of exponential order zero.

Example: The function t^n is of exponential order x, $(k > 0, n \in N)$

Solution: Since

$$\lim_{t\to\infty} t^n e^{-kt} = \lim_{t\to\infty} \frac{t^n}{e^{kt}} = \frac{n!}{k^n e^{kt}} = 0$$

Note:

(i) t^2 is of exponential order 3

As $|t^2| < e^{3t} \;\forall\; t > 0$

(ii) e^{t^3} is not of exponential order as

$$|e^{-kt} e^{t^2}| = e^{t^3 - kt}$$

can be made larger than for any given constant by changing the value of t.

Existence of Laplace Transform: If $F(t)$ is piecewise continuous in every finite interval $[0, t_0]$ and is of exponential order k as $t \to \infty$, then $L\{F(t); p\}$ exists $\forall\ p > k$.

Proof: Let us assume $t_0 > 0$, then

$$f(p) = L\{F(t); p\} = \int_0^\infty e^{-pt} F(t)\, dt$$

$$= \int_0^{t_0} e^{-pt} F(t)dt + \int_{t_0}^\infty e^{-pt} F(t)dt \qquad ...(i)$$

The integral $\int_0^{t_0} e^{-pt} F(t)\, dt$ exists since $F(t)$ is sectionally continuous in each finite interval $[0, t_0]$. Now, for the second integral

$$\left|\int_{t_0}^\infty e^{-pt} F(t)dt\right| \le \int_{t_0}^\infty |e^{-pt} F(t)|\, dt \le \int_{t_0}^\infty e^{-pt} M e^{\alpha t}\, dt$$

Since $|F(t)| \le Me^{\alpha t}$

$$= \int_{t_0}^\infty e^{-(p-\alpha)t} Mdt = \frac{Me^{-(p-\alpha)t_0}}{p-\alpha}, \; p > \alpha$$

$$\Rightarrow \quad \left|\int_{t_0}^\infty e^{-pt} F(t)dt\right| \le \frac{Me^{-(p-\alpha)t_0}}{p-\alpha}, \; p > \alpha$$

$\dfrac{Me^{-(p-\alpha)t_0}}{p-\alpha}$ can be assumed as small as we please by choosing to sufficiently large.

$$\Rightarrow \quad \int_{t_0}^\infty e^{-pt} F(t)dt$$

exists for all values of $p > \alpha$

or $\quad L\{F(t); p\}$ exists for $p > \alpha$.

1.5.1 Laplace Transform of Some Elementary Functions

(a) $$F(t) = t^n = \int_0^\infty e^{-pt} t^n\, dt$$

$$= -\left[\frac{t^n e^{-bt}}{p}\right]_0^\infty + \frac{1}{p}\int_0^\infty n\, t^{n-1} e^{-pt}\, dt$$

$$= -\frac{1}{p}\lim_{t\to\infty}\frac{t^n}{e^{pt}} + \frac{n}{p}\int_0^\infty t^{n-1} e^{-pt}\, dt$$

$$= \frac{n}{p}\int_0^\infty t^{n-1} e^{-pt}\, dt. \text{ As } \lim_{t\to\infty}\frac{t^n}{e^{pt}} = 0,\ p > 0$$

$$= \frac{n(n-1)}{p^2}\int_0^\infty t^{n-2} e^{-pt}\, dt,\quad p > 0$$

(Repeating the steps)

$$= \frac{n!}{p^n}\int_0^\infty e^{-pt}\, dt = \frac{n!}{p^{n+1}}$$

$$\Rightarrow \qquad L\{t^n; p\} = \frac{n!}{p^{n+1}},\ p > 0$$

The above result can be generalised for $n \in C$ and complex variable, p. Then,

$$L\{t^n; p\} = \frac{\Gamma(n+1)}{p^{n+1}}, \qquad \operatorname{Re}(p) > 0,\ \operatorname{Re}(n+1) > 0$$

(b) For $n = 0$, $t_0 = 1$

$$L\{1\} = \int_0^\infty e^{-pt}.1\, dt = \int_0^\infty e^{-pt}\, dt = \frac{1}{p},\ p > 0$$

$\Rightarrow$ $L\{1\} = 1/p$.

(c) For $n = 1$, $F(t) = t$

$$L\{t; p\} = \int_0^\infty t\,.\, e^{-pt}\, dt$$

$$= \left(t\cdot\frac{e^{-pt}}{-p}\right)_0^\infty - \int_0^\infty .\ \frac{e^{-pt}}{-p}\, dt$$

$$= \frac{1}{p}\int_0^\infty e^{-pt}\, dt, \qquad \left(\frac{-t}{p}e^{-pt}\right)_0^\infty = 0$$

$$= \frac{1}{p}\cdot\frac{1}{p}\left[\int_0^\infty e^{-pt}\, dt = 1/p\right] = \frac{1}{p^2} \Rightarrow L\{t\} = 1/p^2.$$

(d) $$F(t) = e^{at}$$

$$L(e^{at}) = \int_0^\infty e^{-pt} \cdot e^{at}\, dt = \int_0^\infty e^{-(p-a)t}\, dt = \frac{1}{p-a},\ p > a$$

$$\Rightarrow L\{e^{at}\} = 1/p - a \text{ for } p > a$$

(e) $$F(t) = \sin at \text{ and } F(t) = \cos at$$

We know that

$$L\{e^{at}\} = \frac{1}{p-a},\ p > a$$

$$\Rightarrow \quad L\{e^{iat}\} = \frac{1}{p-ia} = \frac{p+ia}{p^2+a^2}$$

But $$e^{iat} = \cos at + i \sin at$$

$$\Rightarrow \quad L\{e^{iat}\} = L(\cos at + i \sin at)$$

or $$L(\cos at + i \sin at) = \frac{p+ia}{p^2+a^2}$$

On separating real and imaginary parts, we have

$$L\{\cos at\} = \frac{p}{p^2+a^2},\ p > 0$$

and $$L\{\sin at\} = \frac{a}{p^2+a^2},\ p > 0$$

(f) $$F(t) = \sinh at$$

$$L\{\sinh at\} = L\left\{\frac{1}{2} e^{at} - e^{-at}\right\} = \frac{1}{2}[L\{e^{at}\} - L\{e^{-at}\}]$$

$$= \frac{1}{2}\left[\frac{1}{p-a} - \frac{1}{p+a}\right],\ p > |a|$$

$$= \frac{a}{p^2+a^2},\quad p > |a|$$

(g) $$F(t) = \cosh at$$

$$L\{\cosh at\} = L\left\{\frac{1}{2}(e^{at} + e^{-at})\right\} = \frac{1}{2}(L\{e^{at}\} + L\{e^{-at}\})$$

$$= \frac{1}{2}\left[\frac{1}{p-a} - \frac{1}{p+a}\right],\ p > |a|$$

$$= \frac{1}{2}\left[\frac{2p}{p^2-a^2}\right]$$

$$\Rightarrow \quad L\{\cosh at\} = \frac{p}{p^2 - a^2}$$

Note: $$L\{t^a\} = \frac{\Gamma(a+1)}{p^{a+1}}, \; p > 0, \; a > -1$$

Important Properties of Laplace Transforms: In the following cases, we are assuming that all functions given below satisfy existance of Laplace transforms.

1. **Linear Property:** Let $F_1(t)$ and $F_2(t)$ be two functions; and c_1 and c_2 are two constants, then

$$L\{c_1 F_1(t) + c_2 F_2(t)\} = c_1 L\{F_1(t)\} + c_2 L\{F_2(t)\}$$

or, in general for n constants and n functions of t, we have

$$L\{C_1F_1(t) + C_2 F_2(t) + \ldots + C_n F_n(t)\}$$
$$= C_1 L[F_1(t)] + C_2 L\{F_2(t)\} + C_3L\{F_3(+)\} + \ldots + C_n L\{F_n(t)\}$$

Example: $L\{4t^2 - 3 \cos 2t + 5e^{-t}\}$

$$= 4L\{t^2\} - 3L\{\cos 2t\} + 5L\{e^{-t}\}$$

$$= 4 \cdot \left(\frac{2!}{p^3}\right) - 3\left(\frac{p}{p^2+4}\right) + 5 \cdot \left(\frac{1}{p+1}\right)$$

2. **First Shifting or Translation Property:** If $L[F(t); p] = f(p)$, then $L\{e^{-at} F(t); p\} = f(p + a)$

or $f(p)$ is the Laplace transform of $F(t)$, then $f(p + a)$ will be the Laplace transform of $e^{-at} F(t)$. Here a is any real or complex number.

Proof: We know that

$$L[F(t); p] = f(p)$$

$$\Rightarrow \quad L[e^{-at} F(t); p] = \int_0^\infty e^{-pt} \cdot e^{at} F(t)dt$$

$$= \int_0^\infty e^{-(p+a)t} F(t)\, dt$$

$$= L[F(t); p + a] = f(p + a)$$

3. **Second Shifting or Translation Property:** Let

$$L[F(t); p] = f(p), \text{ and}$$

$$g(t) = \begin{bmatrix} F(t-a), & t > a \\ 0, & t < a \end{bmatrix}$$

Then,

$$L[g(t); p] = e^{-ap} f(p)$$

Proof: We know that

$$L[g(t); p] = \int_0^\infty e^{-pt} g(t)dt$$

$$= \int_0^a e^{-pt} g(t)dt + \int_a^\infty e^{-pt} g(t)\, dt$$

$$= \int_0^a e^{-pt} \cdot 0 \cdot dt + \int_0^\infty e^{-pt} F(t-a)dt$$

Let $t - a = u$

$dt = du$

$$= \int_0^\infty e^{-p(u+a)} F(u)du = e^{-ap} \int_0^\infty e^{-pu} F(u)du$$

$$= e^{-ap} \int_0^\infty e^{-pt} F(t)\, dt \qquad \text{(Change of variable)}$$

$$= e^{-ap} f(p)$$

4. Change of Scale Property: If $L[F(t); p] = f(p)$,

then $$L[F(at); p] = \frac{1}{a} f(p/a),\ a \quad a \neq 0$$

Proof: We know that

$$L\{F(at); p\} = \int_0^\infty e^{-pt} F(at)\, dt$$

Let $at = u \Rightarrow adt = du$ or $dt = \frac{1}{a} du$

$$= \frac{1}{a} \int_0^\infty e^{-\left(\frac{u}{a}\right)p} \cdot F(u)du$$

$$= \frac{1}{a} \int_0^\infty e^{-\left(\frac{p}{a}\right)u} \cdot F(u)du$$

$$\Rightarrow \quad L[F(at); p] = \frac{1}{a} F(p/a)$$

Illustrative Examples

Example 1 Evaluate Laplace transform of the following functions

(i) $e^{-2t}\cos^2 t$

(ii) $(t+2)^2 e^t$

(iii) $e^{-2t}(3\cos 6t - 5\sin 6t)$

(iv) $e^{-2t}\sin 4t$

Solution:

(i) We have

$$L[\cos^2 t, p] = L\left\{\frac{1+\cos 2t}{2}; p\right\} = \frac{1}{2}[L\{1\} + L\{\cos 2t\}, p]$$

$$= \frac{1}{2}\left[\frac{1}{p} + \frac{p}{p^2+4}\right] = \frac{p^2+2}{p(p^4+4)}, \quad L\{\cos at\} = \frac{p}{p^2+a^2}$$

$\Rightarrow$ $L[e^{-2t}\cos^2 t; p] = L[\cos^2 t; p+2]$, on applying the first shifting property

$$= \frac{(p+2)^2+2}{(p+2)[(p+2)^2+4]} = \frac{p^2+4p+6}{(p+2)(p^2+4p+8)}$$

(ii) $$F(t) = (t+2)^2 e^t$$

We have

$$L[(t+2)^2; p] = L[t^2+4t+4; p]$$

$$= L[t^2] + 4L[t] + 4L[1] = \frac{2}{p^3} + \frac{4}{p^2} + \frac{4}{p}$$

$$= \frac{4p^2+4p+2}{p^3}$$

By the first shifting property, we get

$$L[(t+2)^2 e^t; p] = L[(t+2)^2; p-1]$$

$$= \frac{4(p-1)^2+4(p-1)+2}{(p-1)^3}$$

$$= \frac{4p^2-4p+2}{(p-1)^3}$$

(iii) $$F(t) = e^{-2t}(3\cos 6t - 5\sin 6t)$$

We have

$$L[(3\cos 6t - 5\sin 6t); p]$$

$$= 3L[(\cos 6t); p] - 5L[(\sin 6t); p]$$

$$= \frac{3p}{p^2+36} - \frac{5(6p)}{p^2+36}$$

$$= \frac{3p-30}{p^2+36}$$

By the first shifting property, we get

$L[e^{-2t}(3 \cos 6t - 5 \sin 6t); p] = L[(3 \cos 6t - 5 \sin 6t); (p + 2)]$

$$= \frac{3(p+2)-30}{(p+2)^2+36}$$

$$= \frac{3p-24}{p^2+4p+40}$$

(iv) $L[(e^{-2t} \sin 4t); p] = f(p)$

We have $L[(\sin 4t); p] = \dfrac{4}{p^2+16}$

By first shifting property, we get

$L[(e^{-2t} \sin 4t); p] = L[(\sin 4t); p + 2]$

$$= \frac{4}{(p+2)^2+16}$$

$$= \frac{4}{p^2+4p+20}$$

Example 2 Find Laplace transform of the function

$$f(t) = \begin{cases} \sin(t - \pi/3); & t \geq \pi/3 \\ 0 \quad ; & t < \pi/3 \end{cases}$$

Solution: We know that

$$L\{f(t), p\} = \int_0^\infty e^{-pt} F(t)\,dt$$

Then, $$L[f(t); p] = \int_0^\infty e^{-pt} \sin(t - \pi/3)dt$$

$$= \int_0^{\pi/3} e^{-pt} \cdot 0 \cdot dt + \int_{\pi/3}^\infty e^{-pt} \sin(t - \pi/3)dt$$

[Now putting $t - \dfrac{\pi}{3} = x$ (in second integral) $\Rightarrow dt = dx$]

$$= 0 + \int_0^{\infty} e^{-p(x + \pi/3)} \sin x \, dx$$

$$= e^{-p\pi/3} \, L[(\sin x); p]$$

$$= e^{-p\pi/3} \cdot \frac{1}{p^2 + 1}$$

Note: We can solve this example by using the second shifting property,

$$L[(\sin t); p] = \frac{1}{p^2 + 1}$$

Using the second shifting property,

$$L[F(t); p] = e^{-p\pi/3} \cdot \frac{1}{p^2 + 1}$$

Example 3 Find the Laplace transform of the function:

$$F(t) = \begin{cases} \cos (t - 2^{\pi/3}), & t \geq 2\pi/3) \\ 0 & , \quad t < 2\pi/3 \end{cases}$$

Solution: We know that

$$L\{F(t); p\} = \int_0^{\infty} e^{-pt} F(t) dt$$

$$= \int_0^{2\pi/3} e^{-pt} \cdot 0 \, dt + \int_{2\pi/3}^{\infty} e^{-pt} \cos(t - 2\pi/3) \, dt$$

Let $\quad t - 2\pi/3 = x$

$$dt = dx \text{ [in the second integral]}$$

$$= e^{-2\pi p/3} \int_0^{\infty} e^{-px} f(x) \, dx$$

Here, $\quad f(x) = \cos x$

$$= e^{-2\pi p/3} \cdot \frac{p}{p^2 + 1}$$

Examples 4 Find the Laplace transform of the function

$$F(t) = \begin{cases} (t - 1)^2 , & t > 1 \\ 0 & , t < 1 \end{cases}$$

Solution: We have

$$L\{F(t); p\} = \int_0^\infty e^{-pt} F(t)dt$$

$$= \int_0^1 e^{-pt} \cdot 0 \cdot dt + \int_1^\infty (t-1)^2 e^{-pt}\, dt$$

$$= \int_1^\infty e^{-pt} \cdot (t-1)^2\, dt \qquad [(t-1 = x \Rightarrow dt = dx)]$$

$$= \int_0^\infty e^{-p(x+1)} \cdot x^2 dx = e^{-p} \int_0^\infty e^{-px} x^2 dx$$

$$= e^{-p} \cdot \frac{2!}{p^3}, p > 0 = \frac{-2e^{-p}}{p^3}, p > 0$$

EXERCISE

Find the Laplace transform of the following functions:

1. $2t^3 - 6t + 0$

$$\left[\textbf{Ans. } \frac{12}{p^4} - \frac{6}{p^2} + \frac{8}{p}\, p > 0 \right]$$

2. $7e^{2t} + 9e^{-2t} + 5 \cos t + 7t^3 + 5 \sin 3t + 2.$

$$\left[\textbf{Ans. } \frac{16p-4}{p^2-4} + \frac{5p}{p^2+1} + \frac{2p^3+4^2}{p^4} + \frac{15}{p^2+9}, p > 2 \right]$$

3. $2e^{3t} - e^{-3t}.$

$$\left[\textbf{Ans. } \frac{p+9}{p^2-9}, p > 3 \right]$$

4. $\cos^3 t + \cosh^3 3t$

$$\left[\textbf{Ans. } -e^{-\pi p} \log \pi + \frac{e^{-p}}{p^2}(p^2 + p + 1) \right]$$

5. $e^{-2t} \sin 4t.$

$$\left[\textbf{Ans. } \frac{4}{(p^2+4p+20)} \right]$$

6. $e^{-t}(3 \sinh 2t - 5 \cosh 2t).$

$$\left[\textbf{Ans. } \frac{(1-5p)}{(p^2+2p-3)} \right]$$

7. Prove that the following function is sectionally continuous in [0, 3]

$$f(t) = \begin{cases} at+b & , \quad 0 \le t < 1 \\ e^{-t} & , \quad 1 \le t < 2 \\ t^2 & , \quad 1 \le t \le 3 \end{cases}$$

8. Prove that $f(t) = t^3$ is of exponential function of order 4.

9. Show that $L\{e^{at}\, F(dt)\text{: } p\} = \dfrac{1}{b} f\left(\dfrac{p-a}{b}\right)$,

where, $L\{F(t);\, p\} = f(p)$

10. Find Laplace transforms

(i) $$F(t) = \begin{cases} \cos t, & 0 < t < \pi \\ \sin t, & t > \pi \end{cases}$$

$$\left[\textbf{Ans. } \frac{p + (p-1)\, e^{\pi p}}{(p^2+1)}\right]$$

(ii) $$F(t) = \begin{cases} \sin 2t\ , & 0 < t < \pi \\ 0\ , & t > \pi \end{cases}$$

[**Ans.** $2(1 - e^{-\pi p})/(p^2 + 4)$]

(iii) $$F(t) = \begin{cases} 2t\ , & 0 \le t \le 5 \\ 0\ , & t > 5 \end{cases}$$

$$\left[\textbf{Ans. } \frac{2}{p^2}(1 - e^{-5b}) - \frac{9}{p} e^{-5p}\right]$$

(iv) $$F(t) = \begin{cases} e^{t-a}, & t > a \\ 0\ , & t < a \end{cases}$$

[**Ans.** $e^{-ap}/(p - 1)$, $p > 1$]

(v) $$F(t) = ce^{-at} \frac{t^{n-1}}{n!}.$$

$$\left[\textbf{Ans. } \frac{c}{(p+a)^n}\right]$$

(vi) $$F(t) = \begin{cases} 0, & 0 < t < 1 \\ t, & 1 < t < 2 \\ 0, & t > 2 \end{cases}$$

$$\left[\textbf{Ans. } -\left(\frac{2}{p}+\frac{1}{p^2}\right)e^{-2p}+\left(\frac{1}{p}+\frac{1}{p^2}\right)e^{-p}, p \neq 0\right]$$

11. $$F(t) = \begin{cases} t, & 0<t<4 \\ s, & t>4 \end{cases}$$

$$\left[\textbf{Ans. } \frac{1+(p-1)e^{-4p}}{p^2}, p>0\right]$$

12. $$F(t) = \begin{cases} \sin t, & 0<t\,\pi \\ 0\;, & t>\pi \end{cases}$$

$$\left[\textbf{Ans. } \frac{e^{-p\pi}+1}{p^2+1}\right]$$

13. Prove that

$$L\left(\frac{\cos\sqrt{t}}{\sqrt{t}}\right) = e^{-1/4p}\sqrt{\left(\frac{\pi}{p}\right)}.$$

14. Prove that

$$L\left(\frac{1}{\sqrt{\pi t}}\right) = \frac{1}{\sqrt{p}}.$$

Laplace Transform of Derivatives of Function $F(t)$:

Let $F(t)$ be

(i) continuous for $0 \le t \le N$, (ii) of exponential order k for $t > N$, and (iii) sectionally continuous for $0 \le t \le N$ and $L\{F(t); p\} = f(p)$

Then, $\quad L\{F'(t); p\} = pf(p) - F(0)$ for $p > k$

Proof:

Case 1: When $F'(t)$ is continuous for all $t \ge 0$, then

$$L\{F'(t); p\} = \int_0^\infty e^{-pt}F'(t)dt \qquad \text{...(i)}$$

$$= \left[e^{-pt}F(t)\right]_0^\infty + p\int_0^\infty e^{-pt}F(t)dt$$

$$= \lim_{t\to\infty} e^{pt}F(t) - F(0) + p\,L\{F(t); p\} \qquad \text{...(ii)}$$

But $|F(t)| \leq M\, e^{ct}\ \forall\ t > 0$ and for some constant c and M

$$\Rightarrow \qquad |e^{-pt}\, F(t)| = e^{-pt}\, |F(t)| \leq e^{-pt}\, M\, e^{ct}$$

$$= M\, e^{-(p-c)t} \text{ tends to zero as } t \to \infty \text{ and if } p > 0$$

or
$$\lim_{t\to\infty} e^{-pt} F(t) = 0. \text{ For } p > a$$

$$\Rightarrow \qquad L\{F'(t); p\} = pL\{F(t); p\} - F(0)$$

$$= p\, f(p) - F(0)$$

Case 2: Let us assume that $F(t)$ has a finite discontinuity, $x = a$, otherwise continuous in $\{0, \infty\}$.

Then, $\quad L\{F'(t);\ p\} = pf(p) - F(0) - e^{-ap}\ [f(a^+) - f(a^-)],\ p > k$

Proof: We have

$$L\{F'(t); p\} = \int_0^{a^-} e^{-pt} F'(t)dt + \int_{a^+}^{\infty} e^{-pt} F(t)dt$$

$$= \left[e^{-pt} F(t)\right]_0^{a^-} + p\int_0^{a^-} e^{-pt} F(t)dt + \left[e^{-pt} F(t)\right]_{a^+}^{\infty} + p\int_{a^+}^{\infty} e^{-pt} F(t)dt$$

$$= e^{-ap}\ F(a^-) - F(0) - F(a^+)\ e^{-ap} + p\, f(p)$$

$$\Rightarrow \qquad L\{F'(t); p\} = p\, f(p) - F(0) - e^{-ap}\ [F(a^+) - F(a^-)]$$

It can be written as

$$= p\, f(p) - F(0) - e^{ap}\ [F]_a,\ p > k$$

If $F(t)$ has a finite number of discontinuous points, then

$$L\{F'(t); p\} = p\, f(p) - F(0) - \sum_{i=1}^{n} e^{-aip} [F]_{ai}$$

where, $a_1, a_2, \ldots a_n$ are points of discontinuities of the given function, F.

Laplace Transform of the *n*th Order Derivative of *F*(*t*)

Let $F(t), F'(t), F''(t), \ldots, F^{n-1}(t)$ be continuous function for all $t \geq 0$ and be of exponential orders as $t \to \infty$, suppose $F^n(t)$ is the function of class A, and $L\{F^n(t); p\} = f(p)$.

Then, $\quad L\{F^n(t); p\} = p^n f(p) - p^{n-1}\ F(0) - p^{n-1}\ F'(0) \ldots - pF^{n-2}\ (0) - F^{n-1}\ (0),\ p > k.$

Proof: We have

$$L\{F'(t); p\} = p\, f(p) - F(0) \qquad \ldots(i)$$

Using equation (i),

$$L\{F''(t); p\} = p\,L\{F'(t); p\} - F'(0)$$
$$= p[pf(p) - F(0)] - F'(0)$$
$$= p^2 f(p) - pF(0) - F'(0) \qquad ...(ii)$$

Using equations (i) and (ii), we have

$$L\{F'''(t); p\} = p\,L\{F''(t); p\} - F''(0)$$
$$= p[p^2 f(p) - pF(0)] - F'(0)] - F''(0)$$
$$= p^3 f(p) - p^2 F(0) - pF'(0) - F''(0)$$

Similarly, we can find

$$L\{F^n(t); p\} = p^n f(p) - p^{n-1} F(0) - p^{n-2} F'(0) \;...\; - F^{n-1}(0) \text{ (For } p > k)$$

Initial Value Theorem: Suppose function $F(t)$ is continuous for all $t \geq 0$ and is of exponential order as $t \to \infty$, and further suppose that $F'(t)$ is a function of class A. Then

$$\lim_{t \to 0} F(t) = \lim_{p \to \infty} p\,L\{F(t); p\}$$

Proof: We know that

$$L\{F'(t); p\} = p\int_0^\infty e^{-pt} F(t)dt - F(0)$$

$$\Rightarrow \quad \int_0^\infty e^{-pt} F'(t)dt = p\int_0^\infty e^{-pt} F(t)dt - F(0) \qquad ...(i)$$

It is given that $F'(t)$ is a function of class A, therefore, $F'(t)$ is sectionally continuous and of exponential order. Thus,

$$\lim_{p \to \infty} \int_0^\infty e^{-pt} F'(t)dt = 0$$

By equation (i), as $p \to \infty$, we have

$$p \lim_{p \to \infty} \int_0^\infty e^{-pt} F(t)dt - F(0) = 0$$

or

$$F(0) = \lim_{p \to \infty} p\,f(p)$$

$$\Rightarrow \quad \lim_{t \to \infty} F(t) = \lim_{p \to \infty} p\,\{L\{F(t); p\}\}$$

Final Value Theorem: Let $F(t)$ be continuous for all $t \geq 0$ and be of exponential order as $t \to \infty$ and also if $F'(t)$ is a function of class A. Then

$$\lim_{t \to \infty} F(t) = \lim_{p \to 0} p\,f(p)$$

Proof: We have

$$\int_0^\infty e^{-pt} F'(t)dt = p\int_0^\infty e^{-pt} F(t)dt - F(0) \qquad \text{...(i)}$$

By equation (i), taking limit on $p \to 0$, in L.H.S.

$$\lim_{p\to 0} \int_0^\infty e^{-pt} F'(t)dt = \int_0^\infty F'(t)dt$$

$$= \lim_{L\to\infty} \int_0^L F'(t)dt$$

$$= \lim_{L\to\infty} [F(L) - F(0)]$$

$$= \lim_{t\to\infty} [F(t) - F(0)]$$

But from R.H.S. we have

$$\lim_{p\to 0} p\int_0^\infty e^{-pt} F(t)dt - F(0)$$

$$\Rightarrow \quad \lim_{t\to\infty} F(t) - F(0) = \lim_{p\to\infty} p\int_0^\infty e^{-pt} F(t)dt - F(0)$$

$$\Rightarrow \quad \lim_{t\to\infty} F(t) = \lim_{p\to\infty} p\int_0^\infty e^{-pt} F(t)dt$$

Laplace Transform of Integrals: If $L\{F(t); p\} = f(p)$, then

$$L\left(\int_0^t F(u)du; p\right) = \frac{f(p)}{p}$$

Proof: Suppose, $P(t) = \int_0^t F(u)du$

and $P'(t) = F(u)$ and $P(0) = 0$

We know that $L\{P'(u); p\} = p\,L\{P(t); p\} - P(0)$

$$\Rightarrow \quad L\{F(u); p\} = p\,L\left\{\int_0^t F(u)du; p\right\}$$

or $$L\left(\int_0^t F(u)du; p\right) = \frac{f(p)}{p}$$

$$L\left\{\int_0^t \int_0^{u_n} \ldots \int_0^{u_2} F(u_1)\, du_i \ldots du_n; p\right\}$$

$$= p^{1-n} f(p)$$

where $\quad f(p) = L\{F(t); p\}$

Multiplication and Division by Powers of *t*:

Let $F(t)$ be a function of class A and if $L\{F(t); p\} = f(p)$, then

$$L\{t^n F(t); p\} = (-1)^n \frac{d^n}{dp^n} f(p),\ n = 1, 2, \ldots$$

Proof: We know that

$$L\{F(t); p\} = f(p) = \int_0^\infty e^{-pt} F(t)dt \quad \ldots(i)$$

Differentiating equation (i) with respect to p, as differentiating under integration sign,

$$\frac{d}{dp} f(p) = \int_0^\infty \frac{\partial}{\partial p} \{e^{-pt} F(t)\}\, dt$$

$$= \int_0^\infty - te^{-pt} F(t)\}\, dt$$

$$= -\int_0^\infty t\{e^{-pt} F(t)\}\, dt$$

$$= -L\{tF(t); p\}$$

$$\Rightarrow \quad L\{tF(t); p\} = \frac{d}{dp} f(p) = -f'(p) \quad \ldots(ii)$$

Let the statement be true for $n = 1$.

We establish the statement with the help of principle of mathematical induction, by equation (i) statement is true for $n = 1$. Let the statement be true for $n = m$ (m is a finite positive integer),

then, $\quad L\{t^m F(t); p\} = (-1)^m \dfrac{d^m}{dp^m} f(p)$

$$\Rightarrow \quad \int_0^\infty e^{-pt} \{t^m F(t)\}\, dt = (-1)^m \frac{d^m}{dp^m} f(p) \quad \ldots(iii)$$

Differentiating both sides of equation (iii) with respect to p, under integration sign

$$= \int_0^\infty \frac{\partial}{\partial p} \{e^{-pt} \{t^m F(t)\}\, dt = (-1)^m \frac{d^{m+1}}{dp^{m+1}} f(p)$$

$$= \int_0^\infty e^{-pt} \{t^{m+1} F(t)\}\, dt = (-1)^m \frac{d^{m+1}}{dp^{m+1}} f(p)$$

$\Rightarrow$ $$L\{t^{m+1} F(t); p\} = (-1)^{m+1} \frac{d^{m+1}}{dp^{m+1}} f(p)$$

or $$L\{t^{m+1} F(t); p\} = (-1)^{m+1} \frac{d^{m+1}}{dp^{m+1}} f(p)$$

$$= (-1)^{m+1} f^{m+1}(p)$$

Hence, the statement is true for $n = m + 1$.

Thus, statement established by mathematical induction,

$$L\{t^n F(t); p\} = (-1)^n \frac{d^n}{dp^n} f(p)$$

Statement: Let $L\{F(t); p\} = f(p)$, then,

$$L\left[\frac{f(t)}{t}; p\right] = \int_p^\infty F(u)du, \text{ provided both integrals exist}$$

Proof: Let us consider

$$P(t) = \frac{F(t)}{t}$$

$\Rightarrow$ $$F(t) = t\, P(t) \quad \text{...(i)}$$

Applying Laplace transform both sides on equation (i),

$$L\{F(t); p\} = L\{t\, P(t); p\}$$

$$= -\frac{d}{dp} L\{P(t); p\}$$

$$f(p) = -\frac{d}{dp} L\{P(t); p\} \quad \text{...(ii)}$$

Integrating both sides of equation (ii) with respect to p under the limits p to ∞, we get

$$-[L\{P(t); p\}]_p^\infty = \int_p^\infty f(p)dp = \int_p^\infty f(u)du$$

$$\Rightarrow \qquad -\lim_{p\to\infty} L\{p(t)\} + L\{P(t)\} = \int_p^\infty f(p)dp$$

$$\Rightarrow \qquad 0 + L\{P(t)\} = \int_p^\infty f(p)dt$$

$$\Rightarrow \qquad L\left[\frac{1}{t}F(t)\right] = \int_p^\infty f(t)dp$$

Example 5 Find $L\{e^{at}\}$.

Solution: We know that

$$L(F'(t)) = p\,L\{F(t)\} - F(0) \qquad \text{...(i)}$$

Here,

$$F(t) = e^{at}$$

$$F'(t) = ae^{at}$$

$$F(0) = 1$$

From equation (i),

$$L\{ae^{at}\} = p\,L\{e^{at}\} - F(0)$$

$$\Rightarrow \qquad aL(e^{at}) = p\,L\{e^{at}\} - 1$$

$$\Rightarrow \qquad (p-a)\,L\{e^{at}\} = 1$$

$$\Rightarrow \qquad L\{e^{at}\} = \frac{1}{p-a}$$

Example 6 Find $L\{te^{-t}\sin t\}$.

Solution: We have

$$L\{t\,F(t)\} = -f'(p) \qquad \text{...(i)}$$

We know that

$$L\{\sin t\} = \frac{1}{1+p^2}$$

$$L\{t\sin t\} = -\frac{d}{dp}\left(\frac{1}{1+p^2}\right) = \frac{2p}{(1+p^2)^2}$$

$$\therefore \qquad L(e^{-t}\,t\sin t) = \frac{2(p+1)}{[1+(p+1)^2]} = \frac{2p+2}{(p^2+2p+2)^2}$$

Example 7 Find $L\{-a \sin at\}$.

Solution: We know that

Given that,

$$F(t) = -a \sin at,$$

$$\Rightarrow \quad F'(t) = -a^2 \cos at$$

$$F''(t) = 0^3 \sin at$$

We know that

$$L\{F''(t)\} = p^2 L\{F(t)\} - p F(0) - F'(0)$$

$$\Rightarrow \quad L\{a^3 \sin at\} = p^2 L\{-a \sin at\} - 0 - (-a^2)$$

$$\Rightarrow \quad (p^2 + a^2) L\{-a \sin at\} = -a^2$$

$$L\{-a \sin at\} = -\frac{a^2}{p^2 + a^2}$$

Example 8 $L\{(t^2 - 3t + 2) \sin 3t\}$

$$= \frac{6p^4 - 18p^3 + 126p^2 - 16^2 p + 432}{(p^2 + 9)^3}.$$

Solution: We have

$$L\{(t^2 - 3t + 2) \sin 3t\} = L\{t^2 \sin 3t\} - 3 L\{t \sin 3t\} + 2 L \{(\sin 3t)\}$$

$$= (-1)^2 \frac{d^2}{dp^2} L\{\sin 3t\} - 3\left[-\frac{d}{dp} L\{\sin 3t\}\right] + 2L\{\sin 3t\}$$

$$= \frac{d^2}{dp^2}\left(\frac{3}{p^2+9}\right) + 3\frac{d}{dp}\left(\frac{3}{p^2+9}\right) + 2 \cdot \frac{3}{p^2+9}$$

$$= \frac{18p^2 - 54}{(p^2+9)^3} + 3\left(\frac{-6p}{(p^2+9)^2}\right) + \frac{6}{p^2+9}$$

$$= \frac{6p^4 - 18p^3 + 126p^2 - 162p + 432}{(p^2+9)^3}$$

Example 9 Show that $L\left\{\frac{\sin t}{t}\right\} = \tan^{-1}\frac{1}{p}$ and hence find $L\left(\frac{\sin at}{t}\right)$. Prove or disprove $L\left(\frac{\cos at}{t}\right)$ exist?

Solution: Suppose

$$F(t) = \sin t$$

Then, $$\lim_{t\to 0} \frac{\sin t}{t} = 1$$

Now, we have

$$L(\sin t) = \frac{1}{1+p^2} = f(p) \text{ (Assume)}$$

Then, $$L\left\{\frac{\sin t}{t}\right\} = \int_p^\infty f(u)du = \int_p^\infty \frac{du}{u^2+1}$$

$$= \left(\tan^{-1} u\right)_p^\infty$$

$$= \frac{t}{2} - \tan^{-1} p$$

$$= \cot^{-1} p$$

$$= \tan^{-1} \frac{1}{p}$$

$$\Rightarrow \quad L\left\{\frac{\sin at}{t}\right\} = a\, L\left(\sin \frac{dt}{at}\right)$$

$$= a - \frac{1}{a} \tan^{-1} \frac{1}{p/a} \quad \{L\{F(at)\}\}$$

$$= \frac{1}{a} f(p/a)$$

Now,

$$L(\cos at) = \frac{p}{p^2+a^2} = f(p) \text{ (Assume)}$$

$$L\left(\frac{\sin at}{t}\right) = \int_p^\infty \frac{u}{u^2+a^2} du = \left[\frac{1}{2}\log(u^2+a^4)\right]_p^\infty$$

$$= \frac{1}{2} \lim_{u\to\infty} \log(u^2+a^2) - \frac{1}{2} \log(u^2+a^2)$$

But

$$\lim_{u\to\infty} \log(u^2+a^2) \to \infty \text{ on } u \to \infty$$

$$\Rightarrow \qquad L\left(\frac{\cos at}{t}\right) \text{ does not exist.}$$

EXERCISE

Evaluate Laplace transform of the following functions:

1. (i) $t^2 \cos t$ $\left[\textbf{Ans. } \dfrac{2p(p^2-3)}{(p^2+1)^3}\right]$

(ii) $t^3 \cos t$ $\left[\textbf{Ans. } \dfrac{6p^4-36p^2+6}{(p^2+1)^2}\right]$

(iii) $t\, e^{at} \sin bt$ $\left[\textbf{Ans. } \dfrac{2b(p-a)}{(p^2-2ap+a^2+b^2)^2}\right]$

(iv) $t^2 e^t \sin 4t$ $\left[\textbf{Ans. } \dfrac{8[3(p-1)^2-16]}{[(p-1)^2+16]^3}\right]$

(v) $te^{-t} \cosh t$ $\left[\textbf{Ans. } \dfrac{p^2+2p+2}{p^2(p+2)^2}\right]$

2. Prove that $L\left\{\dfrac{\sin t}{t}; p\right\} = \tan^{-1}\dfrac{1}{p}$ and hence find $L\left\{\dfrac{\sin at}{t}; p\right\}$. Does the Laplace transform of cost *at*/*t* exists?

$\left[\textbf{Ans. } \tan^{-1}\left(\dfrac{a}{p}\right), \text{No}\right]$

3. Evaluate $L\left\{\dfrac{e^{-at}-e^{-bt}}{t}\right\}$ and

deduce $\displaystyle\int_0^\infty \frac{e^{-3t}-e^{-6t}}{t}\, dt = \log 2$ $\left[\textbf{Ans. } \log\dfrac{p+b}{p+a}\right]$

Chapter 2

Periodic Function

2.1 PERIODIC FUNCTION

A function $F(t)$ is said to be periodic function if

$$F(t + \alpha) = F(t), \text{ for all } t$$

where α is a least positive constant number and known as period of the function $F(t)$.

Theorem 1. Let $F(t)$ be a periodic function with period α,

such that $F(t + n\alpha) = F(t)$. For n 1, 2,...

Then,

$$L\{F(t); p\} = \frac{1}{1-e^{-\alpha p}} \int_0^{\alpha} e^{-pt} F(t)dt$$

Solution: We have

$$L\{F(t); p\} = \int_0^{\infty} e^{-pt} F(t)dt$$

$$= \int_0^{\alpha} e^{-pt} F(t)dt + \int_{\alpha}^{2\alpha} e^{-pt} F(t)dt + \int_{2\alpha}^{3\alpha} e^{-pt} F(t)dt + ...$$

On replacing, $u = u + \alpha$, $u = u + 2\alpha$, $u = u + 3\alpha$. etc., in second, third, fourth i..., etc., integrals is right-hand side, then

$$L\{F(u); p\} = \int_0^{\alpha} e^{-pu} F(u)du + \int_0^{\alpha} e^{-p(u+\alpha)} F(u+\alpha) + \int_0^{\alpha} e^{-p(u+2\alpha)} F(u+2\alpha)du + ...$$

$$= \int_0^{\alpha} e^{-pu} F(u)du + \int_0^{\alpha} e^{-p\alpha} \cdot e^{-pu} F(u)du + \int_0^{\alpha} e^{-2p\alpha} e^{-pu} F(u)du + ...$$

$$= \int_0^{\alpha} e^{-pu} F(u)du + e^{-p\alpha} \int_0^{\alpha} e^{-pu} F(u)du + e^{-2p\alpha} \int_0^{\alpha} e^{-pt} F(u)du + ...$$

$$= (1 + e^{-p\alpha} + e^{-2p\alpha} + \ldots) \int_0^\alpha e^{-pu} F(u) du$$

Then,

But $1 + x + x^2 + \ldots = \dfrac{1}{1-x}$, $|x| < 1$

$$L\{F(t); p\} = \frac{1}{1 - e^{-\alpha p}} \int_0^\alpha e^{-pu} F(u) du$$

Example 1 Evaluate $L\{F(t); p\}$, where,

$$f(t) = \begin{cases} \sin wt, & \dfrac{2n\pi}{w} < t < \dfrac{(2n+1)\pi}{w} \\ 0\ , & \dfrac{(2n+1)\pi}{w} < t < \dfrac{(2n+2)\pi}{w} \end{cases}$$

(For $n = 0, 1, 2$)

Solution: Given function $F(t)$ is piecewise continuous, bounded and has period $2\pi/w$.

Function $F(t)$ is known and half wave rectified side function.

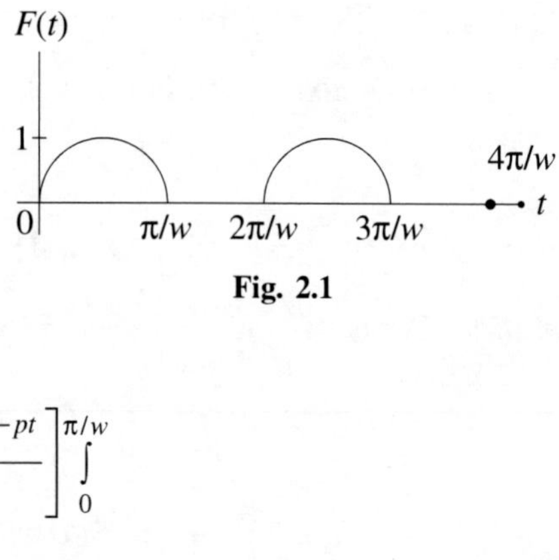

Fig. 2.1

$$L\{F(t); p\} = \frac{1}{(2 - e^{-2\pi p/w})} \int_0^{\pi/w} e^{-pt} \cos wt\, dt$$

$$= \frac{1}{(2 - e^{-2\pi p/w})} \left[\frac{(-p \sin wt - w \cos wt) e^{-pt}}{p^2 + w^2} \right]_0^{\pi/w}$$

$$= \frac{(1 + e^{-\pi p/w})}{(1 - e^{-2\pi p/w})(p^2 + w^2)} = \frac{w}{(1 - e^{-\pi p/w})(p^2 + w^2)}$$

and if $F(t) = |\sin wt|$ which is known as fall wave rectified sin function period with period π/w

Then, $$L\{F(t); p\} = \frac{1}{(1 - e^{-\pi p/w})} \int_0^{\pi/w} e^{-pt} \cos wt\, dt$$

$$= \frac{w}{p^2 + w^2} \left(\frac{1 + e^{-\pi p/w}}{1 - e^{-\pi p/w}} \right)$$

$$= \frac{w}{p^2 + w^2} \coth \frac{\pi p}{w}$$

Example 2 Find $L\{F(t); p\}$; if $F(t)$ has a period 2π and is given by

$$F(t) = \begin{cases} \sin t & , \quad 0 < t < \pi \\ 0 & , \quad \pi < t < 2\pi \end{cases}$$

Solution: The given function is periodic with period 2π. Then by theorem on periodic function,

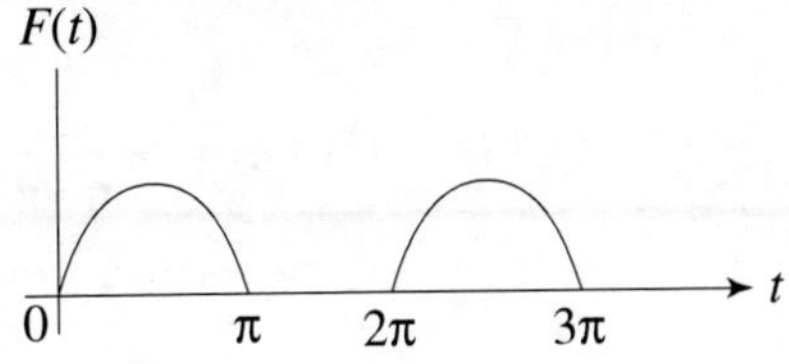

Fig. 2.2

$$L\{F(t); p\} = \frac{1}{2 + e^{-2\pi p}} \int_0^{2\pi} e^{-pt} F(t) dt$$

But

$$\int_0^{2\pi} e^{-pt} F(t) dt = \int_0^{\pi} e^{-pt} F(t) dt + \int_{\pi}^{2\pi} e^{-pt} F(t) dt$$

$$= \int_0^{\pi} e^{-pt} \sin t \, dt + 0$$

$$= \left[\frac{(-p \sin t - \cos t) e^{-pt}}{1 + p^2} \right]_0^{\pi}$$

$$= \frac{e^{-\pi p}}{1 + p^2} + \frac{1}{1 + p^2} = \frac{1 + e^{-\pi p}}{1 + p^2}$$

Hence,

$$L\{F(t); p\} = \frac{1}{1 - e^{-2\pi p}} \int_0^{2\pi} e^{-pt} F(t) dt$$

$$= \frac{1}{1 - e^{-2\pi p}} \cdot \frac{1}{1 + p^2}$$

A. Some Special Functions

(i) Beta functions and Gamma functions:

Beta functions can be defined as

$$B(m, n) = \int_0^1 t^{m-1} (1 - t)^{n-1} dt$$

and Gamma functions can be defined as

$$\Gamma(n) = \int_0^{\infty} e^{-t} x^{n-1} \, dt, \; n > 0$$

Important Results from Beta and Gamma Functions

(a) $\Gamma(n + 1) = n\Gamma(n), n > 0$

(b) $\Gamma(n + 1) = n!$, For 'n' is a positive integer

(c) $\Gamma(1/2) = \sqrt{x}$

(d) $\dfrac{\Gamma(n)}{a^n} = \int\limits_0^\infty e^{-at}\, t^{n-1}\, dt$

(e) $\Gamma(i) = 1$

(f) $\Gamma(n)\,\Gamma(1 - n) = \dfrac{\pi}{\sin n\pi}$

(g) $\Gamma(2n) = \dfrac{\Gamma(n)\,\Gamma(n+1/2)}{\sqrt{\pi}} \cdot 2^{2n-1}$

(h) $B(m, n) = \dfrac{\Gamma(m)\,\Gamma(n)}{\Gamma(m+n)}$

(i) $B(m, n) = B(n, m)$ [Symmetric property]

(j) $\int\limits_0^{\pi/2} \sin^m\theta \cos^n \theta dv = \dfrac{\Gamma\left(\dfrac{m+1}{2}\right)\Gamma\left(\dfrac{n+1}{2}\right)}{2\Gamma\left(\dfrac{m+n+2}{2}\right)}$

B. Bessel Function

Bessel function of the first kind of order n can be given as

$$J_n(t) = \frac{t^n}{2^n\,\Gamma(n+1)}\left[1 - \frac{t^2}{2(2n+2)} + \frac{t^4}{2.4(2n+2)\,(2n+4)} - \ldots\right]$$

$$= \sum_{r=0}^{\infty} \frac{(-1)^r}{r!\,\Gamma(n+r+1)} \cdot \left(\frac{t}{2}\right)^{n+2r}$$

Important Results

(i) $J_n(t) = (-1)^n\, J_n(-t)$

(ii) $J_{n+1}(t) = \dfrac{2n}{t} J_n(t) - J_{n-1}(t)$

(iii) $\dfrac{d}{dt}\, [t^{-n}\, J_n(t)] = -\, t^{-1} J_{n+1}(t)$

(iv) $t^n\, J_{n-1}(t) = \dfrac{d}{dt}[t^n\, J_n(t)]$

For $n = 0$

$-J_1(t) = J_0'(t)$

(v) Generating function for the Bessel function

$$\sum_{n=-\infty}^{\infty} u^n J_n(t) = \exp\left[\frac{t}{2}(u - u^{-1})\right]$$

(Here t is finite and $u \neq 0$)

(vi) $J_{-1/2}(t) = \sqrt{2/\pi t}\ \cos t$

(vii) $J_{1/2}(t) = \sqrt{2/\pi t}\ \sin t$

(viii) $-nJ_n(t) + tJ_{n-1}(t) = tJ_n'(t)$

(ix) $J_{n-1}(t) - J_{n+1}(t) = 2J_n'(t)$

(x) $nJ_n(t) - tJ_{n+1}(t) - tJ_n'(t)$

(xi) The differential equation

$$t^2y'' + ty' + (t^2 - n^2)y = 0$$

is an identity for $y = J_n(t)$

C. Hypergeometric Functions

Factorial function $(a)_n$ can be defined as

$$(a)_n = a(a + 1)\ (a + 2)\ ...\ (a + n - 1)$$

$$= \frac{\Gamma(a+n)}{\Gamma(a)},\ \begin{matrix}(n \geq 1 \\ a \pm 0)\end{matrix}$$

The notation $= F_1(a, b, y, x)$, known as Gauss' hypergeometric function or hypergeometric function, which can be defined as

$$= F_1(a,\ b,\ y,\ x) = \sum_{n=0}^{\infty} \frac{(a)_n\ (b)_n}{n!\,(y)_n} x^n$$

Note: (i) $\sin^{-1}x = x\cdot{}_2F_1\left(\frac{1}{2}\cdot\frac{1}{2}\cdot\frac{3}{2}\cdot x^2\right)$

(ii) $\tan^{-1}x = x\cdot{}_2F_1\left(\frac{1}{2}, 1, \frac{3}{2}, x^2\right)$

D. Error Function

$$\text{erf}(x) = \frac{2}{\sqrt{\pi}}\int_0^x e^{-t^2}\,dt$$

Complimentary error function presented by

$$\text{erfc}(x) = -\ 1 - \text{erf}(x)$$

$$= \frac{2}{\sqrt{\pi}} \int_0^{\infty} e^{-t^2} dt = 1$$

E. Sine and Cosine Integrals

$$Si(x) = \int_0^x \frac{\sin t}{t} dt$$

$$Ci(x) = \int_x^{\infty} \frac{\cos t}{t} dt$$

F. Laguerre Polynomial

Laguerre polynomial can be written as

$$L_n(x) = \frac{e^x}{n!} \frac{d^n}{dx^n}(e^{-x}x^n) \text{ © } n = 0, 1, 2, \ldots$$

which can be put in series form as given,

$$L_n(x) = \sum_{k=0}^{\infty} \frac{(-1)^k n! \, x^k}{k!(n-k)! \, k!}$$

G. Unit Step Function

The unit step function, which is known as Heaviside's unit function, denoted by $H(t - a)$, can be defined as

$$H(t-a) = \begin{cases} 0, & t < a \\ 1, & t > a \end{cases}$$

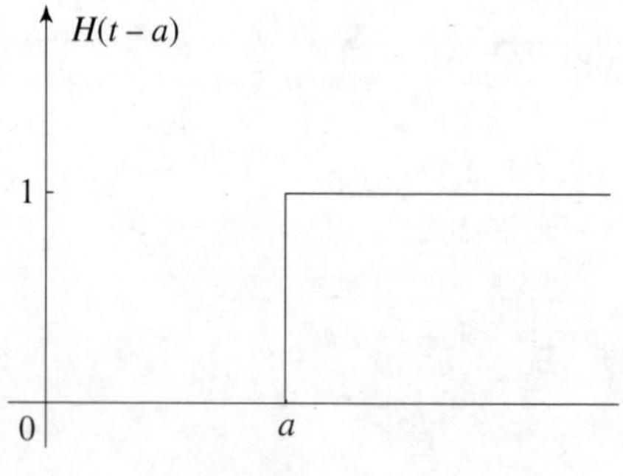

Fig. 2.3

H. Dirac Delta functions

Let us cansider a function $F_a(*)$,

$$F_a(t) = \begin{matrix} 1/a, & 0 \le t \le a \\ 0\;, & t > a \end{matrix}$$

and unit step function can be defined as

$$\delta(t) = \lim_{a \to 0} F_a(t)$$

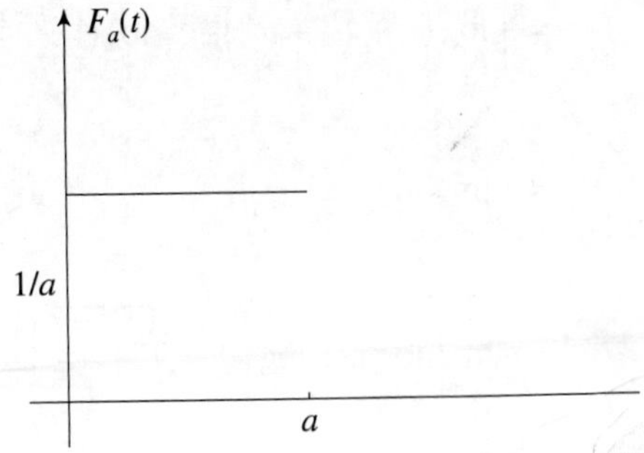

Fig. 2.4

Thus,

(i) $\int_0^\infty \delta(t)\, G(t)\, dt = G(0)$

(ii) $\int_0^\infty \delta(t)\, dt = 1$

(iii) $\int_0^\infty \delta(t - a)\, G(t) dt$

$= G(a)$, $G(0)$ is a continuous function.

Illustrative Examples

Example 1 Prove that $\int_0^\infty \frac{J_0(t) - \cos t}{t} dt = \log 2$

Solution: Assume $F(t) = J_0(t) - \cos t$

such that

$$L\{F(t)\} = L\{J_0(t) - \cos t\}$$

$$= L\{J_0(t)\} - L\{\cos t\}$$

$$= \frac{1}{(1 + p^2)^{1/2}} - \frac{p}{1 + p^2}$$

$$= F(p) \text{ (Assume)}$$

We know that $\int_0^\infty \frac{F(t)}{t} dt = \int_0^\infty f(u)\, du$

provided that the integral converges.

Then,
$$\int_0^\infty \frac{J_0(t)-\cos t}{t}\,dt = \int_0^\infty\left(\frac{1}{(1+u^2)^{1/2}}-\frac{u}{1+u^2}\right)du$$
$$= \left[\log\left(u+\sqrt{1+u^2}\right)-\frac{1}{2}\log(1+u^2)\right]_0^\infty$$
$$= \lim_{u\to\infty}\left[\log u\left\{1+\sqrt{1+\frac{1}{u^2}}\right\}-\frac{1}{2}\log u^2\left(1+\frac{1}{u^2}\right)\right]$$
$$= \lim_{u\to\infty}\left[\log u+\log\left(1+\sqrt{1+\frac{1}{u^2}}\right)-\frac{1}{2}\log u^2-\frac{1}{2}\log\left(1+\frac{1}{u^2}\right)\right]$$
$$= \lim_{u\to\infty}\left[\log\left(1+\sqrt{1+\frac{1}{u^2}}\right)-\log\left(1+\frac{1}{u^2}\right)^{1/2}\right]$$
$$= \log 2.$$

Example 2 $L\{F(t)\}$ where
$$F(t) = \int_0^t \frac{\sin x}{x}\,dx$$

Solution: We know that
$$L\{\sin t\} = \frac{1}{1+p^2}$$
and
$$L\left\{\frac{\sin t}{t}\right\} = \int_p^\infty \frac{1}{1+u^2}\,du$$
$$= (\tan^{-1} u)_p^\infty$$
$$= \frac{\pi}{2}-\tan^{-1} p$$
$$= \tan^{-1} 1/p$$

Thus,
$$L\left\{\int_0^t \frac{\sin x}{x}\,dx\right\} = \frac{1}{p}L\left\{\frac{\sin t}{t}\right\}$$
$$= \frac{1}{p}\tan^{-1} 1/p$$

Example 3 Prove that $L\{H(t-a)\} = \dfrac{e^{-ap}}{p}$

where $H(t-a)$ is Heaviside's unit step function.

Solution: Heaviside's unit step function

$$H(t-a) = \begin{cases} 0, & t < a \\ 1, & t > a \end{cases}$$

$$\begin{aligned} L\{H(t-a)\} &= \int_0^\infty e^{-pt} H(t-a)dt \\ &= \int_0^a e^{-pt} H(t-a)dt + \int_a^\infty e^{-pt} H(t-a)dt \\ &= \int_0^a e^{-pt} \cdot 0 \cdot dt + \int_a^\infty e^{-pt} \cdot 1 \cdot dt \\ &= \int_a^\infty e^{-pt} \cdot dt \\ &= \left[\frac{e^{-pt}}{-p}\right]_0^\infty \\ &= \frac{e^{-ap}}{p} \end{aligned}$$

Example 4 Prove that $L\{\delta(t-a)\} = e^{-ap}$ where $\delta(t)$ is Dirac-delta function.

Solution: We know that

$$\delta(t-a) = \lim_{b \to 0} F_b(t-a)$$

Here,
$$F_b(t-a) = \begin{bmatrix} 0 & t < a \\ 1/b & a \le t \le a+b \\ 0 & t > a+b \end{bmatrix}$$

Then,
$$\begin{aligned} L\{F_b(t-a)\} &= \int_0^\infty e^{-pt} F_b(t-a)dt \\ &= \int_0^\infty e^{-pt} F_b(t-a)dt + \int_a^{a+b} e^{-pt} F_b(t-a)dt \\ &\quad + \int_{a+b}^\infty e^{-pt} F_b(t-a)dt \\ &= \int_a^{a+b} e^{-pt} \cdot \frac{1}{b} dt \\ &= \frac{1}{b}\left[-\frac{e^{-pt}}{p}\right]_0^{a+b} \end{aligned}$$

$$= \frac{e^{-ap}}{p}\left[\frac{1-e^{-pb}}{b}\right]$$

Applying limit $b \to 0$ both sides, we get $\int_0^\infty e^{-pt}\left\{\lim_{b\to 0} F_b(t-a)\right\}dx = \frac{e^{-ap}}{p}\lim_{b\to 0}\frac{1-e^{-pb}}{p}$

or $$\int_0^a e^{-pt}\,\delta(t-a)\,dt = \frac{e^{-ap}}{p}\lim_{b\to 0}\frac{pe^{-pb}}{1}$$

$\Rightarrow$ $$L\,\delta(t-a) = e^{-ap}$$

Example 5 If $$P_n(x) = \frac{e^x}{n!}\frac{d^n}{dx^n}(e^{-x}x^n),$$

Then, $$L\{P_n(x)\} = \frac{(p-1)^n}{p^{n+1}},\ p>1$$

Solution: $$L\{P_n(x)\} = \int_0^\infty e^{-bx}\frac{e^x}{n!}\frac{d^n}{dx}(e^{-x}x^n)dx$$

$$= \frac{1}{n!}\int_0^\infty e^{-(p-1)x}\cdot\frac{d^n}{dx^n}(x^{-x}\cdot x^n)dx$$

$$= \frac{1}{n!}\left[e^{-(p-1)x}\frac{d^{n-1}}{dx^{n-1}}(e^{-x}x^n)\right]_0^\infty$$

$$-\frac{1}{n!}\left[-(p-1)\int_0^\infty e^{-(p-1)x}\frac{d^{n-1}}{dx^{n-1}}(e^{-x}x^n)\right]dx$$

$$= \frac{(p-1)}{n!}\int_0^\infty e^{-(p-1)x}\frac{d^{n-1}}{dx^{n-1}}(e^{-x}x^n)\,dx,\ p>1$$

Processing in similar manner, we get

$$L\{P_n(x)\} = \frac{(p-1)^n}{n!}\int_0^\infty e^{-(p-1)x}e^{-x}\cdot x^n dx$$

$$= \frac{(p-1)^n}{n!}\int_0^\infty e^{-px}\cdot x^n dx$$

$$= \frac{(p-1)^n}{n!}L\{x^n\}$$

$$= \frac{(p-1)^n}{n!}\frac{n!}{p^{n+1}}$$

$$= \frac{(p-1)^n}{p^{n+1}}.$$

⇒ Two different functions may have some Laplace transform.

Example 3

$$F_1(t) = e^{-at}$$

and
$$F_2(t) = \begin{bmatrix} 0 & , & t=1 \\ e^{-at} & , & \text{Otherwise} \end{bmatrix}$$

Have the same Laplace transform $1/(p + a)$

Now, if we not consider the appearance of null function, then inverse of Laplace transform is unique.

3.3 LERCH'S THEOREM

For a function $F(t)$, which is sectionally continuous in every finite interval $0 \le t \le N$ and of exponential order, then the inverse Laplace transform of $f(p)$ is unique.

Standard inverse Laplace transforms

	$f(p)$	$L^{-1} f\{p\} = F(t)$
(i)	$\frac{1}{p}$	1
(ii)	$\frac{1}{p^{n+1}}$, n is a +ve integer	$\frac{t^n}{n!}$
(iii)	$\frac{1}{p^{n+1}} \cdot n > -1$	$\frac{t^n}{\Gamma(n+1)}$
(iv)	$\frac{1}{p-a}$	e^{at}
(v)	$\frac{p}{p^2+a^2}$	$\cos at$
(vi)	$\frac{a}{p^2+a^2}$	$\sin at$
(vii)	$\frac{p}{p^2-a^2}$	$\cosh at$
(viii)	$\frac{a}{p^2-a^2}$	$\sinh at$

3.4 LINEAR PROPERTY

Suppose, $L^{-1}(f_1(p)) = F_1(t)$ and $L^{-1}(f_2(p)) = F_2(t)$. If c_1 and c_2 are two contants, then

$$L^{-1}[c_1 f_1(p) + c_2 f_2(p)] = c_1 L^{-1} f_1(p) + c_2 L^{-1} f_2(p)$$
$$= c_1 F_1(t) + c_2 F_2(t)$$

Proof: We know that

$$L\{c_1 F_1(t) + c_2 F_2(t)\}$$
$$= c_1 L\{F_1(t)\} + c_2 L\{F_2(t)\}$$
$$= c_1 L F_1(t) + c_2 L F_2(t)$$
$$= c_1 f_1(p) + c_2 f_2(p)$$

$$\Rightarrow \quad L^{-1}\{c_1 f_1(p) + c_2 f_2(p)\} = c_1 F_1(t) + c_2 F_2(t)$$
$$= c_1 L^{-1} f_1(p) + c_2 L^{-1} f_2(p)$$

3.5 FIRST TRANSLATION OR SHIFTING PROPERTY

If $\quad L^{-1}\{f(p); t\} = F(t)$

Then,

$$L^{-1}\{f(p-a); t\} = e^{at} F(t)$$
$$= e^{at} L^{-1}\{f(p); t\}$$

Proof:

We know that

$$f(p) = \int_0^\infty e^{-pt} F(t) dt$$

$$\Rightarrow \quad f(p-a) = \int_0^\infty e^{-(p-a)t} F(t) dt$$
$$= \int_0^\infty e^{at} \cdot e^{-pt} F(t) dt$$
$$= \int_0^\infty e^{at} \cdot (e^{-pt} F(t) dt)$$
$$= L\{e^{at} \cdot (e^{-pt} F(t))\}$$

$$\Rightarrow \quad L^{-1} f(p-a) = e^{at} F(t)$$
$$= e^{at} L^{-1} f(p)$$

$$\Rightarrow \quad L^{-1} f(p-a) = e^{at} L^{-1} f(p)$$

3.6 SECOND TRANSLATION OR SHIFTING PROPERTY

If $$L^{-1}[f(p); t] = F(t)$$

and $$G(t) = \begin{cases} 0, & t < a \\ F(t-a), & t > a \end{cases}$$

Then, $$L^{-1}[e^{-ap} f(p); t] = G(t)$$

i.e., $$L^{-1}[e^{-ap} f(p); t] = F(t-a)\, H(t-a)$$

Here, $H(t-a)$ is Heaviside unit function.

Proof: We know that

$$L[G(t); p] = \int_0^\infty e^{-pt}\, G(t) dt$$

$$= \int_0^a e^{-pt}\, G(t) dt + \int_a^\infty e^{-pt}\, G(t) dt$$

$$= \int_0^a e^{-pt} \cdot 0 \cdot dt + \int_a^\infty e^{-pt}\, F(t-a) dt$$

$$= \int_0^\infty e^{-pt}\, F(t-a) dt$$

$$= \int_0^\infty e^{-p(t+a)}\, F(t) dt$$

$$= e^{-ap} \int_0^\infty e^{-pt}\, F(t); p$$

$$L[G(t); p] = e^{-ap}\, L[F(t); p]$$

$$= e^{-ap} f(p)$$

or $$L^{-1}[e^{-ap} f(p); t] = G(t)$$

Illustrative Examples

Example 1 Find $L^{-1}\left(\dfrac{3p+7}{p^2-2p-3}\right)$

Solution: We have, $L^{-1}\ \dfrac{3p+7}{p^2-2p-3}$

$$= L^{-1} \frac{3(p-1)+10}{(p-1)^2-4}$$

$$= L^{-1} \left[\frac{3(p-1)}{(p-1)^2-4} + \frac{10}{(p-1)^2-4} \right]$$

$$= 3L^{-1} \frac{p-1}{(p-1)^2-4} + 10\, L^{-1} \frac{1}{(p-1)^2-4}$$

$$= 3 \cdot e^t \cdot L^{-1} \frac{p}{p^2-2^2} + 10\, e^t \cdot L^{-1} \frac{1}{p^2-2^2}$$

$$= 3e^t \cdot \cosh 2t + 5e^t \sinh 2t$$

$$= 3e^t \cdot \frac{e^{2t}+e^{-2t}}{2} + 5 \cdot e^t \cdot \frac{e^{2t}-e^{-2t}}{2}$$

$$= 4e^{3t} - e^t$$

Ans.

Example 2 Find $L^{-1} \dfrac{p+1}{p(p^2+4p+8)}$

Solution: We have to find $L^{-1} \dfrac{p+1}{p(p^2+4p+8)}$

Now, $$\frac{p+1}{p(p^2+4p+8)} = \frac{1}{8}\left[\frac{1}{p} - \frac{p-4}{p^2+4p+8} \right]$$

$$\Rightarrow \quad L^{-1} \frac{p+1}{p(p^2+4p+8)} = L^{-1} \left[\frac{1}{8}\left(\frac{1}{p} - \frac{p-4}{p^2+4p+8} \right) \right]$$

$$= \frac{1}{8} L^{-1}\left(\frac{1}{p}\right) - \frac{1}{8} L^{-1}\left(\frac{p-4}{p^2+4p+8}\right)$$

$$= \frac{1}{8} L^{-1}\left(\frac{1}{p}\right) - \frac{1}{8} L^{-1}\left(\frac{(p+2)-6}{(p+2)^2+4}\right)$$

$$= \frac{1}{8} L^{-1}\left(\frac{1}{p}\right) - \frac{1}{8} e^{-2t} L^{-1} \frac{p-6}{p^2+4}$$

$$= \frac{1}{8} L^{-1}\left(\frac{1}{p}\right) - \frac{1}{8} e^{-2t} L^{-1} \frac{p}{p^2+4} + \frac{1}{8} e^{-2t} L^{-1} \frac{6}{p^2+4}$$

$$= \frac{1}{8} L^{-1}\left(\frac{1}{p}\right) - \frac{1}{8} e^{-2t} L^{-1} \frac{p}{p^2+4} + \frac{6}{8} e^{-2t} L^{-1} \frac{1}{p^2+4}$$

$$= \frac{1}{8} \cdot 1 - \frac{1}{8} e^{-2t} \cdot \cos 2t + \frac{3}{8} \sin 2t$$

$$= \frac{1}{8} [1 - e^{-2t} \cos 2t + 3 \sin 2t]$$

$$= \frac{1}{8} [1 - e^{-2t} (\cos 2t - 3 \sin 2t)]$$ **Ans.**

Example 3 Find $L^{-1} \dfrac{1}{(p^2+4)(p+1)^2}$

Solution: We have to find $L^{-1}\left(\dfrac{1}{(p^2+4)(p+1)^2}\right)$

We know that $\dfrac{1}{(p^2+4)(p+1)^2}$

$$= \frac{2}{25(p+1)} + \frac{1}{5(p+1)^2} - \frac{2p+3}{25(p^2+4)}$$

$$\Rightarrow L^{-1}\left(\frac{1}{(p^2+4)(p+1)^2}\right)$$

$$= \frac{1}{25}\left[2 \cdot L^{-1}\left(\frac{1}{p+1}\right) + \frac{1}{5} L^{-1}\left(\frac{1}{(p+1)^2}\right) - 2L^{-1}\left(\frac{p}{p^2+4}\right) - 3L^{-1}\left(\frac{1}{p^2+4}\right)\right]$$

$$= \frac{1}{25}\left[2e^{-t} + 5e^{-t} L^{-1} \frac{1}{p^2} - 2\cos 2t - \frac{3}{2} \sin 2t\right]$$

$$= \frac{1}{25} [e^{-t}(2 + 5t) - 2 \cos 2t - 3/2 \sin 2t]$$ **Ans.**

Example 4 Prove that

$$L^{-1} \frac{p^2}{p^4 + 4a^4} = \frac{1}{2a}[\cosh at \sin at + \sinh at \cos at]$$

Solution: We have to prove

$$L^{-1}\left(\frac{p^2}{p^4 + 4a^4}\right) = \frac{1}{2a}[\cosh at \sin at + \sinh at \cos at]$$

Now, we know that

$$\frac{p^2}{p^4+4a^4} = \frac{p^2}{(p^2+2a^2)^2-4a^2p^2}$$

$$= \frac{p^2}{(p^2-2ap+2a^2)(p^2+2ap+2a^2)}$$

$$= \frac{1}{4a}\cdot\frac{p(p^2+2ap+2a^2)-p(p^2-2ap+2a^2)}{(p^2-2ap+2a^2)(p^2-2ap+2a^2)}$$

$$= \frac{1}{4a}\frac{p}{p^2-2ap+2a^2}-\frac{1}{4a}\cdot\frac{p}{p^2+2ap+2a^2}$$

$$= \frac{1}{4a}\cdot\frac{(p-a)+a}{(p-a)^2+a^2}-\frac{1}{4a}\cdot\frac{(p+a)-a}{(p+a)^2+a^2}$$

Then, $$L^{-1}\left(\frac{p^2}{p^4+4a^4}\right) = \frac{1}{4a}L^{-1}\left(\frac{(p-a)+a}{(p-a)^2+a^2}\right)-\frac{1}{da}L^{-1}\left(\frac{(p+a)-a}{(p+a)^2+a^2}\right)$$

$$= \frac{1}{4a}e^{at}\,L^{-1}\left(\frac{p+a}{p^2+a^2}\right)-\frac{1}{4a}e^{-at}\,L^{-1}\left(\frac{p-a}{p^2+a^2}\right)$$

$$= \frac{1}{4a}e^{at}\left[L^{-1}\frac{p}{p^2+a^2}+a.L^{-1}\frac{1}{p^2+a^2}\right]$$

$$= \frac{1}{4a}\,e^{at}\,(\cos at+\sin at)-\frac{1}{4a}\,e^{at}\,(\cos at-\sin at)$$

$$= \frac{1}{2a}\,\sin at\,\frac{e^{at}+e^{-at}}{2}+\cos at\,\frac{e^{at}-e^{-at}}{2}$$

$$= \frac{1}{2a}\,\cosh at\sin at+\sinh at\cos at$$

Example 5 Find $L^{-1}\left[\frac{6}{2p-3}-\frac{3+4p}{9p^2-16}+\frac{8-6p}{16p^2+9}\right]$.

Solution: We have to find $L^{-1}\left[\frac{6}{2p-3}-\frac{3+4p}{9p^2-16}+\frac{8-6p}{16p^2+9}\right]$

$$= 6L^{-1}\frac{1}{2p-3}-3L^{-1}\frac{1}{9p^2-16}-4L^{-1}\frac{p}{9p^2-16}+8L^{-1}\frac{1}{16p^2+9}-6L^{-1}\frac{p}{16p^2+9}$$

$$= 3L^{-1}\frac{1}{p-3/2} - \frac{1}{3}L^{-1}\frac{1}{p^2-\left(\frac{4}{3}\right)^2} - \frac{4}{9}L^{-1}\frac{p}{p^2+\left(\frac{4}{3}\right)^2} + \frac{1}{2}L^{-1}\frac{1}{p^2-\left(\frac{3}{4}\right)^2} - \frac{3}{8}L^{-1}\frac{p}{p^2+\left(\frac{3}{4}\right)^2}$$

$$= 3e^{\frac{3}{2}t} - \frac{1}{3}\frac{1}{(4/3)}\sinh\frac{4t}{3} - \frac{4}{9}\cosh\frac{4t}{3} + \frac{1}{2}.\frac{1}{3/4}\sin\frac{3t}{4} - \frac{3}{8}\cos\frac{3t}{4}$$

$$= 3e^{3t/2} - \frac{1}{4}\sinh\frac{4t}{3} - \frac{4}{9}\cosh + \frac{2}{3}\sin\frac{3t}{4} - \frac{3}{8}\cos\frac{8t}{4}$$

Example 6 Prove that

$$L^{-1}\left(\frac{1}{p}\sin\frac{1}{p}\right) = t - \frac{t^3}{(3!)^2} + \frac{t^5}{(5!)^2} - \frac{t^7}{(7!)^2} + \ldots$$

Solution: Given that

$$L^{-1}\left(\frac{1}{p}\sin\frac{1}{p}\right) = t - \frac{t^3}{(3!)^2} + \frac{t^5}{(5!)^2} - \frac{t^7}{(7!)^2} + \ldots$$

We know that, $\frac{1}{p}\sin\frac{1}{p} = \frac{1}{p}\left(\frac{1}{p} - \frac{(1/p)^3}{3!} + \frac{(1/p)^5}{5!} - \frac{(1/p)^7}{7!} + \ldots\right)$

$$= \left(\frac{1}{p^2} - \frac{1}{3!}\frac{1}{p^4} + \frac{1}{5!}\frac{1}{p^6} - \frac{1}{7!}\frac{1}{p^8} + \ldots\right)$$

Taking inverse Laplace both sides, and applying the standard result:

$$L^{-1}\frac{n!}{p^{n+1}} = t^n$$

We have $L^{-1}\left(\frac{1}{p}\sin\frac{1}{p}\right) = L^{-1}\left(\frac{1}{p^2}\right) - \frac{1}{3!}L^{-1}\left(\frac{1}{p^4}\right) + \frac{1}{5!}L^{-1}\left(\frac{1}{p^6}\right) - \frac{1}{7!}L^{-1}\left(\frac{1}{p^8}\right) + \ldots$

$$= t - \frac{1}{3!}\cdot\frac{t^3}{3!} + \frac{1}{5!}\frac{t^5}{5!} - \frac{1}{7!}\cdot\frac{t^7}{7!} + \ldots$$

$$= t - \frac{1}{(3!)^2}t^3 + \frac{1}{(5!)^2}t^5 - \frac{1}{(7!)^2}t^7 + \ldots$$ ***Ans.***

Example 7 If $L^{-1}\left[\frac{p^2-1}{(p^2+1)^2}\right] = t\cos t$, then find $L^{-1}\left(\frac{9p^2-1}{(9p^2+1)^2}\right)$.

Solution: Given that $L^{-1}\left[\dfrac{p^2-1}{(p^2+1)^2}\right] = t\cos t$

Replacing p by ap using change of scale property, we have,

$$L^{-1}\left[\frac{a^2p^2-1}{(a^2p^2+1)^2}\right] = \frac{1}{a}\cdot\frac{t}{a}\cos\frac{t}{a}$$

Putting $a = 3$

$$L^{-1}\left[\frac{9p^2-1}{(9p^2+1)^2}\right] = \frac{t}{3.3}\cos\frac{t}{3}$$

$$L^{-1}\left[\frac{9p^2-1}{(9p^2+1)^2}\right] = \frac{t}{9}\cos\frac{t}{3}$$ ***Ans.***

Change of Scale Property:

If $\qquad L^{-1}[f(p)] = F(t)$, then

$$L^{-1}[f(ap)] = \frac{1}{a}F(t/a), a>0.$$

Proof: We have

$$f(p) = \int_0^\infty e^{-px}F(t)dt$$

Then,

$$f(ap) = \int_0^\infty e^{-apt}F(t)dt$$

$$= \int_0^\infty e^{-pt}F(t/a)\cdot\frac{du}{0} \qquad \left\{\begin{matrix} at-u \\ adt = du \\ t = u/a \end{matrix}\right\}$$

$$= \int_0^\infty e^{-pt}\left(\frac{1}{a}F(t/a)\right)dt \quad \text{(Change of variable)}$$

$$\Rightarrow \qquad f(ap) = L\left[\frac{1}{a}F(t/a)\right]$$

or

$$L^{-1}f(ap) = \frac{1}{a}F(t/a)$$

Example 8 Show that

$$L^{-1}\left(\frac{e^{-1/p}}{\sqrt{p}}\right) = \frac{\cos 2\sqrt{t}}{\sqrt{\pi t}}$$

hence, find the value of

$$L^{-1}\left(\frac{e^{-a/p}}{\sqrt{p}}\right).$$

Solution: We know that

$$e^{-1/p} = 1-\frac{1}{p}+\frac{1}{2!p^2}-\frac{1}{3!}\frac{1}{p^3}+...$$

$$\Rightarrow \qquad \frac{e^{-1/p}}{\sqrt{p}} = \frac{1}{p^{1/2}}-\frac{1}{p^{3/2}}+\frac{1}{2!}\frac{1}{p^{5/2}}-\frac{1}{3!}\frac{1}{p^{7/2}}+...$$

$$\Rightarrow \qquad L^{-1}\left(\frac{e^{-1/p}}{\sqrt{p}}\right) = L^{-1}(1/p^{1/2})-L^{-1}\left(\frac{1}{p^{3/2}}\right)+\frac{1}{2!}L^{-1}\left(\frac{1}{p^{5/2}}\right)-\frac{1}{3!}L^{-1}\left(\frac{1}{p^{7/2}}\right)+...$$

$$= \frac{1}{\sqrt{\pi t}}-\frac{2t^{1/2}}{\sqrt{\pi}}+\frac{2t^{3/2}}{3\sqrt{\pi}}-\frac{4t^{5/2}}{45\sqrt{\pi}}+...$$

$$= \frac{1}{\sqrt{\pi t}}\left[1-\frac{(2\sqrt{t})^2}{2!}+\frac{(2\sqrt{t})^4}{4!}-\frac{(2\sqrt{t})^6}{6!}+...\right]$$

$$= \frac{\cos 2\sqrt{t}}{\sqrt{\pi t}}$$

Since, $\qquad f(p) = \dfrac{e^{-1/p}}{\sqrt{p}}$

then, by change of scale property,

$$L^{-1}\left(\frac{e^{-1/np}}{\sqrt{np}}\right) = \frac{1}{n}\frac{\cos 2\sqrt{t/n}}{\sqrt{\pi(t/n)}}$$

$$= \frac{1}{\sqrt{n}}\frac{\cos 2\sqrt{t/n}}{\sqrt{\pi t}}$$

$$\Rightarrow \qquad L^{-1}\left(\frac{e^{-1/np}}{\sqrt{p}}\right) = \frac{\cos 2\sqrt{t/n}}{\sqrt{\pi t}}$$

Now, putting the value of $n = 1/a$, we have

$$L^{-1}\left(\frac{e^{-a/p}}{\sqrt{p}}\right) = \frac{\cos 2\sqrt{at}}{\sqrt{\pi t}}$$

Example 9 Find the inverse Laplace of

$$f(p) = \frac{6p-4}{p^2-4p+20}$$

Solution: We have $$f(p) = \frac{6p-4}{p^2-4p+20}$$

$$\Rightarrow \quad L^{-1}(f(p)) = L^{-1}\left(\frac{6p-4}{p^2-4p+20}\right) = L^{-1}\left(\frac{6(p-2)+8}{(p-2)^2+16}\right)$$

$$= 6L^{-1}\left(\frac{p-2}{p^2-4p+20}\right) + 2L^{-1}\left(\frac{4}{(p-2)^2+16}\right)$$

(Using linear property)

$$= 6e^{2t}\,L^{-1}\left(\frac{p}{p^2+4^2}\right) + 2e^{2t}\,L^{-1}\left(\frac{4}{p^2+4^2}\right)$$

$$= 6e^{2t}\cos 4t + 2e^{2t}\sin 4t$$

$$= 2e^{2t}(3\cos 4t + \sin 4t)$$ ***Ans.***

Example 10 Find the inverse Laplace of $f(p) = \dfrac{p}{(p+3)^{7/2}}$.

Solution: Given that $$f(p) = \frac{p}{(p+3)^{7/2}}$$

Then, $$L^{-1} f(p) = L^{-1} \frac{p}{(p+3)^{7/2}}$$

$$= e^{-3t} \cdot L^{-1} \frac{p-3}{p^{7/2}}$$

$$= e^{-3t} L^{-1}\left(\frac{p}{p^{7/2}} - \frac{3}{p^{7/2}}\right)$$

$$= e^{-3t} L^{-1}\left(\frac{p}{p^{7/2}}\right) - 3e^{-3t}L^{-1}\frac{1}{p^{7/2}}$$

$$= e^{-3t}\frac{t^{3/2}}{\Gamma 5/2} - 3.e^{-3t}\frac{t^{5/2}}{\Gamma 7/2}$$

$$= \frac{e^{-3t}}{\sqrt{\pi}}\left[\frac{4}{3}t^{3/2} - \frac{8}{5}t^{5/2}\right]$$

$$= \frac{4t^{3/2}}{15\sqrt{\pi}}(5-6t)$$ ***Ans.***

Example 11 Find inverse Laplace transform of

$$f(p) = \frac{1}{\sqrt{p^2 - 4p + 20}}$$

Solution: Given that

$$f(p) = \frac{1}{\sqrt{p^2 - 4p + 20}}$$

Then,

$$L^{-1}f(p) = L^{-1}\frac{1}{\sqrt{p^2 - 4p + 20}}$$

$$= L^{-1}\frac{1}{\sqrt{(p-2)^2 + 16}}$$

$$= e^{2x}L^{-1}\frac{1}{\sqrt{p^2 + 4^2}}$$

$$= e^{2x}\, J_o(4x)$$

Example 12 Find inverse Laplace transform of

$$f(p) = \frac{1}{\sqrt{2p+5}}$$

Solution: Given that

$$f(p) = L^{-1}\frac{1}{\sqrt{2p+5}}$$

$$L^{-1} f(p) = L^{-1}\frac{1}{\sqrt{2p+5}}$$

$$= \frac{1}{\sqrt{2}} L^{-1}\frac{1}{\sqrt{p+5/2}}$$

$$= \frac{1}{\sqrt{2}} e^{-5t/2} L^{-1}\frac{1}{p^{1/2}}$$

$$= \frac{1}{\sqrt{2}} e^{-5t/2} \frac{t^{-1/2}}{\sqrt{1/2}}$$

$$= \frac{1}{\sqrt{2}} e^{5t/2} \cdot \frac{1}{\sqrt{\pi}} t^{-1/2}$$

$$= \frac{1}{\sqrt{2\pi}} e^{-5t/2} \cdot t^{-1/2}$$ ***Ans.***

Example 13 Find the inverse Laplace transform of

$$f(p) = \frac{4p+15}{p^2+8p+16}$$

Solution: Given that $f(p) = \dfrac{4p+15}{p^2+8p+16}$

Then,

$$L^{-1} f(p) = L^{-1} \frac{4p+15}{p^2+8p+16}$$

$$= L^{-1} \frac{4(p+4)-1}{(p+4^2)}$$

$$= 4L^{-1} \frac{1}{(p+4)} - L^{-1} \frac{1}{(p+4)^2}$$

$$= 4e^{-4t} - L^{-1} \frac{1}{(p+4)^2}$$

$$= 4e^{-4t} - e^{-4t} L^{-1} \frac{1}{p^2}$$

$$= 4e^{-4t} - e^{-4t} \cdot t$$

$$= 4e^{-4t} - t \cdot e^{-4t}$$

$$= 4e^{-4t} (4 - t)$$ ***Ans.***

Example 14 Find inverse Laplace transform of

$$f(p) = \frac{p}{(p+1)^5}$$

Solution: Given that $f(p) = \dfrac{p}{(p+1)^5}$

Then,

$$L^{-1} f(p) = L^{-1} \frac{p}{(p+1)^5}$$

$$= L^{-1} \frac{p+1-1}{(p+1)^5} = L^{-1} \frac{1}{(p+1)^4} - L^{-1} \frac{1}{(p+1)^5}$$

$$= e^{-t} L^{-1} \frac{1}{p^4} - e^{-t} \frac{1}{p^5}$$

$$= e^{-t} \cdot \frac{t^3}{3!} - e^{-t} \cdot \frac{t^4}{4!} = e^{-t} \left(\frac{t^3}{3!} - \frac{t^4}{4!} \right)$$ ***Ans.***

Example 15 Find $L^{-1}\dfrac{1}{(p^2+a^2)^{3/2}}$ and, hence, find $L^{-1}\dfrac{1}{(p^2+2p+5)}$

Solution: We have to find $L^{-1}\dfrac{1}{(p^2+a^2)^{3/2}}$

when $$f(p) = \frac{1}{(p^2+a^2)^{3/2}}$$

We know that $$L[J_0(at)] = \frac{1}{(p^2+a^2)^{1/2}} \qquad \text{...(i)}$$

Differentiating with respect to a, we get

$$\frac{d}{da}(L\,J_0(at)) = \frac{d}{da}\left(\frac{1}{(p^2+a^2)^{1/2}}\right) = -\frac{a}{(p^2+a^2)^{3/2}}$$

$$\Rightarrow \quad L\frac{d}{da}(J_0(at)) = L\,t\,J_0'(at) = -\frac{a}{(p^2+a^2)^{3/2}}$$

$$\Rightarrow \quad L^{-1}\frac{1}{(p^2+2p+5)} = -\frac{t}{a}\,J_0'(at)$$

$$= \frac{t}{a}\,J_1(at)$$

Since $J_0'(at) = -J_1(at)$

Now, $$L^{-1}\frac{1}{(p^2+2p+5)^{3/2}} = L^{-1}\frac{1}{((p+1)^2+2^2)^{3/2}}$$

$$= e^{-t}L^{-1}\frac{1}{(p^2+2p+5)}$$

$$= e^{-t}\frac{t}{2}\,J_1(2t)$$

$$= \frac{-te^{-t}}{2}\,J_1(2t)$$

Ans.

Example 16 Find $L^{-1}\dfrac{(p+1)\,e^{-\pi p}}{p^2+p+1}$.

Solution: We have $L^{-1}\dfrac{p+1}{p^2+p+1}$

$$= L^{-1}\frac{(p+1/2)+1/2}{(p+1/2)^2+\left(\frac{\sqrt{3}}{2}\right)^2}$$

$$= e^{-t/2} L^{-1} \frac{p + 1/2}{p^2 + \left(\frac{\sqrt{3}}{2}\right)^2}$$

$$= e^{-t/2} L^{-1} \frac{p}{p^2 + \left(\frac{\sqrt{3}}{2}\right)^2} + e^{-t/2} \frac{1}{2} L^{-1} \frac{1}{p^2 + \left(\frac{\sqrt{3}}{2}\right)^2}$$

$$= e^{-t/2} \cos \frac{t\sqrt{3}}{2} + \frac{1}{2} e^{-t/2} \cdot \frac{1}{\sqrt{3/2}} \sin \frac{t\sqrt{3}}{2}$$

$$= \frac{e^{-t/2}}{\sqrt{3}} \left(\sqrt{3} \cos \frac{\sqrt{3}t}{2} + \sin \frac{\sqrt{3}t}{2}\right)$$

Hence, $L^{-1} \frac{(p+1)\, e^{-\pi p}}{p^2 + p + 1}$

$$= \begin{cases} \frac{e^{-(t-\pi)/2}}{\sqrt{3}} \left[\sqrt{3} \cos \frac{\sqrt{3}}{2}(t-\pi) + \sin \frac{\sqrt{3}}{2}(t-\pi)\right] & t > \pi \\ 0 \quad , & t < \pi \end{cases}$$

$$= \frac{e^{-(t-\pi)/2}}{\sqrt{3}} \left[\sqrt{3} \cos \frac{\sqrt{3}}{2}(t-\pi) + \sin \frac{\sqrt{3}}{2}(t-\pi)\right] H(t-\pi)$$

Example 17 Find $L^{-1} \frac{e^{-4p}}{(p+2)^3}$.

Solution: We know that

$$L^{-1} \frac{1}{(p+2)^3} = e^{-2t} L^{-1} \frac{1}{p^3}$$

$$= e^{-2t} \cdot \frac{t^2}{2!}$$

$$= \frac{t^2}{2} e^{-2t}$$

$$\Rightarrow \quad L^{-1}\frac{e^{-4p}}{(p+2)^3} = \begin{cases} \frac{1}{2}(t-4)^2 \cdot e^{-2(t-4)} & , t > 4 \\ 0 & , t < 4 \end{cases}$$

$$= \frac{1}{2}(t-4)^2 e^{-2(t-4)} H(t-4)$$

Example 18 Evaluate $L^{-1}\dfrac{e^{-2p}}{\sqrt{p^2+9}}$.

Solution: We know that

$$L^{-1}\frac{1}{(p^2+9)^{1/2}} = J_0(3x)$$

$$\Rightarrow \quad L^{-1}\frac{e^{-2p}}{(p^2+9)^{1/2}} = \begin{cases} J_0(3x-6), & x>2 \\ 0 \quad , & x<2 \end{cases}$$

$$= J_0(3x-6)\, H(x-2)$$

Example 19 Find the inverse Laplace transform of

$$f(p) = \frac{p\, e^{-4\pi\pi/p}}{p^2+25}$$

Solution: We know that $L^{-1}\dfrac{p}{p^2+25} = \cos 5t$

$$L^{-1}\frac{p\, e^{-4\pi/p}}{p^2+25} = \begin{cases} \cos 5(t-4\pi/5), & t>4\pi/5 \\ 0 \quad , & t<4\pi/5 \end{cases}$$

$$= \cos 5t\; H(x-4\pi/5)$$

Example 20 Evaluate $L^{-1}\dfrac{1}{p^3(p-e^{-p})}$

hence, find $F(\pi)$ and $F(4)$.

Solution: We know that

$$f(p) = \frac{1}{p^3(p-e^{-p})} = \frac{1}{p^4}\left(1+\frac{e^{-p}}{p}+\frac{e^{-2p}}{p^2}+\frac{e^{-3p}}{p^3}+\ldots\right)$$

$$= \frac{1}{p^4}+\frac{e^{-p}}{p^5}+\frac{e^{-2p}}{p^6}+\frac{e^{-3p}}{p^7}+\ldots$$

$$= \sum_{k=0}^{\infty} \frac{e^{-kp}}{p^{k+4}}$$

$$\Rightarrow \qquad L^{-1} \frac{1}{p^3(p-e^{-p})} = \sum_{k=0}^{[t]} \frac{(t-k)^{k+3}}{(k+3)!}$$

$$= F(t)$$

Now,
$$F(\pi) = \sum_{k=0}^{3} \frac{(t-k)^{k+3}}{(k+3)!}$$

$$= \frac{\pi^3}{3!} + \frac{(\pi-1)^3}{4!} + \frac{(\pi-2)^3}{5!} + \frac{(\pi-3)^3}{6!}$$

$$= 6.06 \text{ (Approx)}$$

and
$$F(4) = \sum_{k=0}^{4} \frac{(4-k)^{k+3}}{(k+3)!} = \frac{4^3}{3!} + \frac{3^4}{4!} + \frac{2^5}{5!} + \frac{1^6}{6!}$$

$$= 14.32 \text{ (Approx.)}$$

Example 21 Find $L[F(t)]$, where $F(t) = n$, $n \le t < n+1$, $n = 0, 1, 2, \ldots$. Hence, find

$$L^{-1}\left[\frac{e^{-p}}{p(1-re^{-p})}\right], \quad \text{For } r = 1 \text{ and } r \neq 1.$$

Solution: We know that,

$$L[F(t)] = \int_0^{\infty} e^{-pt} F(t)dt \qquad \ldots\text{(i)}$$

$$= \int_0^1 e^{-pt} \cdot 0 \cdot dt + \int_1^2 e^{-pt} \cdot 1 \cdot dt + \int_2^3 e^{-pt} \cdot 2 \cdot dt + \ldots$$

$$= \frac{e^{-p} - e^{-2p}}{p} + 2 \cdot \frac{e^{-2p} - e^{-3p}}{p} + 3\frac{e^{-3p} - e^{-4p}}{p} + \ldots$$

$$= \frac{e^{-p}(1-e^{-p})}{p}(1 + 2e^{-p} + 3e^{-2p} + \ldots) \qquad \ldots\text{(ii)}$$

For a geometric progression,

$$1 + t + t^2 + \ldots = \frac{1}{1-t}, \text{ if } |t| < 1$$

On differentiation,

$$1 + 2t + 3t^2 + ... = \frac{1}{(1-t)^2} \quad ...(iii)$$

by equations (ii) and (iii), we get

$$L[F(t)] = \frac{e^{-p}}{p(1-e^{-p})}$$

and also

$$\frac{e^{-p}}{p(1-re^{-p})} = \frac{e^{-p}}{p}(1 + re^{-p} + r^2 e^{-2p} + ...)$$

$$= \frac{e^{-p}}{p} + \frac{re^{-2p}}{p} + \frac{r^2 e^{-3p}}{p} + ...$$

$$\Rightarrow \quad L^{-1}\left[\frac{e^{-p}}{p(1-re^{-p})}\right] = H(t-1) + rH(t-2) + r^2H(t-3) + ...$$

$$\Rightarrow \quad F(t) = \begin{cases} \sum_{k=1}^{[t]} r^k, & t \geq 1 \\ 0, & t < 1 \end{cases}$$

For $r = 1$

$$L^{-1}\left[\frac{e^{-p}}{p(1-e^{-p})}\right] = n, \quad n \leq t < n + 1, n = 0, 1, ...$$

For $r \neq 1$

$$F(t) = r + r^2 + r^3 + ... = \frac{r(r^n - 1)}{r - 1}, r > 1.$$

Example 22 Evaluate $L^{-1}\left[\dfrac{2p^3 + 10p^2 + 8p + 40}{p^2(p^2+9)}\right]$

Solution: We know that

$$\frac{1}{p^2(p^2+9)} = \frac{1}{9}\left(\frac{1}{p^2} - \frac{1}{p^2+9}\right)$$

$$\Rightarrow \qquad \frac{2p^3+10p^2+8p+40}{p^2(p^2+9)} = \frac{1}{9}\left[\frac{2p^3+10p^2+8p+40}{p^2} - \frac{2p^3+10p^2+8p+40}{p^2+9}\right]$$

$$= \frac{1}{9}\left[2p+10+\frac{8}{p}+\frac{40}{p^2}-\left(2p+10+\frac{-10p-50}{p^2+9}\right)\right]$$

$$= \frac{1}{9}\left[\frac{8}{p}+\frac{40}{p^2}+\frac{10p}{p^2+9}+\frac{50}{p^2+9}\right]$$

$$\Rightarrow \qquad L^{-1}\left[\frac{2p^3+10p^2+8p+40}{p^2(p^2+9)}\right]$$

$$= \frac{1}{9}L^{-1}\left[\frac{8}{p}+\frac{40}{p^2}+\frac{10p}{p^2+9}+\frac{50}{p^2+9}\right]$$

$$= \frac{1}{9}\left[L^{-1}\left(\frac{8}{p}\right)+L^{-1}\left(\frac{40}{p^2}\right)+L^{-1}\left(\frac{10p}{p^2+9}\right)+L^{-1}\left(\frac{50}{p^2+9}\right)\right]$$

$$= \frac{1}{9}\left[8+40t+10\cos 3t+\frac{50}{3}\sin 3t\right]$$ ***Ans.***

Example 23 Evaluate $L^{-1}\left[\log\left(1+\frac{1}{p^2}\right)\right]$.

Solution: Assume $f(p) = \log\left(1+\frac{1}{p^2}\right)$

$$= -\log\frac{p^2}{1+p^2} = -\log p^2 + \log(1+p^2)$$

$$\Rightarrow \qquad f'(p) = -\frac{2}{p}+\frac{2p}{1+p^2}$$

Therefore, $\qquad L^{-1}[f'(p)] = -2 + 2\cos t$

and $\qquad -t\,L^{-1}[f(p)] = -2(1-\cos t)$

$$\Rightarrow \qquad L^{-1}\log\left(1+\frac{1}{p^2}\right) = \frac{2}{t}(1-\cos t)$$ ***Ans.***

Example 24 Show that $L^{-1}\left[\frac{5}{p^2}+\left(\frac{\sqrt{p}-1}{p}\right)^2-\frac{7}{3p+2}\right] = 1+6t-4\sqrt{\frac{t}{\pi}}-\frac{7}{3}e^{-2t/3}$

Solution: We have

$$L^{-1}\left[\frac{5}{p^2}+\left(\frac{\sqrt{p}-1}{p}\right)^2-\frac{7}{3p+2}\right] = L^{-1}\,\frac{5}{p^2}+L^{-1}\left(\frac{p+1-2\sqrt{p}}{p^2}\right)-7L^{-1}\,\frac{1}{3p+2}$$

$$= 6L^{-1}\,\frac{1}{p^2}+L^{-1}\frac{1}{p}-2L^{-1}\,\frac{1}{p^{3/2}}-\frac{7}{3}L^{-1}\,\frac{1}{p+2/3}$$

$$= 6\cdot\frac{t^{2-1}}{1!}+1-2\cdot\frac{t^{3/2-1}}{\Gamma 3/2}-\frac{7}{3}e^{-2t/3}$$

$$= 6t + 1 - 4\,\sqrt{(t/\pi)} - \frac{7}{3}e^{-2t/3}$$

Example 25 Find $L^{-1}\,\dfrac{1}{(p^2+4)}$.

Solution:

$$L^{-1}\left(\frac{1}{p^2+4}\right) = L^{-1}\,\frac{1}{p^2+2^2}$$

$$= \frac{1}{2}\sin 2t$$

Ans.

Example 26 Show that $L^{-1}\left(\dfrac{1}{p}e^{-1/p}\right) = J_0\,(2\sqrt{t})$.

Solution: Given that $f(p) = \dfrac{1}{p}e^{-1/p}$

Using expansion of exponential power, we have

$$\frac{1}{p}e^{-1/p} = \frac{1}{p}\left[1-\frac{1}{p}+\frac{1}{2!}\frac{1}{p^2}-\frac{1}{3!}\frac{1}{p^3}+\ldots\right]$$

$$= \frac{1}{p}-\frac{1}{p^2}+\frac{1}{2!}\frac{1}{p^3}-\frac{1}{3!}\frac{1}{p^4}+\ldots$$

Using linear property or inverting term by term, we have

$$L^{-1}\,\frac{e^{-1/p}}{p} = 1 - t + \frac{1}{2!}\frac{t^2}{2!}-\frac{1}{3!}\frac{t^3}{3!}+\ldots$$

$$= 1 - \frac{(2\sqrt{t})^2}{2^2} + \frac{(2\sqrt{t})^4}{2^2 4^2} - \frac{(2\sqrt{t})^6}{2^2 \cdot 4^2 \cdot 6^2} + \ldots$$

$$= J_0\,(2\sqrt{t})\,.$$

EXERCISE

Find the following

1. $L^{-1}\left(\frac{1}{p^2 - 6p + 10}\right)$. [*Ans.* $e^{3t} \sin t$]

2. $L^{-1}\left[\frac{p}{(p+1)^{3/2}}\right]$. $\left[\textit{Ans.}\quad \frac{2}{3} e^{-t} \sqrt{\frac{t}{\pi}}\,(3 - 2t)\right]$

3. $L^{-1}\left[\frac{p-1}{(p+3)\,(p^2+2p+2)}\right]$. $\left[\textit{Ans.}\quad -\frac{4}{5} e^{-3t} + \frac{1}{5} e^{-t}\,(4\cos t - 3\sin t)\right]$

4. $L^{-1}\left[\frac{1}{(p+a)^n}\right]$. $\left[\textit{Ans.}\quad e^{-at} \cdot \frac{t^{n-1}}{(n-1)!}\right]$

5. $L^{-1}\left(\frac{3p-8}{p^2+4} - \frac{4p-24}{p^2-16}\right)$. [*Ans:* $3\cos 2t - 4\sin 2t - 4\cos h4t + 6\sin h4t$]

6. $L^{-1}\left(\frac{7}{p^2+9} + \frac{5}{p-7} + \frac{1}{2p^{3/2}}\right)$. $\left[\textit{Ans.}\quad \frac{7}{3}\sin 3t + 5e^{7t} + \sqrt{t/\pi}\right]$

7. $L^{-1}\left(\frac{5p+4}{p^3} - \frac{2p-18}{p^2+9} + \frac{24 - 30\sqrt{p}}{p^4}\right)$. $\left[\textit{Ans.}\quad 5t + 2t^2 - 2\cos 3t + 6\sin 3t + 4t^3 - \frac{16}{\sqrt{\pi}} t^{5/2}\right]$

8. Show that $L^{-1}\left[\frac{1}{p}\cos\frac{1}{p}\right] = 1 - \frac{t^2}{(2!)^2} + \frac{t^4}{(4!)^2} - \frac{t^6}{(6!)^2} + \ldots$

9. Show that $L^{-1}\left[\frac{1}{p} J_0\left(\frac{2}{\sqrt{p}}\right)\right] = 1 - \frac{t}{1} + \frac{t^2}{(2!)^3} - \frac{t^3}{(3!)^3} + \ldots.$

10. Find the function $F(t)$ for which

$$F(t) = L^{-1}\left[\frac{3}{p} - \frac{4e^{-p}}{p^2} + \frac{4e^{-3p}}{p^2}\right]$$

$$\left[\textit{Ans.}\quad F(t) = \begin{cases} 3\,, & t < 1 \\ 7 - 4t, & 1 < t < 3 \\ -5\,, & t > 3 \end{cases}\right]$$

11. Prove that

$$L^{-1}\left[\frac{1}{p+e^{-p}}\right] = \sum_{n=0}^{[t]} \frac{(-1)^n\,(t-n)^n}{n!}$$

where $[t]$ denotes the greatest integer less than or equal to t.

12. Find inverse transforms for the given function:

(i) $\dfrac{e^{4-3p}}{(p+4)^{5/2}}$. $\left[\textbf{\textit{Ans.}}\quad \dfrac{4(t-3)^{3/2}\,e^{-4(t-4)}}{3\sqrt{\pi}}\,H(t-3)\right]$

(ii) $\dfrac{p\,e^{-ap}}{p^2-k^2}, a>0$. [***Ans.*** $\cosh k(t-a)\,H(t-a)$]

(iii) $\dfrac{e^{-p}}{\sqrt{p+1}}$. $\left[\textbf{\textit{Ans.}}\quad \dfrac{(t-1)^{-1/2}\,e^{-(t-1)}\,H(t-1)}{\sqrt{\pi}}\,H(t-1)\right]$

(iv) $\dfrac{e^{-5p}}{(p-2)^4}$. $\left[\textbf{\textit{Ans.}}\quad \dfrac{1}{6}(t-5)^3\,e^{2(t-5)}\,H(t-5)\right]$

(v) $\dfrac{1}{(p+a)^{n+1}}, p>-a$. $\left[\textbf{\textit{Ans.}}\quad \dfrac{t^n\,e^{-at}}{\sqrt{(n+1)}}\right]$

(vi) $\dfrac{1}{(8p-27)^{1/3}}$. $\left[\textbf{\textit{Ans.}}\quad \dfrac{t^{-2/3}\,e^{27t/8}}{2\,\Gamma 1/3}\right]$

(vii) $\dfrac{p+1}{2p^2+p+1}$. $\left[\textbf{\textit{Ans.}}\quad \dfrac{e^{-t/4}}{2}\left[\cos\dfrac{\sqrt{7}}{4}t+\dfrac{3}{\sqrt{7}}\sin\dfrac{\sqrt{7}}{4}t\right]\right]$

(viii) $\dfrac{2p-1}{p^2+4p+29}$. [***Ans.*** $e^{-2t}\,(2\cos 5t - \sin 5t)$]

Inverse Laplace Transform of Derivative:

If $$L^{-1}\{f(p)\} = F(t)$$

Then, $$L^{-1}\left(\frac{d^n}{dp^n}f(p)\right) = (-1)^n\,t^n\,F(t)$$

Proof: We have

$$L^{-1}\{f(p)\} = F(t)$$

or $$L\{F(t)\} = f(p)$$

and $$L\{t^n F(t)\} = (-1)^n \frac{d^n}{dp^n} f(p)$$

$$= (-1)^n f^n(p)$$

$\Rightarrow$ $$L^{-1}\{(-1)^n f^n(p)\} = t^n F(t)$$

or $$L^{-1}\{f^n(p)\} = (-1)^n t^n F(t)$$

Inverse Laplace Transform of Integrals:

If $L^{-1}[f(p)] = F(t)$, then

$$L^{-1}\left[\int_p^\infty f(u)du\right] = \frac{F(t)}{t}$$

Proof: We know that

$$L^{-1}[f(p)] = F(t)$$

or $$L[F(t)] = f(p)$$

and $$L\left(\frac{F(t)}{t}\right) = \int_p^\infty f(u)du,$$

if $\lim_{t \to 0}\left(\frac{F(t)}{t}\right)$ exists

$\Rightarrow$ $$L^{-1}\left[\int_p^\infty f(u)du\right] = \frac{F(t)}{t}$$

Multiplication and Division by Powers of *p*:

1. If $L[F(t)] = f(p)$ and $F(0) = 0$, then

$$L^{-1}[f(p)] = f'(t)$$

Proof: We know that

$\Rightarrow$ $$L[F'(t)] = pf(p) - F(0)$$

$$= pf(p) \quad \because F(0) = 0$$

$\Rightarrow$ $$L^{-1}[pf(p)] = F'(t)$$

Note: If $F(0) \neq 0$

Then, $L^{-1}(pf(p) - F(0)) = F'(t)$

$\Rightarrow$ $$L^{-1}[pf(p)] = F'(t) + F(0)\,\delta(t)$$

$\delta(t)$ is the unit impulse function.

(i) $$F^{n-1}(0) = F^{(n-1)}(0) = \dots F''(0) = F'(0) = F(0) = 0$$

Then, $$L^{-1}[p^n f(p)] = F^n(t) = \frac{d^n}{dt^n} F(t)$$

2. If $L^{-1}[f(p)] = F(t)$ and $F(t)$ is sectionally continuous with exponential order K, such that $\lim\limits_{t\to 0} \frac{F(t)}{t}$ exists, then for $p > k$, we have

$$L^{-1}\left[\frac{f(p)}{p}\right] = \int_0^t F(u)du$$

Proof:

Let $$G(t) = \int_0^t F(u)du$$

Then, $$G'(t) = F(t),\ G(0) = 0$$

$$\Rightarrow \quad L(G'(t)) = pL[G(t)] - G(0)$$

$$= pL[G(t)]$$

$$\Rightarrow \quad L[F(t)] = pL(G(t))$$

$$\Rightarrow \quad f(p) = pLG(t)$$

or $$L^{-1}\left[\frac{f(p)}{p}\right] = G(t) = \int_0^t F(u)du.$$

3. If $L^{-1}[f(p)] = F(t)$, then

$$L^{-1}\left[\frac{f(p)}{p}\right] = \int_0^t \int_0^v F(u)du\, dv$$

Proof: Consider

$$G(t) = \int_0^t \int_0^v F(u)du\, dv$$

$$\Rightarrow \quad G'(t) = \int_0^t F(u)du$$

and $$G''(t) = F(t), \text{ if } G(0) = G(0) = 0$$

Now, $$L[G''(t)] = p^2 L[G(t)] - pG(0) - G'(0)$$

$$\Rightarrow \qquad L[G''(t)] = p^2\, L[G(t)], \qquad \because G(0) = G'(0) = 0$$

$$\Rightarrow \qquad L[F(t)] = p^2\, L[G(t)]$$

$$\Rightarrow \qquad F(p) = p^2\, L[G(t)]$$

$$\Rightarrow \qquad \frac{f(p)}{p^2} = L[G(t)]$$

or $$L^{-1}\left[\frac{f(p)}{p^2}\right] = G(t) = \int_0^t \int_0^v F(u)du\, dv \text{ and can be written as}$$

$$L^{-1}\left[\frac{f(p)}{p^2}\right] = \int_0^t \int_0^t F(t)(dt)^2$$

In general,

$$L^{-1}\left[\frac{f(p)}{p^n}\right] = \int_0^t \int_0^t \ldots \int_0^t F(t)(dt)^n$$

Illustrative Examples

Example 1 Find $L^{-1}\left[\dfrac{1}{p^3\,(p^2+1)}\right]$

Solution: We know that

$$L^{-1}\left(\frac{1}{1+p^2}\right) = \sin t$$

Therefore, $$L^{-1}\left[\frac{1}{p}\,\frac{1}{(1+p^2)}\right] = \int_0^t \sin u\, du = 1 - \cos t$$

and $$L^{-1}\left(\frac{1}{p^2\,(1+p^2)}\right) = \int_0^t (1 - \cos u)\, du = t - \sin t$$

$$\Rightarrow \qquad L^{-1}\left(\frac{1}{p^3}\left(\frac{1}{1+p^2}\right)\right) = \int_0^t (u - \sin u)\, du = \frac{1}{2}\, t^2 + \cos t - 1$$

Example 2 Find $L^{-1}\left(\dfrac{1}{p}\log\dfrac{p+2}{p+1}\right)$.

Solution: We know that

$$f(p) = \log\frac{p+2}{p+1}$$

$\therefore$ $$f(p) = \log(p+2) - \log(p+1)$$

$\Rightarrow$ $$f'(p) = \frac{1}{p+2} - \frac{1}{p+1}$$

Now, $$L^{-1}(f'(p)) = e^{-2t} - e^{-t}$$

or $$-t\,L^{-1} f(p) = e^{-2t} - e^{-t}$$

$\Rightarrow$ $$L^{-1} f(p) = \frac{-e^{-2t} + e^{-t}}{t}$$

Hence, $$L^{-1}\left(\frac{1}{p} f(p)\right) = \int_0^t \frac{e^{-u} - e^{-2u}}{u}\,du$$

or $$L^{-1}\left(\frac{1}{p}\log\frac{p+2}{p+1}\right) = \int_0^t \frac{e^{-u} - e^{-2u}}{u}\,du$$ ***Ans.***

Example 3 Find $L^{-1}\left[\dfrac{a^2}{p(p+a)^2}\right]$.

Solution: We know that

$$L^{-1}\left[\frac{a^2}{(p+a)^2}\right] = -a^2 L^{-1} \frac{d}{dp}\left(\frac{1}{p+a}\right)$$

$$= -a^2(-1)\,t\,e^{-at}$$

$$= a^2\,t\,e^{-at}$$

$\Rightarrow$ $$L^{-1}\left[\frac{a^2}{(p+a)^2}\right] = a^2\int_0^t u\,e^{-au}\,du$$

$$= 1 - (at+1)e^{-at}$$ ***Ans.***

Example 4 Find (i) $L^{-1}\log\left(1+\dfrac{1}{p^2}\right)$

(ii) $L^{-1}\left[\dfrac{1}{p}\log\left(1+\dfrac{1}{p^2}\right)\right]$.

Solution: Let us assume $f(p) = \log\left(1+\dfrac{1}{p^2}\right)$

$$= \log \frac{1+p^2}{p^2} = \log(1+p^2) - \log p^2$$

$$\Rightarrow \qquad f'(p) = \frac{2p}{1+p^2} - \frac{2}{p}$$

Therefore, $\quad L^{-1}[f(p)] = -2 + 2\cos t$

and $\quad -t\,L^{-1}[f(p)] = -2(1-\cos t)$

$$\Rightarrow \qquad L^{-1} \log\left(1+\frac{1}{p^2}\right) = \frac{2(1-\cos t)}{t}$$

Since, $\quad L^{-1} \log\left(1+\frac{1}{p^2}\right) = 2(1-\cos t) = F(t)$

$$\Rightarrow \qquad L^{-1}\left[\frac{1}{p}\log\left(1+\frac{1}{p^2}\right)\right] = L^{-1}\left(\frac{1}{p}f(p)\right)$$

$$= \int_0^t F(u)du = 2\int_0^t \frac{(1-\cos t)}{t}dt \qquad \textbf{\textit{Ans.}}$$

Example 5 Find $L^{-1}\left(\frac{p^2}{(p^2+4)^2}\right)$

Solution: Let $\quad f(p) = \frac{p}{(p^2+4)^2}$

Then, $\quad L^{-1}\{f(p)\} = L^{-1}\left\{\frac{p}{(p^2+4)^2}\right\}$

$$= \frac{t}{4}\sin 2t$$

$$= F(t) \text{ (Assume)}$$

Now, $\quad L^{-1}\{p\,f(p)\} = F'(t)$, if $F(0) = 0$

$$\Rightarrow \qquad L^{-1}\left(\frac{p^2}{(p^2+4)^2}\right) = \frac{d}{dt}\left(\frac{t}{4}\sin 2t\right)$$

$$= \frac{1}{4}(\sin 2t + 2t\cos 2t) \qquad \textbf{\textit{Ans.}}$$

Example 6 Find $L^{-1}\dfrac{p+2}{p^2(p+3)}$.

Solution: We know that

$$L^{-1}\frac{p+2}{p^2(p+3)} = L^{-1}\left[\frac{p+3-1}{p^2(p+3)}\right]$$

$$= L^{-1}\left[\frac{1}{p^2}-\frac{1}{p^2(p+3)}\right]$$

$$= L^{-1}\left[\frac{1}{p^2}\right]-L^{-1}\left[\frac{1}{p^2(p+3)}\right]$$

Since, $$L^{-1}\frac{1}{p+3} = e^{-3t} = F(t)$$

$\Rightarrow$ $$L^{-1}\left[\frac{1}{p(p+3)}\right] = \int_0^t e^{-3t}\,dt$$

$$= \frac{1}{3}(1-e^{-3t})$$

and, therefore $$L^{-1}\left[\frac{1}{p^2(p+3)}\right] = \frac{1}{3}\int_0^t (1-e^{-3u})du$$

$$= \frac{1}{3}t+\frac{1}{9}(e^{-3t}-1)$$

Hence, $$L^{-1}\frac{p+2}{p^2(p+3)} = t-\frac{1}{3}t-\frac{1}{9}(e^{-3t}-1)$$

$$= \frac{2}{3}t-\frac{1}{9}e^{-3t}+\frac{1}{9}$$ ***Ans.***

Example 7 Find $L^{-1}\left[\dfrac{1}{p^2(p+1)^2}\right]$.

Solution: We know that

$$L^{-1}\left(\frac{1}{(p+1)^2}\right) = e^{-t}L^{-1}\left(\frac{1}{p^2}\right)$$

$$= t\,e^{-t}$$

Therefore,

$$L^{-1}\left[\frac{1}{p}\cdot\frac{1}{(p+1)^2}\right] = \int_0^t u\,e^{-u}\,du$$

$$= 1 - e^{-t}\,(t + 1)$$

and

$$\left[\frac{1}{p}\cdot\frac{1}{(p+1)^2}\right] = \int_0^t (1 - e^{-u}\,(u - 1))du$$

$$= t\,e^{-t} + 2e^{-t} + t - 2 \qquad \textbf{\textit{Ans.}}$$

Example 8 Find $L^{-1}\left[\log\dfrac{p+\sqrt{1+p^2}}{2p}\right]$.

Solution: We have

$$f(p) = \log\frac{p+\sqrt{1+p^2}}{2p}$$

$$= \log\,(p+\sqrt{1+p^2}) - \log 2p$$

and

$$f'(p) = \frac{1}{p+\sqrt{1+p^2}}\left(1+\frac{1}{2}(1+p^2)^{-1/2}\cdot 2p\right) - \frac{1}{p}$$

$$= \frac{1}{\sqrt{1+p^2}} - \frac{1}{p}$$

Taking inverse Laplace transform,

$$L^{-1}\,[f'(p)] = L^{-1}\,\frac{1}{\sqrt{1+p^2}} - L^{-1}\,\frac{1}{p}$$

$$\Rightarrow \qquad -t\,L^{-1}[f'(p)] = J_0\,(t) - 1$$

$$\Rightarrow \qquad L^{-1}\,\log\frac{p+\sqrt{1+p^2}}{2p} = \frac{1-J_0(t)}{t}$$

Example 9 Find $L^{-1}\left[\dfrac{e^{-p}(1-e^{-p})}{p(1+p^2)}\right]$.

Solution: We have

$$L^{-1}\left[\frac{1}{p^2(p+1)^2}\right] = L^{-1}\left[\frac{1}{p}-\frac{p}{p^2+1}\right]$$

$$= 1 - \cos t$$

Therefore,

$$L^{-1}\left[\frac{e^{-p}(1-e^{-p})}{p(p^2+1)}\right] = L^{-1}\left(\frac{e^{-p}}{p(p^2+1)}-\frac{e^{-2p}}{p(p^2+1)}\right)$$

$$= L^{-1}\left[\frac{e^{-p}}{p(1+p^2)}\right] - L^{-1}\left[\frac{e^{-2p}}{p(1+p^2)}\right]$$

$$= [1 - \cos(t-1)]\, H(t-1) - [1 - \cos(t-2)]\, H(t-2)$$

$$= \begin{cases} 0 & , \quad t<1 \\ 1-\cos(t-1) & , \quad 1<t<2 \\ \cos(t-2)-\cos(t-1), & \quad t>2 \end{cases}$$

EXERCISE

Find the following:

1. $L^{-1}\left[\frac{1}{(p-a)^3}\right]$. $\left[\textbf{\textit{Ans.}}\quad \frac{1}{2}\, e^{at}\, t^2\right]$

2. $L^{-1}\left[\frac{p+1}{(p^2+2p+2)^2}\right]$. $\left[\textbf{\textit{Ans.}}\quad \frac{1}{2}\, te^{-t}\sin t\right]$

3. $L^{-1}\left[\frac{p}{(p^2-a^2)^2}\right]$. $\left[\textbf{\textit{Ans.}}\quad \frac{1}{2a}\, t\sinh at\right]$

4. $L^{-1}\left[\log\left(1-\frac{1}{p^2}\right)\right]$. $\left[\textbf{\textit{Ans.}}\quad \frac{2}{t}\,(1-\cosh t)\right]$

5. $L^{-1}\left[\frac{1}{p(p+1)^3}\right]$. $\left[\textbf{\textit{Ans.}}\quad 1-e^{-t}\left(\frac{1}{2}t^2+t+1\right)\right]$

6. $L^{-1}\left[\frac{1}{(p+1)p^3}\right]$. $\left[\textbf{\textit{Ans.}}\quad \frac{1}{2}t^2-t+1-e^{-t}\right]$

7. $L^{-1}\left[\dfrac{1}{(1+p^2)p^3}\right]$. $\left[\textbf{\textit{Ans.}}\quad -1+\dfrac{1}{2}t^2+\cos t\right]$

8. $L^{-1}\left[\log\dfrac{p+2}{p+1}\right]$. $\left[\textbf{\textit{Ans.}}\quad \dfrac{1}{t}\,(e^{-t}-e^{-2t})\right]$

9. $L^{-1}\left[\log\dfrac{p+3}{p+2}\right]$. $\left[\textbf{\textit{Ans.}}\quad \dfrac{1}{t}\,(e^{-2t}-e^{-3t})\right]$

10. $L^{-1}\left[\dfrac{1}{p}\log\dfrac{p+3}{p+2}\right]$. $\left[\textbf{\textit{Ans.}}\quad \int_0^t \dfrac{1}{u}\,(e^{-2u}-e^{-3u})du\right]$

11. If $L^{-1}\left[\dfrac{p}{(1+p^2)^2}\right]=\dfrac{1}{2}t\sin t$, then find

$L^{-1}\left[\dfrac{1}{(1+p)^2}\right]$. $\left[\textbf{\textit{Ans.}}\quad \dfrac{1}{2}\,(\sin t - t\cos t)\right]$

12. If $L^{-1}\left[\dfrac{1}{(1+p)^2}\right]=t\,e^{-t}$, find $L^{-1}\left[\dfrac{p}{(1+p)^2}\right]$. [***Ans.*** $(1-t)e^{-t}$]

13. Find $L^{-1}\left[\log\left(1+\dfrac{w^2}{p^2}\right)\right]$. $\left[\textbf{\textit{Ans.}}\quad \dfrac{2}{t}\,(1-\cos wt)\right]$

14. Find $L^{-1}\dfrac{1}{p\sqrt{p^2+a^2}}$. $\left[\textbf{\textit{Ans.}}\quad \int_0^t J_0(au)du\right]$

15. If $L^{-1}[f(p)] = F(t)$, prove that,
   (i) $L^{-1}[p\,f'(p)] = -t\,F'(t) - F(t)$
   (ii) $L^{-1}[p\,f''(p)] = t^2\,F'(t) + 2t\,F(t)$
   (iii) $L^{-1}[p^2 f''(p)] = p - t^2\,F''(t) + 4t\,F'(t)$
   (iv) $L^{-1}[p^2 f''(p) + f'(0)] = -t\,F''(t) - 2F'(t)$

3.7 CONVOLUTION OF TWO FUNCTIONS

Let us consider two functions $F(t)$ and $G(t)$ of class A, then the convolution of the two functions, denoted by, $F*G$ defined as (also known as falting of F and G)

$$F*G = \int_0^t F(u)\,G(t-u)du \qquad ...(i)$$

By using the property of definite integral

$$\int_0^x \phi(u)\, du = \int_0^x f(x-u)du \qquad \text{...(ii)}$$

Therefore, equation (i) can be written as

$$F*G = \int_0^t F(t-u)\, G(u)du \qquad \text{...(iii)}$$

Properties of Convolution

(i) Convolution is commutative $\Rightarrow F*G = G*F$

(ii) Convolution is associative $\Rightarrow F*(G*H) = (F*G)*H$

(iii) Convolution is distributive over addition

$\Rightarrow$ (i) $\quad F*(G+H) = F*G + F*H$

Proof:

$$F*G = \int_0^t F(u)\, G(t-u)du$$

Let

$$t-u = y$$

$$-du = dy \Rightarrow du = -dy$$

$$F*G = \int_0^t G(y)\, F(t-y)dy$$

$$= G*F$$

$$\Rightarrow \quad F*G = G*F$$

(ii) $F*G$ is associative, i.e., $(F*G)*H = F*(G*H)$

Suppose

$$f*(g*h)$$

$$= f*\psi$$

Then,

$$\psi = g*h$$

$$= \int_0^t G(u)\, H(t-u)\, du$$

$$= \int_0^t G(t-u)\, H(u)\, du$$

Now,

$$f*\psi = \int_0^t F(v)\, \psi(t-u)\, dv$$

$$= \int_0^t F(v) \left(\int_0^{t-u} G(t-v-u)\, H(u)du \right) dv$$

$$= \int_0^t \int_0^{t-v} F(v)\, G(t-u-v)\, H(u)\, du\, dv$$

$$= \int_0^t \int_0^{t-u} F(v)\, G(t - v - u)\, H(u)\, dv\, du \quad \text{(Change of order)}$$

$$= \int_0^t H(u) \left(\int_0^{t-u} F(v)\, G(t - u - v) dv \right) du$$

$$= h * (f * g)$$

$$= (f * g) * h$$

$$\Rightarrow \qquad f * (g * h) = (f * g) * h$$

(iii) $F*G$ is distributive with respect to addition

$$F*(G + H) = F*G + F*H$$

$$F*(G + H) = \int_0^t F(u)\, [G(t - u) + H(t - u)] du$$

$$= \int_0^t F(u)\, G(t - u) du + \int_0^t F(u)\, H(t - u)] du$$

$$= F*H + F*H$$

Convolution Theorem: For two functions $F(t)$ and $G(t)$ of class A, and $L^{-1}\,[f(p)] = F(t)$, $L^{-1}\,[g(p)] = G(t)$, we have

$$L^{-1}\,[f(p) \cdot g(p)] = \int_0^t F(u)\, G(t - u)] du$$

$$= F*G$$

Proof: First we prove that

$$L\left[\int_0^t F(u)\, G(t-u) du\right] = f(p) g(p) \qquad \text{...(i)}$$

We know that, if

$$H(t) = \int_0^t F(u)\, G(t - u)] du$$

Then $$L[H(t)] = \int_0^t e^{-pt} \left[\int_0^t F(u)\, G(t-u) du\right] dt$$

Using rule for changing integration order,

$$L[H(t)] = \int_0^\infty F(u) \left[\int_0^\infty e^{-pt}\, G(t - u) dt\right] du$$

$$= \int_0^\infty F(u)\left[\int_0^\infty e^{-p(u+y)}\, G(y)dy\right] du$$

$$= \int_0^\infty e^{-pu}\, F(u)\, du \int_0^\infty e^{-py}\, G(y)dy$$

$$= f(p)\, g(p)$$

$\Rightarrow$ $$L^{-1}[f(p)\, g(p)] = H(t)$$

or $$L^{-1}[f(p)\, g(p)] = \int_0^\infty F(u)\, G(t-u)du$$

$$= F*G$$

Heaviside's Expansion Theorem (or Rule)

For any two polynomials in p, $f(p)$ and $g(p)$, where degree of $g(p)$ is more than the degree of $f(p)$. Let $g(p)$ have n distinct roots a_i, $i = 1, 2, 3 \ldots n$, or $g(p) = (p - a_1)(p - a_2) \ldots (p - a_n)$, then,

$$L^{-1}\left[\frac{f(p)}{g(p)}\right] = \sum_{i=1}^{n} \frac{f(a_i)}{g'(a_i)} e^{a_i t}$$

Proof: It is given that degree of $g(p)$ is greater than $f(p)$ and $f(p)$ has n distinct roots, therefore,

$$\frac{f(p)}{g(p)} = \frac{f(p)}{(p-a_1)(p-a_2)\ldots(p-a_n)}$$

$$= \frac{c_1}{p-a_1} + \frac{c_2}{p-a_2} + \ldots \frac{c_i}{p-a_i} + \frac{c_n}{p-a_n}$$

Multiplying both sides by $(p - a_i)$ and applying limit as $p \to a_i$, we get

$$a_i = \lim_{p\to a_i} \frac{f(p).(p-a_i)}{g(p)}$$

$$= f(a_i) \lim_{p\to a_i} \frac{p-a_i}{g(p)}$$

$$= f(a_i) \lim_{p\to a_i} \frac{1}{g'(p)} \quad \text{(using L–H Rule)}$$

Therefore,

$$L^{-1}\left[\frac{f(p)}{g(p)}\right] = \frac{f(a_1)}{g'(a_1)} L^{-1}\left[\frac{1}{p-a_1}\right] + \frac{f(a_2)}{g'(a_2)} L^{-1}\left[\frac{1}{p-a_2}\right]$$

$$+ \ldots + \frac{f(a_i)}{g'(a_i)} L^{-1}\left[\frac{1}{p-a_i}\right] + \ldots + \frac{f(a_n)}{g'(a_n)} L^{-1}\left[\frac{1}{p-a_n}\right]$$

$$= e^{a_1 t}\frac{f(a_1)}{g'(a_1)} + e^{a_2 t}\frac{f(a_2)}{g'(a_2)} + \ldots + e^{a_i t}\frac{f(a_i)}{g'(a_i)} + \ldots + e^{a_n t}\frac{f(a_n)}{g'(a_n)}$$

$$= \sum_{i=1}^{n} \frac{f(a_i)}{g'(a_i)} e^{a_i t}$$

Illustrative Examples

Example 1 Use convolution theorem to find $L^{-1}\left[\dfrac{p^2}{(p^2+a^2)^2}\right]$.

Solution: We know that $L^{-1}\left(\dfrac{p}{p^2+a^2}\right) = \cos at$

Using convolution formula,

$$L^{-1}\left[\frac{p^2}{(p^2+a^2)^2}\right] = L^{-1}\left[\frac{p}{p^2+a^2}\cdot\frac{p}{p^2+a^2}\right]$$

$$= \int_0^t \cos au \cos a\,(t-u)]du$$

$$= \int_0^t \cos au\,(\cos at \cos au - \sin at \sin au)du$$

$$= \cos at \int_0^t \cos^2 au\, du + \sin at \int_0^t \cos au \sin au\, du$$

$$= \frac{1}{2}\cos at \int_0^t (1 + \cos 2au)du + \frac{1}{2}\sin at \int_0^t \sin 2\,au\, du$$

$$= \frac{1}{2}\cos at \left[u + \frac{1}{2a}\sin 2au\right]_0^t + \frac{1}{2}\sin at\left(-\frac{1}{2a}\cos 2au\right)_0^t$$

$$= \frac{1}{2}\cos at\left[t + \frac{1}{2a}\sin 2at\right] + \frac{1}{4a}\sin at\,(1-\cos 2at)$$

$$= \frac{1}{2} t \cos at + \frac{1}{4a} \sin at + \frac{1}{4a} (\sin 2at \cos at - \sin at \cos 2at)$$

$$= \frac{1}{2} t \cos at + \frac{1}{4a} [\sin at + \sin(2at - at)]$$

$$= \frac{1}{2a} [at \cos at + \sin at] \qquad \textit{Ans.}$$

Example 2 Use of convolution formula to find $L^{-1}\left[\frac{1}{p^2 (p+1)^2}\right]$.

Solution: We know that

$$L^{-1}\left(\frac{1}{p^2}\right) = t$$

and $$L^{-1}\left(\frac{1}{(p+1)^2}\right) = t e^{-t}$$

Using convolution formula, we have

$$L^{-1}\left[\frac{1}{p^2} \cdot \frac{1}{(p+1)^2}\right] = \int_0^t (u e^{-u} (t - u) du$$

$$= \int_0^t (ut - u^2) e^{-u} du$$

$$= [(ut - u^2)(-e^{-u}) - (t - 2u)e^{-u} + (-2)(-e^{-u})]_0^t$$

$$= t - 2 + 2 e^{-t} + t e^{-t} \qquad \textit{Ans.}$$

Example 3 Using convolution formula to find $L^{-1}\left[\frac{1}{p(p^2+4)^2}\right]$.

Solution: We know that

$$\frac{1}{p(p^2+4)^2} = \frac{1}{p^2} \cdot \frac{p}{(p^2+4)^2}$$

and $$L^{-1}\left(\frac{1}{p^2}\right) = t$$

$$L^{-1}\left(\frac{p}{(p^2+4)^2}\right) = \frac{t}{4}\sin 2t$$

Now, using convolution formula,

$$L^{-1}\left[\frac{1}{p(p^2+4)^2}\right] = \int_0^t \left(\frac{1}{4}u\sin 2u\right)(t-u)du$$

$$= \frac{t}{4}\int_0^t u\sin 2u\,du - \frac{1}{4}\int_0^t u^2\sin 2u\,du$$

$$= \frac{t}{4}\left[-\frac{u}{2}\cos 2u + \frac{1}{4}\sin 2u\right]_0^t - \frac{1}{4}\left[\frac{-u^2}{2}\cos 2u + \frac{u\sin 2u}{2} + \frac{\cos 2u}{4}\right]_0^t$$

$$= \frac{t}{4}\left[-\frac{t}{2}\cos 2t + \frac{1}{4}\sin 2t\right] - \frac{1}{4}\left[-\frac{t^2}{2}\cos 2t + \frac{t\sin 2t}{2} + \frac{\cos 2t}{4} - \frac{1}{4}\right]$$

$$= \frac{1}{16}[1 - \cos 2t - t\sin 2t]$$ ***Ans.***

Example 4 Apply Heaviside expansion formula to find $L^{-1}\left[\frac{p+1}{(p+1)(p^2+1)}\right]$.

Solution: Given that, on comparison,

$$f(p) = p + 5$$

$$g(p) = (p+1)(p^2+1) = (p+1)(p+i)(p-i)$$

and $$g'(p) = 3p^2 + 2p + 1 \text{ and}$$

$$a_1 = -1$$

$$a_2 = -i$$

$$a_3 = i$$

Then,

$$L^{-1}\left[\frac{p+1}{(p+1)(p^2+1)}\right] = \frac{f(-1)}{g'(-1)}e^{-t} + \frac{f(-i)}{g'(-i)}e^{-it} + \frac{f(i)}{g'(i)}e^{it}$$

$$= \frac{4}{2}e^{-t} + 2\,\mathrm{Re}\left(\frac{f(i)}{g'(i)}.e^{it}\right)$$

$$= 2\,e^{-t} + 2\,\mathrm{Re}\left[\frac{5+i}{2i-2}e^{it}\right]$$

$$= 2e^{-t} - \text{Re}\left[\frac{(5+i)(II+i)}{2}\,e^{it}\right]$$

$$= 2e^{-t} - \frac{1}{2}\,\text{Re}\,[(4 + 6i)(\cos t + i \sin t)]$$

$$= 2e^{-t} + 2 \cos t + 3 \sin t$$

Example 5 Find $L^{-1}\left[\dfrac{19p+37}{(p+1)(p-2)(p+3)}\right]$ by using Heaviside expansion.

Solution: On comparison, we have

$$f(p) = 19p + 37$$

$$g(p) = (p + 1)\,(p - 2)\,(p + 3)$$

$\Rightarrow$ $$g'(p) = 3p^2 + 4p - 5$$

and $$a_1 = -1$$

$$a_2 = 2$$

$$a_3 = -3$$

Using Heaviside expansion formula, we get

$$L^{-1}\left[\frac{19p+37}{(p+1)(p-2)(p+3)}\right] = \frac{f(-1)}{g'(-1)}\,e^{-t} + \frac{f(2)}{g'(2)}\,e^{2t} + \frac{f(-3)}{g'(-3)}\,e^{-3t}$$

$$= \frac{18}{-6}\,e^{-t} + \frac{75}{15}\,e^{2t} + \frac{-20}{10}\,e^{-3t}$$

$$= -3\,e^{-t} + 5\,e^{2t} - 2\,e^{-3t}$$ ***Ans.***

Example 6 Apply Heaviside's expansion to find $L^{-1}\left[\dfrac{p}{(p^2-2p+2)\,(p^2+2p+2)}\right]$.

Solution: On comparison with standard result

$$f(p) = p$$

$$g(p) = (p^2 - 2p + 2)\,(p^2 + 2p + 2)$$

$$= p^4 + 4$$

$\Rightarrow$ $$g'(p) = 4p^3$$

and $$a_1 = 1 + i$$

$$a_2 = 1 - i$$

$$a_3 = -(1 + i)$$

$$a_4 = -(1 - i)$$

Therefore, by Heaviside's expansion, we get

$$L^{-1}\frac{f(1+i)}{g'(1+i)}e^{(1+i)t}+\frac{f(1-i)}{g'(1-i)}e^{(1-i)t}+\frac{f(-1-i)}{g'(-1-i)}e^{-(1+i)t}+\frac{f(-1+i)}{g'(-1+i)}e^{-(-1+i)}$$

$$= 2\text{Re}\left[\frac{f(1+i)}{g'(1+i)}e^{(1+i)t}+\frac{f(-1+i)}{g'(-1+i)}e^{-(1-i)t}\right]$$

$$= 2\text{Re}\left[\frac{1+i}{4(1+i)^3}e^{(1+i)t}+\frac{1-i}{4(1-i)^3}e^{-(1-i)t}\right]$$

$$= \frac{1}{4}\text{ Re }[-i\,e^t\cdot e^{it} + ie^{-t}\,e^{it}]$$

$$= \frac{1}{4}\text{ Re }[ie^{it}\,(e^{-t} - e^t)]$$

$$= -\frac{1}{2}\text{ Re }[i(\cos t + i\sin t)\sinh t]$$

$$= \frac{1}{2}\sinh t\sin t$$

Example 7 Apply Heaviside's expansion formula to find $L^{-1}\left[\dfrac{2p^2-6p+5}{p^3-6p^2+11p-6}\right]$.

Solution: On comparison

$$f(p) = 2p^2 - 6p + 5$$
$$g(p) = p^3 - 6p^2 + 11p - 6$$
$$= (p-1)(p-2)(p-3)$$

$\Rightarrow$ $$g'(p) = 3p^2 - 12p + 11$$

and $$a_1 = 1,\ a_2 = 2,\ a_3 = 3$$

On applying Heaviside's expansion formula, we get

$$L^{-1}\left[\frac{2p^2-6p+5}{p^3-6p^2+11p-6}\right] = \frac{f(1)}{g'(1)}e^t+\frac{f(2)}{g'(2)}e^{2t}+\frac{f(3)}{g'(3)}e^{3t}$$

$$= \frac{1}{3}e^t - e^{2t} + \frac{5}{2}e^{3t}$$ ***Ans.***

Example 8 Apply Heaviside's expansion to find $L^{-1}\left[\dfrac{2p+3}{(p+1)^2\,(p+2)^2}\right]$.

Solution: Using, Heaviside's expansion formula for multiple roots,

$$L^{-1}\left[\frac{2p+3}{(p+1)^2\,(p+2)^2}\right] = (a_1t + a_2)e^{-t} + (b_1t + b_2)e^{-2t}$$

where
$$a_1 = \lim_{p\to-1} \frac{2p+3}{(p+2)^2} = 1$$

$$a_2 = \lim_{p\to-1} \frac{d}{dp}\left(\frac{2p+3}{(p+2)^2}\right) = 0$$

$$b_1 = \lim_{p\to-2} \frac{2p+3}{(p+1)^2} = -1$$

$$b_2 = \lim_{p\to-2} \frac{d}{dp}\left(\frac{2p+3}{(p+1)^2}\right) = 0$$

Hence,

$$L^{-1}\left(\frac{2p+3}{(p+1)^2\,(p+2)^2}\right) = t\,.e^{-t} + (-t)\,e^{-2t} = t(e^{-t} - e^{-2t})$$ ***Ans.***

Example 9 Apply Heaviside's expansion to find $L^{-1}\left(\frac{2p^2-9p+19}{(p-1)^2\,(p+3)}\right)$.

Solution: By using Heaviside's expansion, we have

$$L^{-1}\left(\frac{2p^2-9p+9}{(p-1)^2\,(p+3)}\right) = e^t(a_1t + a_2) + b_1\,e^{-3t}$$

and we can calculate a_1, a_2, b_1 as follows:

$$a_1 = \lim_{p\to1} \frac{2p^2-9p+19}{(p+3)} = 3$$

$$a_2 = \lim_{p\to1} \frac{d}{dp}\left(\frac{2p^2-9p+19}{p+3}\right) = -2$$

$$b_1 = \frac{f(-3)}{g'(-3)}$$

$$= \frac{64}{16} = 4$$

$$L^{-1}\left(\frac{2p^2-9p+19}{(p-1)^2\,(p+3)}\right) = (3t-2)\,e^t + 4e^{-3t}$$ ***Ans.***

Example 10 Using Heaviside's expansion formula to find $L^{-1}\left[\dfrac{2p^2+5p-4}{p^3+p^2-2p}\right]$.

Solution: Given that

$$f(p) = 2p^2 + 5p - 4$$

$$g(p) = p^3 + p^2 - 2p$$

$$= p(p-1)(p+2)$$

$$g'(p) = 3p^2 + 2p - 2$$

Now, $\qquad g(p) = 0 \Rightarrow p = 0, 1, -2$

$$\text{Thus, } L^{-1}\left(\frac{2p^2+5p-4}{p^3+p^2-2p}\right) = \frac{f(0)}{g'(0)}e^{0t} + \frac{f(1)}{g'(1)}e^{t} + \frac{f(-2)}{g'(-2)}e^{-2t}$$

$$= 2 + e^t - e^{-2t} \qquad \textbf{\textit{Ans.}}$$

Example 11 Using Heaviside's expansion formula to find $L^{-1}\left(\dfrac{3p+1}{(p-1)(p^2+1)}\right)$.

Solution: Given that

$$f(p) = 3p + 1$$

$$g(p) = (p-1) + (p^2+1) \Rightarrow p = 1, -i, i$$

$$\Rightarrow \qquad g'(p) = 3p^2 - 2p + 1$$

$$L^{-1}\left(\frac{3p+1}{(p-1)(p^2+1)}\right) = \frac{f(1)}{g'(1)}e^{t} + \frac{f(i)}{g'(i)}e^{it} + \frac{f(-i)}{g'(-i)}e^{-it}$$

$$= \frac{4}{2}e^t + \frac{(3i+1)}{-(2+2i)}e^{it} + \frac{(-3i+1)}{(-2+2i)}e^{-it}$$

$$= 2e^t - \frac{(3i+i)(1-i)}{2(1+i)(1-i)}e^{it} + \frac{(3i-1)(1+i)}{2(1-i)(1+i)}e^{-it}$$

$$= 2e^t - \frac{1}{2}i\,(e^{it} - e^{-it}) - (e^{it} + e^{-it})$$

$$= 2e^t - \sin t - 2\cos t \qquad \textbf{\textit{Ans.}}$$

EXERCISE

1. Using convolution theorem to find the Laplace inverse of the following functions:

(a) $\dfrac{1}{(p-1)(p+3)}$. $\left[\textbf{Ans.}\ \dfrac{1}{4}(e^{t}-e^{-3t})\right]$

(b) $\dfrac{1}{(p-2)(p+1)}$. $\left[\textbf{Ans.}\ \dfrac{1}{3}(e^{2t}-e^{-t})\right]$

(c) $\dfrac{1}{(p-1)(p+2)}$. $\left[\textbf{Ans.}\ \dfrac{1}{3}(e^{t}-e^{-2t})\right]$

(d) $\dfrac{p+2}{p^2-6p+8}$. $\left[\textbf{Ans.}\ 3e^{4t}-2e^{2t}\right]$

(e) $\dfrac{1+2p}{(p+2)^2(p-1)^2}$. $\left[\textbf{Ans.}\ \dfrac{1}{3}(e^{t}-e^{-2t})\right]$

(f) $\dfrac{1}{p^2+ap}$. $\left[\textbf{Ans.}\ \dfrac{1}{a}\ (1-e^{-at})\right]$

(g) $\dfrac{1}{(p+1)^2(p^2+4)}$. $\left[\textbf{Ans.}\ \dfrac{2}{25}e^{-t}+\dfrac{1}{5}\ t\,.\,e^{-t}-\dfrac{2}{25}\cos 2t-\dfrac{3}{50}\sin 2t\right]$

(h) $\dfrac{p^2-6}{p^3+4p^2+3p}$. $\left[\textbf{Ans.}\ -2+\dfrac{5}{2}\ e^{-t}+\dfrac{1}{2}\ e^{-3t}\right]$

(i) $\dfrac{2p^2-1}{(p^2+1)(p^2+4)}$. $\left[\textbf{Ans.}\ -\sin t+\dfrac{3}{2}\ \sin 2t\right]$

(j) $\dfrac{4p+5}{(p-1)^2(p+2)}$. $\left[\textbf{Ans.}\ 3t\ e^{t}+\dfrac{1}{3}e^{t}-\dfrac{1}{3}e^{-2t}\right]$

(k) $\dfrac{p^2}{(p^2+a^2)(p^2+b^2)}, ab\neq 0, a^2\neq b^2$. $\left[\textbf{Ans.}\ \dfrac{a\sin at-b\sin bt}{a^2-b^2}\right]$

(l) $\dfrac{1}{(p-2)(p+2)^2}$. $\left[\textbf{Ans.}\ \dfrac{1}{16}[e^{2t}-(4t+1)\ e^{2t}]\right]$

(m) $\dfrac{16}{p(p^2+4)^2}$. $\left[\textbf{Ans.}\ 1-\cos 2t-t\sin 2t\right]$

(n) $\dfrac{p^2-3}{(p+2)(p-3)(p^2+2p+5)}$.

$$\left[\textbf{Ans.}\quad \frac{3}{50}e^{3t}-\frac{1}{25}e^{-2t}\,\frac{1}{50}e^{-t}\cos 2t+\frac{9}{25}e^{-t}\sin 2t\right]$$

2. Using Heaviside's expansion formula, Find:

(i) $L^{-1}\dfrac{p+5}{(p+1)(p^2+1)}$. $\left[\textbf{Ans.}\quad 2e^{-t}-2\cos t+3\sin t\right]$

(ii) $L^{-1}\dfrac{p^2-6}{p^3+4p^2+3p}$. $\left[\textbf{Ans.}\quad -2+\frac{5}{2}e^{-t}+\frac{1}{2}e^{-3t}\right]$

(iii) $L^{-1}\dfrac{19p+37}{(p+1)(p-2)(p+3)}$. $\left[\textbf{Ans.}\quad -3e^{-t}+5e^{2t}-2e^{-3t}\right]$

(iv) $L^{-1}\dfrac{2p^2-6p+5}{p^3-6p^2+11p-6}$. $\left[\textbf{Ans.}\quad \frac{1}{2}e^{t}-e^{2t}+\frac{5}{2}e^{3t}\right]$

(v) $L^{-1}\dfrac{3p+1}{(p-1)(p^2+1)}$. $\left[\textbf{Ans.}\quad 2e^{t}-2\cos t+\sin t\right]$

(vi) $L^{-1}\dfrac{2p^2-4}{(p+1)(p-2)(p-3)}$. $\left[\textbf{Ans.}\quad -\frac{1}{6}e^{-t}-\frac{4}{3}e^{2t}+\frac{7}{2}e^{3t}\right]$

(vii) $L^{-1}\dfrac{2p-11}{(p+2)(p-3)}$. $\left[\textbf{Ans.}\quad 3e^{-2t}-e^{3t}\right]$

(viii) $L^{-1}\dfrac{p-1}{(p+3)(p^2+2p+2)}$. $\left[\textbf{Ans.}\quad \frac{1}{5}e^{-t}(4\cos t-3\sin t)-\frac{4}{5}e^{-3t}\right]$

(ix) $L^{-1}\left[\dfrac{2p^3-p^2-1}{(p+1)^2(p^2+1)^2}\right]$. $\left[\textbf{Ans.}\quad \frac{1}{2}\sin t+\frac{1}{2}t\cos t-te^{-t}\right]$

(x) $L^{-1}\left[\dfrac{11p^3-47p^2+56p+4}{(p+2)(p-2)^3}\right]$. $\left[\textbf{Ans.}\quad (2t^2-t+5)e^{2t}+6e^{-2t}\right]$

Chapter 4

Laplace Transforms (Complex Variables)

4.1 FOURIER INTEGRAL THEOREM

Let $F(t)$ satisfy the Dirichlet conditions in every finite interval $-\alpha \le t \le \alpha$ and if $\int_{-\infty}^{\infty} |F(t)|\,dt$ converges then at each point of continuity t of $f(t)$.

$$f(t) = \frac{1}{2\pi}\int_{-\infty}^{\infty} du \int_{-\infty}^{\infty} F(v)\cos[u(t-v)]du \qquad ...(i)$$

Let us consider a point of discontinuity t, then left-hand side of equation (i) substituted by $\frac{1}{2}$ $[F(t + 0) + F(t - 0)]$ or the mean value of $F(t)$ at the point of discontinuity. This is sufficient condition.

Now, $\sin[u(t - v)]$ is an odd function of u,

$$\Rightarrow \qquad 0 = \frac{1}{2\pi}\int_{-\infty}^{\infty} du \int_{-\infty}^{\infty} F(v)\sin[u(t-v)]dv \qquad ...(ii)$$

By equations (i) and (ii), we get

$$f(t) = \frac{1}{2\pi}\int_{-\infty}^{\infty} du \int_{-\infty}^{\infty} F(v)\, e^{iu(t-u)} dv$$

$$\Rightarrow \qquad F(t) = \frac{1}{2\pi}\int_{-\infty}^{\infty} e^{iut} du \int_{-\infty}^{\infty} F(v)\, e^{-iuv} dv \qquad ...(iii)$$

Equation (iii) is known as complex form of Fourier integer.

4.2 COMPLEX INVERSION FORMULA

Let us consider a function $F(t)$ has a continuous derivative and is of exponential order k for large positive values of t, where $k > 0$ and if $f(p) = L\{F(t)\}$ then, $L^{-1}\{f(p)\}$ can be written as

$$F(t) = \frac{1}{2\pi i}\int_{k-i\infty}^{k+i\infty} e^{pt}f(p)dp,\ t > 0 \quad \text{...(i)}$$

$$= 0,\ t < 0$$

Solution: If $\int_{-\infty}^{\infty} G(t)dt$ is absolutely convergent and if $G(t)$ has continuous derivative. Then $G(t)$ can be written as

$$G(t) = \frac{1}{\pi}\int_{-\infty}^{\infty} e^{iut}\,du \int_{-\infty}^{\infty} G(v)e^{-iuv}\,dv \quad \text{...(ii)}$$

Assume that
$$G(t) = \begin{cases} e^{-kt}F(t) & , \ t>0 \\ 0 & , \ t<0 \end{cases}$$

Hence, $F(t)$ is of exponential order k and $\int_{-\infty}^{\infty} G(t)dt$ is absolutely convergent. Thus, all the conditions of Fourier's integral theorem are confirmed by equation (ii), we have (for $t > 0$)

$$e^{-kt}F(t) = \frac{1}{2\pi}\int_{-\infty}^{\infty} e^{iut}\left[\int_{-\infty}^{\infty} e^{-kv}F(v)e^{-iuv}dv\right]du,\ k > t$$

$$= \frac{1}{2\pi}\int_{-\infty}^{\infty} e^{iut}\ [e^{-(k+iu)v}F(u)dv]du$$

$$= \frac{1}{2\pi}\int_{-\infty}^{\infty} e^{iut}\,f(k + iu)du \quad [\text{Since } L\{F(t)\} = f(p)]$$

$$= \frac{1}{2\pi}\int_{k-i\infty}^{k+i\infty} \frac{1}{e}e^{t(p-k)}f(p)dp, \text{ by assuming } k + iu = p$$

$$\Rightarrow \quad du = \frac{1}{i}dp$$

$$\Rightarrow \quad F(t) = \frac{1}{2\pi}\int_{k-i\infty}^{k+i\infty} e^{pt}f(p)dp,\ t > 0$$

In this solution, we consider that $e^{-kv}F(v)$ is absolutely integrable in $(0, \infty)$ or $\int_0^\infty e^{-kv}|F(v)|dv$ converges, for applying Fourier's integral theorem.

Note: If $f(p)$ is an analytic function of the complex variable p and of order $0(p^{-n})$ in some half-plane $\text{Re}(p) \geq k$, here k and n are real number constants and $n \geq 1$. Then the integral

$$I = \frac{1}{2\pi i} \lim_{w \to \infty} \int_{c-iw}^{c+iw} e^{px} f(p)dp$$

along any line $\text{Re}(p) = c > k$ converges to a function $f(t)$ which is independent of c and whose Laplace transform is $f(p)$, $\text{Re}(p) > k$.

The function $F(t)$ is continuous for $t > 0$ and $0(e^{ct})$ as $t \to \infty$.

4.3 THE BROMWICH CONTOUR

We know that the integral, evaluated in the section of the complex plane, is evaluated by assuming the contour integral

$$\frac{1}{2\pi i}\int_C e^{pt} f(p)dp$$

C is the contour given in Fig. 4.1, known as Bromwich contour and consists of

(a) line AB, and

(b) an arc $BDEGA$ with radius r and origin at O.

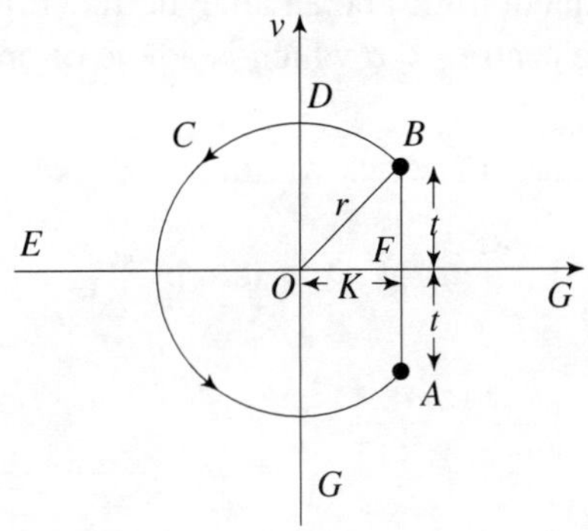

Fig. 4.1

Let us consider the arc $BDEGA$ is denoted by C_1, then we get

$$\oint_C e^{pt} f(p)dp = \int_{k-iL}^{k-iL} e^{pt} f(p)dp + \int_{C_1} f(p)dp$$

$$\Rightarrow \quad \frac{1}{2\pi i}\int_{k-iL}^{k-iL} e^{pt} f(p)dp = \frac{1}{2\pi i}\oint_C e^{pt} f(p)dp - \frac{1}{2\pi i}\oint_{C_1} e^{pt} f(p)dp$$

as $L \to \infty$ or $R \to \infty$, $L^2 + k^2 = r^2$ and using complex inversion formula, we get

$$F(t) = \lim_{r\to\infty}\left[\frac{1}{2\pi i}\oint_C e^{pt} f(p)dp - \frac{1}{2\pi i}\int_{C_1} e^{pt} f(p)dp\right]$$

4.4 FUNDAMENTAL CONCEPTS OF COMPLEX VARIABLE THEORY

(a) Analytic Functions: A function f of the complex variable z is analytic at a point z_0 if it has a derivative at each point in some neighbourhood of z_0. It can be stated as if f is analytic at point z_0, it must be analytic at each point in some neighbourhood of z_0.

(b) Singular Point or Singularity: The point at which function fails to be analytic is called singular point or singularity of $f(z)$. If a function $f(z)$ is analytic everywhere in some region except at an interior point $z = c_1$, the point $z = c_1$ is said to be isolated singularity of $f(z)$.

(c) Poles: Let us consider a function $F(z) = \dfrac{f(z)}{(g-a)^n}$ where $f(z)$ is analytic everywhere in a region also at $z = a$. For positive integer n. $z = a$ is called isolated singularity and also known as pole of order n, for $n = 1$, pole is called simple pole while for $n = 2$, it is called double pole.

For a function $F(z) = \dfrac{3z-1}{(z-2)^2(z+5)(z-1)}$

$z = 1, -5$ is a simple pole, $z = 2$ is called double pole.

Residue at pole: Consider a function $F(z)$ be an analytic function within a circle c of radius r and centre $z = a$, except at the centre $z = a$ which is a pole of order n. Thus, $F(z)$ is analytic within the annulus $0 < 1\ z - a| < r$.

So, it can be expanded as Laurent's series within $0 < |z - a| < r$.

$$F(z) = \sum_{n=0}^{\infty} a_n(z-a)^n + \sum_{n=1}^{\infty} b_n(z-a)^{-n}$$

where $$b_n = \frac{1}{2\pi i}\int_c \frac{F(z)dz}{(z-a)^{-n+1}}$$

and $$b_1 = \frac{1}{2\pi i}\int_c F(t)dz$$

here, b_1 is known as the residue of $F(z)$ at the pole $z = a$.

Residue at a finite pole: Any—

(i) Residue of a given function $f(z)$ at a simple pole $z = a_1$ can be defined as

$$\lim_{z\to a_1} (z - a_1) f(z)$$

If $f(z)$ can be choosen such that

$$f(z) = \frac{F(z)}{\phi(z)}$$

where, $\phi(a_1) \neq 0$ and $F(z)$ has a simple zero at $z = a_1$, then residue of the given function, $f(z)$ at $z = a_1$, is equal to $F(a_1)/\phi'(a_1)$.

(ii) If $f(z)$ has a pole of order k which is greater than one ($k > 1$) at $z = a_1$ the residue of $f(z)$ at $z = a_1$ can be given as

$$\lim_{z \to a_1} \frac{1}{(k-1)!} \frac{d^{k-1}}{d^{k-1}} [(z - a_1)^k F(z)/\phi(z)]$$

Let $f(z) = \dfrac{F(z)}{(z-a)^k}$ and $F(a_1) \neq 0$. Then we can calculate residue of $f(z)$ at $z = a_1$ by $\dfrac{F^{k-1}(a_1)}{(k-1)!}$.

(iii) **Cauchy's Residue Theorem:** Let us consider a function $f(z)$ which is analytic inside and on a simple closed contour (C) except at a finite number of poles $z = c_1, c_2, \ldots c_n$ inside C, having residue $R_1, R_2, \ldots, R_n$ respectively, then

$$\int_C f(z)dz = 2\pi i(R_1, R_2 + \ldots + R_n)$$

Theorem: Suppose L denotes the curved portion '*ABCDEF*' of the Bromwich contour with equation $p = Se^{i\theta}$, $\theta_0 < \theta < 2\pi - \theta_0$ or L is the arc of a circle of radius S with centre at the origin. Let us consider on L, we have

$$|F(p) = \frac{M}{S^m}$$

where $m > 0$ and M are constant, then show that

$$\lim_{S \to \infty} \int e^{bx} F(p)dp = 0$$

Solution: Suppose L_1, L_2, L_3, and L_4 denote arcs AB, BD, DE and EF respectively, then

$$\int_L e^{px} F(p)dp = \int_{L_1} e^{px} F(p)dp + \int_{L_2} e^{px} F(p)dp + \int_{L_3} e^{px} F(p)dp + \int_{L_4} e^{px} F(p)dp$$

Case 1: Along L_1 or AB.

Along arc AB, we have

$$p = Se^{i\theta}, \theta_0 \leq \theta \leq \pi/2$$

Then,

$$I_1 = \int_{C_1} e^{px} F(p)dp$$

$$= \int_{\theta_0}^{\pi/2} e^{se^{i\theta}x} \cdot F(Se^{i\theta})\, iSe^{i\theta}\, d\theta$$

$$\Rightarrow \qquad |I_1| = \int_{\theta_0}^{\pi/2} |e^{(s\cos\theta)x}|\,|e^{i(s\sin\theta)x}|\,|F(Se^{i\theta})|\,|iSe^{i\theta}|d\theta$$

$$\leq \int_{\theta_0}^{\pi/2} |e^{(s\cos\theta)x}|\,|F(Se^{i\theta})|\,sd\theta$$

$$\leq \frac{M}{S^{m-1}} \int_{\theta_0}^{\pi/2} |e^{(s\cos\theta)x}\,d\theta$$

Since, $|f(p)| \leq \dfrac{M}{S^m}$ on $L_1 = \dfrac{M}{S^{m-1}} \displaystyle\int_0^{\alpha_0} |e^{(s\sin\alpha)x}\,d\alpha$

and $\qquad \theta = \pi/2 - \alpha$

Here, $\qquad \alpha_0 = \pi/2 - \theta_0 = \sin^{-1} \dfrac{r}{s})$

$$= \frac{M}{S^{m-1}} \int_0^{\alpha_0} e^{(s\sin\alpha)2}\,d\alpha$$

Then, $\dfrac{M}{S^{m-1}} \displaystyle\int_0^{\alpha_0} |e^{(s\sin\alpha)x}\,d\alpha \leq \dfrac{M}{S^{m-1}} \displaystyle\int_0^{\alpha_0} e^{rx}\,d\alpha$

$$= \frac{Me^{rx}\alpha_0}{S^{m-1}} = \frac{Me^{rx}}{S^{m-1}} \sin^{-1}\left(\frac{r}{R}\right)$$

When $\sin^{-1} \dfrac{r}{S} \simeq \dfrac{r}{S}$, if S is larger $S \to \infty$, then

$$\lim_{S\to\infty} I_1 = 0$$

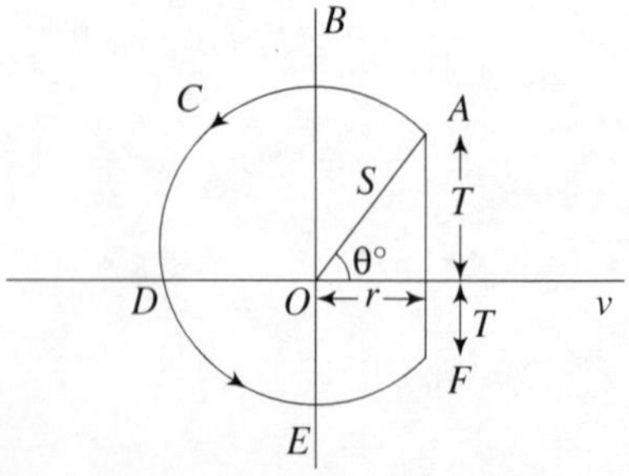

Fig. 4.2

Case 2: Along L_2 (or BCD)

Along BCD $\quad p = Se^{i\theta}, \pi/2 \le \theta \le \pi$

$$I_2 = \int_{L_2} e^{p2} F(p)\, dp$$

$$I_2 = \int_{\pi/2}^{\pi} e^{Se^{i\theta}x} \cdot F(Se^{i\theta})\, i \cdot Se^{i\theta}\, d\theta$$

Then,
$$|I_2| \le \frac{M}{S^{m-1}} \int_{\pi/2}^{\pi} e^{(s\sin\theta)x}\, d\theta$$

$$\le \frac{M}{S^{m-1}} \int_0^{\pi/2} e^{-(s\sin\theta)x}\, d\alpha, \; (0 \le \alpha \le \pi/2 \sin\alpha \ge 2\alpha)$$

$$|I_2| \le \frac{M}{S^{m-1}} \int_0^{\pi/2} e^{-2s\alpha 2/\pi}\, d\alpha$$

$$= \frac{\pi M}{2xS^m}(1 - e^{-Sx})$$

As $S \to \infty$, we have $\lim_{S\to\infty} I_2 = 0$

and similarly, $\quad \lim_{S\to\infty} I_3 = 0$

and $\quad \lim_{S\to\infty} I_4 = 0$

Then $\lim_{S\to\infty} \int_L e^{px} F(p)dp = 0$

Use of residue theorem in obtaining inverse Laplace transform.

Theorem: Assume that the only singularities of $F(p)$ are poles which all lie to the left of the line $\text{Re}(p) = r$ for some real constant, r. Further assume that $F(p)$ satisfies the condition

$$|F(p)| < \frac{M}{S^m}$$

and $\quad \lim_{S\to\infty} \int_L e^{px} F(p)\, dp = 0$

Then prove that the inverse Laplace transform of $F(p)$ can be given as $f(x)$ = Sum of residues of $e^{px}F(p)$ at all the poles of $F(p)$.

Example 1 Find the Laplace transform of the function

$$f(x) = \frac{a^{ax} - 1}{a}$$

and prove that by the method of contour integration that $f(x)$ is itself the inverse Laplace transform of the function arrived at.

Solution: Given that,

$$f(x) = \frac{a^{ax}-1}{a}$$

Then,

$$Lf(x) = L\left[\frac{1}{a}(e^{ax}-1)\right]$$

$$= \frac{1}{a}L(e^{ax}) - \frac{1}{a}L\,(1)$$

$$= \frac{1}{a}\left[\frac{1}{p-a}-\frac{1}{p}\right] = \frac{1}{p(p-a)}$$

Now,

$$L^{-1}\left[\frac{1}{p(p-a)}\right]$$

$$= \frac{1}{2\pi i}\int_{r-i\infty}^{r+i\infty}\frac{e^{px}}{p(p-a)}\,dp$$

$$= \frac{1}{2\pi i}\oint\frac{e^{px}}{p(p-a)}\,dp$$

$$= \text{Sum of residue of } \frac{e^{px}}{p(p-a)} \text{ at poles } p = 0,\ a$$

Now, residue at simple pole $p = 0$

$$\lim_{p\to 0} p\left[\frac{e^{px}}{p(p-a)}\right]$$

$$= \lim_{p\to 0}\frac{e^{px}}{p-a} = -\frac{1}{a}$$

and residue at simple pole $p = a$ is

$$\lim_{p\to a}(p-a)\frac{e^{px}}{p(p-a)}$$

$$= \lim_{p\to a}\frac{e^{px}}{p} = \frac{e^{ax}}{a}$$

$$\Rightarrow \qquad L^{-1}\left[\frac{1}{p(p-a)}\right] = -\frac{1}{a}+\frac{e^{ax}}{a}$$

$$= \frac{e^{ax}-1}{a} = f(x)$$

Example 2 Using the complex inversion formula to calculate

$$L^{-1}\left[\frac{p}{(p-1)^2(p+1)^3}\right].$$

Solution: Given that

$$f(p) = \frac{p}{(p-1)^2(p+1)^3}$$

Let $\qquad p = Se^{i\theta}$

$$|F(p)| = \left|\frac{p}{(p-1)^2(p+1)^3}\right| < \frac{32}{S^4} \quad (S > 2)$$

Then, $\qquad L^{-1}\left[\frac{p}{(p-1)^2(p+1)^3}\right]$

$$= \frac{1}{2\pi i}\int_{r-i\infty}^{r+\infty}\frac{pe^{px}}{(p-1)^2(p+1)^3}\,dp$$

$$= \frac{1}{2\pi i}\oint_C \frac{pe^{px}}{(p-1)^2(p+1)^3}\,dp$$

$$= \text{Sum of residue of } \frac{pe^{px}}{(p+1)^3(p-1)^2}$$

at poles $p = -1, 1$.

(i) Residue at pole $p = -1$ of order 3 is $\lim\limits_{p\to -1}\frac{1}{2!}\frac{d^2}{dp^2}\left[(p+1)^3\frac{pe^{px}}{(p+1)^3(p-1)^2}\right]$

$$= \frac{1}{2}\lim_{p\to -1}\frac{d^2}{dp^2}\left[\frac{pe^{px}}{(p-1)^2}\right]$$

$$= \frac{1}{16}e^{-t}\,(1-2t^2)$$

(ii) Residue at pole $p = 1$ of order 2 is $\lim_{p\to 1} \frac{1}{1!}\frac{d}{dp}\left[(p-1)^2 \frac{pe^{pt}}{(p-1)^2(p+1)^3}\right]$

$$= \lim_{p\to 1} \frac{d}{dp}\left[\frac{pe^{px}}{(p+1)^3}\right]$$

$$= \frac{1}{16} e^t (2t - 1)$$

$$\Rightarrow \quad L^{-1}\left[\frac{p}{(p+1)^3(p-1)^2}\right] = \frac{1}{16} e^{-t} (1 - 2t^2) + \frac{1}{16} e^t (2t - 1)$$

Example 3 Using complex inversion formula to obtain inverse Laplace transform of

$$\frac{p}{(p+1)(p-1)^2}$$

Solution: Given that

$$f(p) = \frac{p}{(p+1)(p-1)^2}$$

Let us consider $p = Se^{i\theta}$, then

$$\left|\frac{p}{(p+1)(p-1)^2}\right| = \frac{|p|}{|p+1||p-1|^2}$$

$$= \frac{|Se^{i\theta}|}{|Se^{i\theta}+1||Se^{i\theta}-1|^2} \le \frac{S}{(|Se^{i\theta}|-1)(|Se^{i\theta}|-1)^2} \le \frac{S}{(S-1)^3}$$

$$= \frac{1}{S^2\left(1-\frac{1}{S^2}\right)} < \frac{8}{S^2}, S > 2$$

Then, we have

$$L^{-1}\left(\frac{p}{(p+1)(p-1)^2}\right) = \frac{1}{2\pi i}\int_{r-i\infty}^{r+i\infty} e^{pt}\frac{p}{(p+1)(p-1)^2}dp$$

$$= \frac{1}{2\pi i}\oint_c \frac{pe^{pt}}{(p+1)(p-1)^2}dp$$

$= \text{Sum of residue of } e^{px}\frac{p}{(p+1)(p-1)^2}$ at the simple pole $p = -1$ and double pole $p = 1$.

(i) Residue at simple pole $p = -1$ is

$$\lim_{p \to -1} (p+1) \frac{pe^{pt}}{(p+1)(p-1)^2}$$

$$= -\frac{1}{9}e^{-t}$$

(ii) Residue at the touble pole $p = 1$ is

$$\lim_{p \to 1} \frac{1}{1!} \frac{d}{dp} \left[(p-1)^2 . \frac{pe^{pt}}{(p-1)^2 (p+1)} \right]$$

$$= \lim_{p \to 1} \frac{d}{dp} \left[\frac{pe^{pt}}{p+1} \right]$$

$$= \frac{(1+2x)e^t}{4}$$

Then, $$L^{-1} \left[\frac{p}{(p-1)^2 (p+1)} \right] = \frac{1}{4}(1+2t)e^t - \frac{1}{9}e^t.$$

EXERCISE

Find the inverse Laplace transform using the complex inversion formula:

1. (i) $\dfrac{p^2+3}{(p^2-2p+5)(p+1)}$. $\left[\textbf{Ans. } \dfrac{1}{2} e^x(\cos 2x + \sin 2x) + \dfrac{1}{2} e^{-x} \right]$

(ii) $\dfrac{p^2}{(p^2+4)^2}$. $\left[\textbf{Ans. } \dfrac{1}{2} x \cos 2x + \dfrac{1}{4} \sin 2x \right]$

(iii) $\dfrac{p}{p^2+a^2}$. [**Ans.** $\cos ax$]

(iv) $\dfrac{1}{(p-2)(p+3)(p-4)}$. $\left[\textbf{Ans. } \dfrac{1}{14}e^{4x} - \dfrac{1}{10}e^{2x} + \dfrac{1}{35}e^{-3x} \right]$

(v) $\dfrac{1}{p^3(p^2+1)}$. $\left[\textbf{Ans. } \cos x - 1 + \dfrac{1}{2}x^2 \right]$

2. (i) $\dfrac{2}{(p-1)^2(1+p)^2}$. [***Ans.*** $\cos x + e^x(x-1)$]

(ii) $\dfrac{3p-1}{p(p+1)(p-1)^2}$. [***Ans.*** $x\,e^x - 1 - e^{-x}$]

(iii) $\dfrac{1}{(p+a)(p-b)^2}$. $\left[\textbf{\textit{Ans.}}\ \dfrac{e^{-ax}}{(a+b)^2} - \dfrac{e^{bx}}{(a+b)^2} + \dfrac{xe^{bx}}{a+b}\right]$

(iv) $\dfrac{1}{(p+1)(p-2)^2}$. $\left[\textbf{\textit{Ans.}}\ \dfrac{1}{3}xe^{2x} - \dfrac{1}{9}e^{2x} + \dfrac{1}{9}e^{-x}\right]$

Chapter 5

Applications of Laplace Transform to Differential Equations

5.1 INTRODUCTION

Laplace transform is an important tool to solve various types of differential equations developed while solving scientific and technical problems by using their mathematical modelling. The process of getting solution of a differential equation has three main sleps:

(i) Taking Laplace transform of the given differential equation.

(ii) Convert resultant into algebraic or simple equations and applying initial/boundary conditions.

(iii) Applying inverse Laplace transform to get the solutions.

5.2 SOLUTION OF LINEAR DIFFERENTIAL EQUATIONS WITH CONSTANT COEFFICIENTS

Let us consider a linear differential equation with constant coefficient

$$\frac{d^n y}{dx^n} + a_1 \frac{d^{n-1} y}{dx^{n-1}} + a_2 \frac{d^{n-2} y}{dx^{n-2}} + \ldots + a_{n-1} \frac{dy}{dx} + a_n y = f(x) \qquad \ldots\text{(i)}$$

and satisfying conditions

$$y = y_0,\ \frac{dy}{dx} = b_1,\ \frac{d^2 y}{dx^2} = b_2 \ldots \text{ etc. at } x = 0 \qquad \ldots\text{(ii)}$$

Now, applying Laplace transform on equation (i) both sides, we have

$$L\left\{\frac{d^n y}{dx^n}\right\} + a_1 L\left\{\frac{d^{n-y}}{dx^{n-1}}\right\} + \ldots + a_{n-1} L\left\{\frac{dy}{dx}\right\} + a_n L\{y\} = L\{F(x)\} \qquad \ldots\text{(iii)}$$

Let us assume $L\{F(x)\} = f(p)$

and $L\{y\} = Y$

Then, we have

$$p^n y - [p^{n-1}y_0 + p^{n-2}b_1 + ... + pb_{n-2} + b_{n-1}]$$

$$+ a_1[p^{n-1}y - (p^{n-2}y_0 + p^{n-3}b_1 + ... + b_{n-2})]$$

$$+ ... + a_{n-1}(pY - y_0) + a_n y = f(p)$$

or $$(p^n + a, p^{n-1} + ... + a_{n-1}\, p + a_n)\, y$$

$$= f(p) + a_{n-1}\, y_0 + a_{n-2}(py_0 + b_1)$$

$$+ ... + (p^{n-1}y_0 + p^{n-2}b_1 + ... + b_{n-1}) \qquad ...(iv)$$

Equation (iv) is known as subsidiary equation from which we can find y and the required solution can be obtained by taking inverse Laplace transform.

Example 1 Solve $(D^2 + 9) = \cos 2t$, if $y(0) = 1$, $y(\pi/2) = -1$

Solution: Given differential equation is

$$(D^2 + 9)\, y = \cos 2t$$

or $$y'' + 9y = \cos 2t \qquad ...(i)$$

Applying Laplace transform on both sides of the given equation,

$$L(y'') + 9L(y) = L \cos 2t$$

$$p^2 L(y) - py(0) - y'(0) + 9L(y) = \frac{p}{p^2+4}$$

$$(p^2 + 9)\, L(y) = \frac{p}{p^2+4} + p + y'(0)$$

$$L(y) = \frac{p+y'(0)}{p^2+9} + \frac{p}{(p^2+9)(p^2+4)}$$

$$L(y) = \frac{p}{p^2+9} + \frac{y'(0)}{p^2+9} + \frac{p}{5(p^2+4)} - \frac{p}{5(p^2+9)}$$

Taking Laplace inverse and replace $y'(0) = c_1$

$$y = L^{-1}\left(\frac{p}{p^2+9}\right) + c_1 L^{-1}\left(\frac{1}{p^2+9}\right) + \frac{1}{5}L^{-1}\left(\frac{p}{p^2+4}\right) - \frac{1}{5}L^{-1}\left(\frac{p}{p^2+9}\right)$$

$$= \cos 3t + \frac{1}{3}c_1 \sin 3t + \frac{1}{5}\cos 2t - \frac{1}{5}\cos 3t$$

$$= \frac{4}{5}\cos 3t + \frac{1}{3}c_1 \sin 3t + \frac{1}{5}\cos 2t$$

but, $y(\pi/2) = -1$ (given)

Then, $$-1 = \frac{4}{5}\cos\frac{3\pi}{2} + \frac{1}{3}c_1 \sin\frac{3\pi}{2} + \frac{1}{5}\cos\pi$$

or $$-1 = -\frac{1}{3}c_1 - \frac{1}{5} \Rightarrow c_1 = \frac{12}{5}$$

Therefore, $$y = \frac{4}{5}\cos 3t + \frac{4}{5}\sin 3t + \frac{1}{5}\cos 2t$$ ***Ans.***

Example 2 Solve $y''(t) + y(t) = t$,

Given that $y'(0) = 1$, $y(\pi) = 0$.

Solution: Given differential equation

$$y''(t) + y(t) = t, \quad \text{...(i)}$$

Applying Laplace transform both sides,

$$L\, y''(t) + L\, y(t) = L\{t\},$$

$$\Rightarrow p^2 L(y) - p\, y(0) - y'(0) + L\,(y) = \frac{1}{p^2}$$

$$(p^2 + 1)\, L(y) = \frac{1}{p^2} + p\,(y(0)) + y'(0)$$

$$(p^2 + 1)\, L(y) = \frac{1}{p^2} + p\, y(0) + y'(0)$$

$$= \frac{1}{p^2} + 1 + p\, y(0)$$

$$= \frac{1}{p^2}\; 1 + pc = \frac{1+p^2}{p^2} + pc$$

$$L(y) = \frac{1}{p^2} + \frac{p}{p^2+1}c$$

Taking Laplace inverse,

$$y = L^{-1}(1/p^2) + cL^{-1}\left(\frac{p}{p^2+1}\right)$$

$$y = t + c\cos t$$

Now, $y(\pi) = 0$, then

$$0 = \pi + c\cos\pi \quad \Rightarrow \quad c = \pi$$

$\Rightarrow$
$$y = t + \pi\cos t$$

Example 3 Solve $(D^2 + 1)\, y = t\cos 2t,\ y(0) = 0$

$$y'(0) = 0$$

Solution: Given differential equation,

$$(D^2 + 1)\, y = t\cos 2t$$

or
$$y'' + y = t\cos 2t \qquad \text{...(i)}$$

Taking Laplace transform both sides,

$$L(y'') + L(y) = L(t\cos 2t)$$

$$\Rightarrow\ p^2L(y) - py(0) - y'(0) + L(y) = -\frac{d}{dp}L(\cos 2t)$$

$$(p^2 + 1)\, L(y) - py(0) - y'(0) = -\frac{d}{dp}\left(\frac{p}{p^2+4}\right)$$

$$(p^2 + 1)\, L(y) - 0 - 0 = -\frac{1}{p^2+4} + \frac{2p^2}{(p^2+4)^2}$$

or
$$L(y) = \frac{p^2-4}{(p^2+1)(p^2+4)^2}$$

$$= -\frac{5}{9}\cdot\frac{1}{p^2+1} + \frac{5}{9}\cdot\frac{1}{(p^2+4)} + \frac{8}{3}\cdot\frac{1}{(p^2+4)^2} \qquad \text{...(ii)}$$

Taking inverse Laplace transform of equation (ii), we get

$$y = -\frac{5}{9}L^{-1}\left(\frac{1}{p^2+1}\right) + \frac{5}{9}L^{-1}\left(\frac{1}{(p^2+4)}\right) + \frac{8}{3}L^{-1}\left(\frac{1}{(p^2+4)^2}\right)$$

$$= -\frac{5}{9}\sin t + \frac{5}{18}\sin 2t + \frac{8}{3}\int_0^t \frac{1}{2}(\sin 2x)\frac{1}{2}\sin 2(t-x)\,dx$$

(Using convolution theorem for third term in right hand side)

$$= -\frac{5}{9}\sin t + \frac{5}{18}\sin 2t + \frac{1}{3}\left[-\frac{1}{4}\sin(2t-4x) - x\cos 2t\right]_0^t$$

$$= -\frac{5}{9}\sin t + \frac{5}{18}\sin 2t - \frac{1}{3}t\cos 2t + \frac{1}{12}\sin 2t + \frac{1}{12}\sin 2t$$

$$= -\frac{5}{9}\sin t + \frac{4}{9}\sin 2t - \frac{1}{3}t\cos 2t \qquad \textbf{\textit{Ans.}}$$

Example 4 Solve $(D^2 + m^2)y = a \cos nt$, $t > 0$ y, Dy equal to y_0 and y_1 when y $t = 0$, $n = m$

Solution: Given differential equation

$$(D^2 + m^2)y = a \cos nt \quad \text{...(i)}$$

Taking Laplace transform both sides of equation (i), we get

$$L(y'') + m^2 L(y) = a L \cos nt$$

$$\Rightarrow \quad p^2L(y) - py(0) - y'(0) + m^2 L(y) = \frac{ap}{p^2+n^2}$$

$$(p^2 + m^2) L(y) = py_0 + y_1 + \frac{ap}{p^2+n^2}$$

$$L(y) = \frac{p}{p^2+m^2}y_0 + \frac{1}{p^2+m^2}y_1 + \frac{ap}{(p^2-n^2)}\left(\frac{p}{p^2+n^2} - \frac{p}{p^2+m^2}\right) \quad \text{...(ii)}$$

Applying inverse Laplace of equation (ii), we get

$$y = y_0L^{-1}\left\{\frac{p}{(p^2+m^2)}\right\} + y_1L^{-1}\frac{1}{(p^2+m^2)} + a\cdot\frac{1}{(m^2-n^2)}\left[L^{-1}\left\{\frac{p}{p^2+m^2}\right\} - L^{-1}\frac{p}{p^2+m^2}\right]$$

or
$$y = y_0 \cos mt + \frac{y_1}{m}\sin mt + \frac{a}{m^2-n^2}(\cos nt - \cos mt) \quad \textit{Ans.}$$

Example 5 Solve $(D + 1)^2 y = t$. Given that $y(0) = -3$, $y(0) = -1$.

Solution: Given differential equation

$$(D + 1)^2y = t$$

or
$$(D^2 + 2D + 1)y = t \quad \text{...(i)}$$

Applying Laplace transform on equation (i), we get

$$L(y'') + 2L(y') + L(1) = L(t)$$

$$p^2L(y) - py(0) - y'(0) + 2[pL(y) - y(0)] + L(y) = \frac{1}{p^2}$$

$$\Rightarrow \quad (p^2 + 2p + 1) L(y) - p(-3) - c_1 - 2(-3) = \frac{1}{p^2}, \quad y'(0) = c_1 \text{ (Assume)}$$

$$(p+1)^2L(y) = \frac{1}{p^2} - 3p + c_1 - 6$$

$$L(y) = \frac{1}{p^2(p+1)^2} - \frac{3p+6}{(p+1)^2} + \frac{c_1}{(p+1)^2}$$

$$= -\frac{2}{p} + \frac{1}{p^2} + \frac{2}{p+1} + \frac{1}{(p+1)^2} - \frac{3}{p+1} - \frac{3}{(p+1)^2} + \frac{c_1}{(p+1)^2}$$

$$= -\frac{2}{p} - \frac{1}{(p+1)} + \frac{1}{p^2} + \frac{c_1 - 2}{(p+1)^2} \qquad \text{...(ii)}$$

Applying inverse Laplace on equation (ii), we get

$$y = -2 \cdot L^{-1}\left(\frac{1}{p}\right) - L^{-1}\left(\frac{1}{(p+1)}\right) + L^{-1}\left(\frac{1}{p^2}\right) + (c_1 - 2)L^{-1}\left(\frac{1}{(p+1)^2}\right)$$

$$= -2(1) - e^{-t} + t + (c_1 - 2)\, e^{-t}\, L^{-1}\left(\frac{1}{p^2}\right)$$

$$= -2 - e^{-t} + t + (c_1 - 2)t \cdot e^{-t}$$

Now, putting $y(1) = -1$, we get

$$-1 = -2 - e^{-1} + 1 + (c_1 - 2) \cdot 1 \cdot e^{-1}$$

$$\Rightarrow \qquad c_1 = 3$$

Then, $\qquad y = -2 + t - e^{-t} + t \cdot e^{-t}$ ***Ans.***

Example 6 Solve $(D^4 + 2D^2 + 1)y = 0$, Given that $y(0) = 0$, $y'(0) = 1$, $y''(0) = 2$, $y'''(0) = -3$

Solution: Given differential equation,

$$(D^4 + 2D^2 + 1)y = 0 \qquad \text{...(i)}$$

Applying Laplace transform on equation (i), we get

$$L(y^{\text{iv}}) + 2L(y'') + L(y) = 0$$

$$\Rightarrow \quad p^4 L(y) - p^3 y(0) - p^2 y'(0) - py''(0) - y'''(0) + 2\,[p^2 L(y) - py(0) - y'(0)] + L(y) = 0$$

$$\Rightarrow \quad (p^4 + 2p^2 + 1)\, L(y) = p^2 + 2p - 3 + 2\,(+1)$$

$$= p^2 + 2p - 1$$

or

$$L(y) = \frac{p^2 + 2p - 1}{(p^2+1)^2}$$

$$= \frac{1}{(p^2+1)} + \frac{2p - 1}{(p^2+1)^2}$$

$$\Rightarrow \qquad L(y) = \frac{1}{(p^2+1)} - \frac{2}{(p^2+1)^2} + \frac{2p}{(p^2+1)^2} \qquad \text{...(ii)}$$

Applying inverse Laplace transform on equation (ii), we get

$$y = L^{-1}\left(\frac{1}{(p^2+1)}\right) - 2 \cdot L^{-1}\left(\frac{1}{(p^2+1)^2}\right) + 2 \cdot L^{-1}\left(\frac{p}{(p^2+1)^2}\right) \qquad \text{...(iii)}$$

But $$L^{-1}\frac{1}{(p^2+1)^2}=L^{-1}\left(\frac{1}{p^2+1}\cdot\frac{1}{p+1}\right)=F(t)*F(t)$$

Here, $$F(t)=L^{-1}\left(\frac{1}{p^2+1}\right)=\sin t$$

$$=\int_0^t \sin x \sin(t-x)dx$$

$$=\frac{1}{2}\int_0^t ((\cos 2x-t)-\cos t)dx$$

$$=\frac{1}{2}\left[\frac{1}{2}\sin(2x-t)-x\cos t\right]_0^t$$

$$=\frac{1}{2}(\sin t-t\cos t)$$

and

$$L^{-1}\left(\frac{2p}{p^2+1}\right)=-L^{-1}\left[\frac{d}{dp}\left(\frac{1}{p^2+1}\right)\right]$$

$$=-(-t)L^{-1}\left(\frac{1}{p^2+1}\right)$$

$$=t\sin t$$

Therefore, solution to the equation (i), is

$$y=\sin t-(\sin t-t\cos t)+t\sin t$$

$$\Rightarrow \qquad y=t(\sin t+\cos t)$$

Example 7 Solve $(D^4-1)y=1$.

Given that $y(0)=y'(0)=y''(0)=y'''(0)=0$.

Solution: Given that

$$(D^4-1)y=1. \qquad \text{...(i)}$$

Applying Laplace transform on equation (i), we get

$$L(y^{iv})-L(y)=1.$$

$$\Rightarrow \qquad p^4L(y)-p^3y(0)-p^2y'(0)-py''(0)-y'''(0)-L(y)=\frac{1}{p}$$

On putting values of $y'(0)=y''(0)=y'''(0)=0$,

We get

$$(p^4 - 1)\, L(y) = \frac{1}{p}$$

$$L(y) = \frac{1}{p(p^4-1)}$$

$$= \frac{1}{p(p^2-1)(p^2+1)}$$

$$\Rightarrow \qquad L(y) = -\frac{1}{p} + \frac{p}{2(p^2-1)} + \frac{p}{2(p^2+1)}$$

Applying Laplace inverse transformations, we get

$$y = -L^{-1}\left(\frac{1}{p}\right) + \frac{1}{2}L\left(\frac{p}{p^2-1}\right) + \frac{1}{2}L^{-1}\left(\frac{p}{p^2+1}\right)$$

$$\Rightarrow \qquad y = -1 + \frac{1}{2}(\cosh t + \cos t).$$ ***Ans.***

Example 8 Solve $(D^3 - 2D^2 + 5D)y = 0$

Given that $y'(0) = 0$, $y(\pi/8) = 1$.

Solution: Given that $(D^3 - 2D^2 + 5D)y = 0$ ...(i)

Applying Laplace transform both sides of equation (i), we get

$L(y''') - 2\, L(y'') + 5\, L(y') = 0$

$[p^3 L(y) - p^2 y(0) - py''(0) - y'''(0)] + 2[p^2 L(y) - py(0) - y'(0)] + 5[pL(y) - y(0)] = 0$

Putting the values of $y'(0) = 1$, $y'(0) = 0$ and $y''(0) = c_1$, we get

$p^3 L(y) - p - c_1 - 2(p^2 L(y) - 1) + 5p\, L(y) = 0$

$\Rightarrow \quad (p^3 - 2p^2 + 5p)\, L(y) = c_1 - 2 + p$

or

$$L(y) = \frac{c_1 - 2}{p(p^2-2p+5)} + \frac{p}{p(p^2-2p+5)}$$

$$= \frac{c_1 - 2}{p(p^2-2p+5)} + \frac{1}{p^2-2p+5}$$

$$= \frac{c_1 - 2}{5}\left[\frac{1}{p} - \frac{p-2}{p^2-2p+5}\right] + \frac{1}{(p-1)^2+2^2} \qquad \text{...(ii)}$$

Applying inverse Laplace transform, we get

$$y = \frac{c_1 - 2}{5}\left[L^{-1}\left(\frac{1}{p}\right) - L^{-1}\left(\frac{p-2}{p^2-2p+5}\right)\right] + L^{-1}\frac{1}{(p-1)^2+2^2}$$

$$= \frac{c_1 - 2}{5}\left[L^{-1}\left(\frac{1}{p}\right) - L^{-1}\left(\frac{(p-1)-1}{(p-1)^2+2^2}\right)\right] + e^t L^{-1}\frac{1}{p^2+2^2}$$

$$= \frac{c_1 - 2}{5}\left[1 - e^t L^{-1}\left(\frac{p-1}{p^2+2^2}\right)\right] + e^t \frac{1}{2}\sin 2t$$

$$= \frac{c_1 - 2}{5}\left[1 - e^t\left(\cos 2t + \frac{1}{2}\sin 2t\right)\right] + e^t \cdot \frac{1}{2}\sin 2t$$

$$= \frac{c_1 - 2}{5} - e^t\left[\frac{c_1 - 2}{5}\cos 2t - \frac{c_1 + 3}{10}\sin 2t\right]$$

Putting $y(\pi/8) = 1$, we get

$$L = \frac{c_1 - 2}{5} - e^{\pi/8}\left[\frac{1}{\sqrt{2}}\left(\frac{c_1 - 2}{5} - \frac{c_1 + 3}{10}\right)\right]$$

$\Rightarrow \qquad c_1 = 7$ **Ans.**

$\therefore \qquad y = 1 - e^t(\cos 2t - \sin 2t).$

Example 9 Solve $(D^2 + n^2)y = a \sin(nt + \alpha)$, $y'(0) = y(0) = 0$.

Solution: Given differential equation

$$(D^2 + n^2)y = a \sin(nt + \alpha) \qquad \text{...(i)}$$

Applying Laplace transform on the equation (i), we get

$$L(y'') + n^2 L(y) = a L \sin(nt + \alpha)$$

$$\Rightarrow \quad p^2 L(y) - py(0) - y'(0) + n^2 L(y) = a L(\sin nt \cos\alpha + \cos nt \sin\alpha)$$

$$= a \cos\alpha\, L(\sin nt) + a \sin\alpha\, L(\cos nt)$$

$$\Rightarrow \quad (p^2 + n^2) L(y) = (a\cos\alpha)\frac{n}{p^2+n^2} + (a\sin\alpha)\frac{p}{p^2+n^2}$$

$$\Rightarrow \quad L(y) = \frac{an\cos\alpha}{(p^2+n^2)^2} + \frac{(a\sin\alpha)p}{(p^2+n^2)^2} \qquad \text{...(ii)}$$

Applying inverse Laplace transform on equation (ii), we get

$$y = (an\cos\alpha)L^{-1}\left(\frac{1}{(p^2+n^2)^2}\right) + (a\sin\alpha)L^{-1}\left(\frac{p}{(p^2+n^2)^2}\right) \qquad \text{...(iii)}$$

But, $L^{-1}\left(\frac{p}{(p^2+n^2)^2}\right) = \frac{t}{2n}\sin nt$

and $$L^{-1}\left(\frac{p}{(p^2+n^2)^2}\right) = \frac{1}{2n}\left[\frac{-t}{n}\cos nt + \frac{1}{n^2}\sin nt\right]$$

Then by equation (iii), we have

$$y = \frac{an\cos\alpha}{2n}\left[-\frac{t}{n}\cos nt + \frac{1}{n^2}\sin nt\right] + (a\sin\alpha)\frac{t}{2n}\sin nt$$

or $$y = \frac{a}{2n^2}[\sin nt\cos\alpha - nt\cos(nt+\alpha)]$$ ***Ans.***

EXERCISE

Solve the following differential equations by using Laplace transform:

1. $(D^2 + 4)y = 9t$; $y(0) = 0$, $y'(0) = 7$. $\left[\textbf{Ans.}\quad y = \frac{9}{4}t + \frac{19}{8}\sin t\right]$
2. $(D^2 + 1)y = 0$; $y(0) = 1 = y'(0)$. [***Ans.*** $y = \sin t + \cos t$]
3. $(D^2 + 2D + 5)y = e^{-t}\sin t$; $y(0) = 0$, $y'(0) = 1$. $\left[\textbf{Ans.}\quad y = \frac{1}{3}e^{-t}(\sin 2t + \sin t)\right]$
4. $(D^2 - 3D + 2)y = 1 - e^{2t}$; $y(0) = 1$, $y'(0) = 0$. $\left[\textbf{Ans.}\quad y = \frac{1}{2} + \frac{1}{2}e^{2t} - t\,e^{2t}\right]$
5. $(D^2 + D + 1)y = 3\,e^t$, $y(0) = 0 = y'(0)$. $\left[\textbf{Ans.}\quad e^t - e^{-t/2}\left(\cos\frac{t\sqrt{3}}{2} + \sqrt{3}\sin\frac{t\sqrt{3}}{2}\right)\right]$
6. $(2D^2 + 3D - 2)y = 0$; $y(0) = 1$ and $y(t) \to 0$ as $t \to \infty$. [***Ans.*** $y = e^{-2t}$]
7. $(D^3 - D)y = 2\cos t$; $y(0) = 3$, $y''(0) = 1$, $y'(0) = 2$.
 [***Ans.*** $y = 3\sinh t + \cosh t - \sin t + 2$]
8. $(D^2 + 2D + 2)y = 5\sin t$; $y(0) = 0 = y'(0)$.
 [***Ans.*** $y = -2\cos t + \sin t + 2e^{-t}\cos t + e^{-t}\sin t$]
9. $(D^2 + 1)^2 y = t$; $y(0) = 1$, $y'(0) = -2$. [***Ans.*** $y = t + \cos t - 3\sin t$]
10. $(D^2 - 3D + 2)y = 4t + e^{2t}$, $y(0) = 1$, $y'(0) = -1$.
 $\left[\textbf{Ans.}\quad y = t\,e^{2t} - \frac{11}{4}e^{2t} + 3e^t + \frac{t}{2} + \frac{3}{4}\right]$
11. $(D + 1)^2 y = 3t\,e^{-t}$; $y(0) = 4$, $y'(0) = 0$ $\left[\textbf{Ans.}\quad y = e^{-t}(4 + 4t + \frac{1}{2}t^3)\right]$
12. Find the general solution of the differential equation

 $(D^3 - 3D^2 + 3D - 1)y = t^2e^t$ $\left[\textbf{Ans.}\quad c_1t^2e^t + c_2t\,e^t + c_3e^t + \frac{t^5e^t}{60}\right]$

5.3 SOLUTION OF ORDINARY DIFFERENTIAL EQUATIONS WITH VARIABLE COEFFICIENTS

It is very useful to find the solution of ordinary differential equation with variable coefficient using Laplace transform. In this method, we generally use

$$L\left[t^n f^n(t)\right] = (-1)^n \frac{d^n}{dp^n} L\left[f^n(t)\right].$$

Illustrative Examples

Example 1 Solve $xy'' + (x - 1)y' - y = 0$; $y(0) = 5$, $y(\infty) = 0$.

Solution: Given differential equation

$$xy'' + (x - 1)y' - y = 0$$

or $$xy'' + xy' - y' - y = 0 \quad \text{...(i)}$$

Applying Laplace transform to the equation (i), we get

$$L(xy'') + L(xy') - L(y') - L(y) = 0$$

$$\Rightarrow -\frac{d}{dp}L(y'') - \frac{d}{dp}L(y') - L(y') - L(y) = 0$$

$$\Rightarrow -\frac{d}{dp}\left(p^2 L(y) - py(0) - y'(0)\right) - \frac{d}{dp}(p\,L(y) - y(0)) - (p\,L(y) - y(0)) - L(y) = 0$$

$$\Rightarrow -\frac{d}{dp}\left(p^2 L(y) - 5p - a\right) - \frac{d}{dp}(p \cdot L(y) - 5) - (p\,L(y) - 5) - L(y) = 0, \quad \text{Let } y'(0) = a$$

$$\Rightarrow (p^2 + p)\frac{d}{dp}Y + (3p + 2)Y = 10. \quad \text{Let } L(y) = Y$$

$$\Rightarrow \frac{dY}{dp} + \frac{3p+2}{p(p+1)}Y = \frac{10}{p(p+1)} \quad \text{...(ii)}$$

Equation (ii) is a linear equation of first order and first digree, therefore, the solution to equation (ii) is

$$Y(p + 1)p^2 = \int \frac{10}{p(p+1)} \cdot (p+1)p^2 dp + c_1$$

$$= 5\,p^2 + c_1$$

$$\Rightarrow \quad Y = \frac{5}{p+1} + \frac{c_1}{p^2(p+1)}$$

or $$L(y) = \frac{5}{p+1} + c_1 \cdot \frac{1}{p^2(p+1)} \quad \text{...(iii)}$$

Applying inverse Laplace transform to equation (iii), we get

$$y = 5L^{-1}\frac{1}{p+1} + c_1 L^{-1}\frac{1}{p^2(p+1)}$$

But $\quad L^{-1}\left(\frac{1}{p+1}\right) = e^{-t}$

$$L^{-1}\frac{1}{p^2(p+1)} = \int_0^t e^{-u}du = t - e^{-t}$$

and $\quad L^{-1}\frac{1}{p^2(p+1)} = \int_0^t (1-e^{-u})du = t + e^{-t} - 1$

Then, $\quad y = 5 \cdot e^{-t} + c_1 (t + e^{-t} - 1)$

Given $\quad y(\infty) = 0$

$\Rightarrow \quad c_1 = 0$

Therefore, $y = 5 \cdot e^{-t}$. **_Ans._**

Example 2 Solve $ty'' + y' + 4ty = 0$

If, $\quad y(0) = 3, y'(0) = 0$

Solution: Given that

$$ty'' + y' + 4ty = 0 \qquad \text{...(i)}$$

Applying Laplace transform to the equation (i), we get

$$L(ty'') + L(y') + 4L(ty) = 0$$

$$\Rightarrow -\frac{d}{dp}L(y'') + L(y') - 4\frac{d}{dp}L(y) = 0$$

$$\Rightarrow -\frac{d}{dp}[p^2L(y) - py(0) - y'(0)] + [p\,L(y) - y(0)] - 4\frac{d}{dp}(L(y)) = 0$$

$$\Rightarrow -\frac{d}{dp}[p^2Y - 3p] + [p\,Y - 3] - 4\frac{dY}{dp} = 0$$

$$\Rightarrow pY + [p^2 + 4]\frac{dY}{dp} = 0$$

$$\Rightarrow \frac{dY}{Y} + \frac{p}{p^2+4}dp = 0 \qquad \text{...(ii)}$$

On integration equation (ii), we get

$$\log Y + \frac{1}{2}\log(p^2 + 4) = \log c$$

$$\log Y^2 (p^2 + 4) = \log c$$

$$Y = \frac{c}{\sqrt{p^2+4}}$$

or $$L(y) = \frac{c}{\sqrt{p^2+4}}$$

or $$y = cL^{-1}\frac{1}{\sqrt{p^2+4}}$$

$\Rightarrow$ $$y = c\,J_0(2t) \quad \text{(On taking inverse Laplace transfer)}$$

but $$y(0) = 3$$

$\Rightarrow$ $$c\,J_0(0) = 3 \quad \text{as} \quad J_0(0) = 1$$

$\therefore$ $$y = 3\,J_0(2t)$$

Ans.

Example 3 Solve $ty'' + y' + 2y = 0, \quad y(0) = 1.$

Solution: Given differential equation

$$ty'' + y' + 2y = 0 \qquad \text{...(i)}$$

Applying Laplace transform to the equation (i), we get

$$-\frac{d}{dp}[p^2L(y) - py(0) - y'(0)] + [p\,L(y) - y(0)] + 2L(y) = 0$$

$$\Rightarrow \quad -p^2\frac{dY}{dp} - pY + 2Y = 0$$

$$\Rightarrow \quad \frac{dY}{dp} + \left(\frac{1}{p} - \frac{2}{p^2}\right)Y = 0$$

$$\Rightarrow \quad L(y) = \frac{k}{p}e^{-2/p}$$

$$= k\sum_{n=0}^{\infty}\frac{(-1)^n\,2^n}{n!\,p^{n+1}}$$

$$\Rightarrow \quad y = k\sum_{n=0}^{\infty}\frac{(-1)^n\,(2t)^n}{(n!)^2}$$

But $$y(0) = 1 \Rightarrow k = 1$$

Therefore, $$y = \sum_{n=0}^{\infty}\frac{(-1)^n\,(2t)^n}{(n!)^2} = J_0\,(2\sqrt{2t})$$

Note: An unbounded solution of this differential equation also exists but cannot be evaluated by this method.

Example 4 Solve $y'' - ty' + y = 1$; $y(0) = 1$, $y'(0) = 2$.

Solution: Given differential equation

$$y'' - ty' + y = 1 \quad \text{...(i)}$$

Applying Laplace transform both sides equation (i), we get

$$L(y'') - L(ty') + L(y) = L(1)$$

$$\Rightarrow \quad p^2 L(y) - p\,y(0) - y'(0) + \frac{d}{dp}[p\,L(y) - y(0)] + L(y) = 1/p \quad \text{...(ii)}$$

On applying given conditions, we get

$$p\frac{dY}{dp} + (p^2 + 2)Y = \frac{1}{p} + p + 2, \quad L(y) = Y$$

$$\frac{dY}{dp} + \left(p + \frac{2}{p}\right)Y = \frac{1}{p^2} + \frac{2}{p} + 1 \quad \text{...(iii)}$$

Equation (iii) is linear in Y, therefore, solution of the equation (iii), we get

$$p^2 \cdot e^{p^2/2}\, Y = \int e^{p^2/2}\,(p^2 + 2p + 1)dp + c_1$$

$$= \int p^2 e^{p^2/2}\,dp + 2\int e^{p^2/2}\,p\,dp + \int e^{p^2/2}\,dp + c_1$$

$$= (p + 2)\,e^{p^2/2} + c_1$$

But $Y = L(y)$, then

$$p^2 \cdot e^{p^2/2}\, L(y) = (p + 2)\,e^{p^2/2} + c_1$$

$$\Rightarrow \quad L(y) = \frac{p+2}{p^2} + \frac{c_1}{p_2}\,e^{-p^2/2}$$

$$= \frac{1}{p} + \frac{2}{p^2} + \frac{c_1}{p^2}\left(1 - \frac{p^2}{2} + \frac{1}{2!} + \frac{p^4}{4} + \ldots\right)$$

$$= \frac{1}{p} + \frac{c_1 + 2}{p^2} - \frac{c_1}{2} + \frac{c_1}{8}\,p^2 + \ldots$$

Since, the solution is bounded, therefore, $L(y)$ should be finite as $p \to \infty$, and this is possible when $c_1 = 0$.

Then, $$L(y) = \frac{1}{p} + \frac{2}{p^2} \quad \text{...(iv)}$$

Applying inverse Laplace transform equation (iv); we get

$$y = 1 + 2t$$ ***Ans.***

EXERCISE

Solve the following differential equation using Laplace transform:

1. $t\,y'' + (1 - 2t)y' - 2y = 0, \quad y(0) = 1, \quad y'(0) = 2$ [*Ans.* $y = 2e^{2t}$]

2. $y'' + ty' - y = 0, \quad y(0) = 0, \quad y'(0) = 1$ [*Ans.* $y = t$]

3. $ty'' + y' + ty = 0, \quad y(0) = 1, \quad y'(0) = 0$ [*Ans.* $y = J_0(t)$]

4. $y'' + aty' - 2ay = 1, \quad y(0) = y'(0) = 0, \quad a > 0$ $\left[\textbf{Ans.} \quad y = \frac{1}{2}t^2\right]$

5. $ty'' + y' + a^2\,ty = 0, \quad y(0) = 1$ [*Ans.* $y = J_0(at)$]

6. $y'' + ty' - 2y = 4, \quad y(0) = -1, \quad y'(0) = 0$ [*Ans.* $y = t^2 - 1$]

5.4 SOLUTION OF SIMULTANEOUS ORDINARY DIFFERENTIAL EQUATIONS

In this section, we are illustrating the solution method of simultaneous ordinary differential equations by using Laplace transforms.

Example 1 Solve $\dfrac{d^2x}{dt^2} + 5\cdot\dfrac{dy}{dt} - x = t$

$$2\cdot\frac{dx}{dt} - \frac{d^2y}{dt^2} + 4y = 2$$

and given that, $x(0) = 0 = y(0)$, $x'(0) = 0 = y'(0)$.

Solution: Given differential equation

$$\frac{d^2x}{dt^2} + 5\cdot\frac{dy}{dt} - x = t$$

or $$x'' + 5y' - x = t \qquad ...(i)$$

and $$2\cdot\frac{dx}{dt} - \frac{d^2y}{dt^2} + 4y = 2$$

or $$2x' - y'' + 4y = 2 \qquad ...(ii)$$

Applying Laplace transform to equations (i) and (ii), we get

$$[p^2X - px(0) - x'(0)] + 5[pY - y(0)] - X = \frac{-1}{p^2} \qquad ...(iii)$$

$$2[pX - x(0)] + 5[p^2Y - y(0) - y'(0)] - 4Y = \frac{2}{p} \qquad ...(iv)$$

$$\Rightarrow \qquad (p^2 - 1)X + 5pY = \frac{1}{p^2} \qquad ...(v)$$

$$2pX - (p^2 - 4)Y = \frac{2}{p} \qquad \text{...(vi)}$$

On solving equations (v) and (vi)

$$X = \frac{11p^2 - 4}{p^2(p^2+1)(p^2+4)}$$

$$Y = \frac{4 - 2p^2}{p(p^2+1)(p^2+4)}$$

$$\Rightarrow \qquad X = \frac{-1}{p^2} + \frac{5}{p^2+1} - \frac{4}{p^2+4}$$

Applying inverse Laplace transform, we have

$$x = -L^{-1}\left(\frac{1}{p^2}\right) + 5L^{-1}\left(\frac{1}{p^2+1}\right) - 4L^{-1}\left(\frac{1}{p^2+4}\right), \; L^{-1}(X) = x$$

$$\Rightarrow \qquad x = -t + 5\sin t - 2\sin 2t$$

and

$$y = \frac{1}{p} - \frac{2p}{p^2+1} + \frac{p}{p^2+4}$$

Applying inverse Laplace transform, we have

$$y = L^{-1}\left(\frac{1}{p}\right) - 2L^{-1}\left(\frac{p}{p^2+1}\right) + L^{-1}\left(\frac{p}{p^2+4}\right)$$

$$\Rightarrow \qquad y = 1 - 2\cos t + \cos 2t$$

Hence,

$$x = -t + 5\sin t - 2\sin 2t$$

$$y = 1 - 2\cos t + \cos 2t$$

Ans.

Example 2 Solve: $(D^2 - 3)x - 4y = 0, \quad x + (D^2 - 1)y = 0, \; t > 0$

Given that $x(0) = 0 = y(0) = y'(0), \; x'(0) = 2.$

Solution: Given differential equations:

$$(D^2 - 3)x - 4y = 0$$

or

$$x'' - 3x - 4y = 0 \qquad \text{...(i)}$$

and

$$x + (D^2 - 1)y = 0$$

or

$$x + y'' - y = 0 \qquad \text{...(ii)}$$

Applying Laplace transform to equations (i) and (ii), we get

$$L(x'') - 3L(x) - 4L(y) = 0$$

$$p^2X - px(0) - x'(0) - 3X - 4Y = 0, \text{ taking } \; L(x) = X, L(y) = Y$$

and

$$L(x) + L(y'') + L(y) = 0$$

$\Rightarrow \qquad p^2X - px(0) - x'(0) - 3X - 4Y = 0$

and $\qquad X + p^2Y - py(0) - y'(0) + Y = 0$

On putting values, we get

$$(p^2 - 3)X - 4Y = 2$$

and $\qquad X + (p^2 + 1)Y = 0$

On solving for X and Y, we have

$$X = \frac{2(p^2+1)}{(p^2-1)^2} = \frac{1}{(p-1)^2} + \frac{1}{(p+1)^2}$$

Applying inverse Laplace transform, we have

$$x = L^{-1}\left(\frac{1}{(p-1)^2}\right) + L^{-1}\left(\frac{1}{(p+1)^2}\right)$$

$$= e^t\, L^{-1}\left(\frac{1}{p^2}\right) + e^{-t}\, L^{-1}\left(\frac{1}{p^2}\right)$$

$$= e^t \cdot t + e^{-t} \cdot t = t(e^t + e^{-t})$$

and $$Y = \frac{-2}{(p+1)^2\,(p-1)^2}$$

$\Rightarrow$ $$Y = \frac{1}{2}\left[-\frac{1}{p+1} + \frac{1}{p-1} - \frac{1}{(p+1)^2} - \frac{1}{(p-1)^2}\right]$$

Applying inverse Laplace transform, we have

$$y = \frac{1}{2}\left[-L^{-1}\left(\frac{1}{p+1}\right) + L^{-1}\left(\frac{1}{p-1}\right) - L^{-1}\left(\frac{1}{(p+1)^2}\right) - L^{-1}\left(\frac{1}{(p-1)^2}\right)\right]$$

$$= \frac{1}{2}[-e^{-t} + e^t - t\,e^{-t} - t\,e^t]$$

$\Rightarrow$ $$y = \frac{1}{2}[(1-t)\,e^t] - \frac{1}{2}[(1+t)\,e^t]$$

Hence, $\qquad x = t(e^t + e^{-t}),\ y = \frac{1}{2}(1-t)\,e^t - \frac{1}{2}(1+t)\,e^{-t}$ ***Ans.***

Example 3 Solve

$$\frac{d^2x}{dt^2} + \frac{dy}{dt} + 3x = 15\,e^{-t}$$

$$\frac{d^2y}{dt^2} - 4\frac{dx}{dt} + 3y = 15\sin 2t$$

If, $x(0) = 35$, $x'(0) = -48$, $y(0) = 27$, $y'(0) = -55$.

Solution: Given that

$$\frac{d^2x}{dt^2} + \frac{dy}{dt} + 3x = 15e^{-t}$$

$$\frac{d^2y}{dt^2} - 4\frac{dx}{dt} + 3y = 15\sin 2t$$

or $$x'' + y' + 3x = 15e^{-t} \qquad \text{...(i)}$$

and $$-4x' + y'' + 3y = 15\sin 2t \qquad \text{...(ii)}$$

Applying Laplace transform on equations (i) and (ii), we get

$$L(x'') + L(y') + 3L(x) = 15\,L(e^{-t})$$

and $$-4L(x') + L(y'') + 3L(y) = 15\,L(\sin 2t)$$

On putting $L(x) = X$, $L(y) = Y$, we have,

$$\Rightarrow \quad p^2X - px(0) - x'(0) + pY - y(0) + 3X = \frac{15}{p+1}$$

and $$-4[pX - x(0)] + p^2y - py(0) - y'(0) + 3Y = \frac{30}{p^2+4}$$

On putting initial conditions, we get

$$(p^2 + 3)X + pY = 35p - 21 + \frac{15}{p+1}$$

$$-4pX + (p^2 + 3)Y = 27p - 195 + \frac{30}{p^2+4}$$

Solving for X and Y, we get

$$X = \frac{15(p^2+3)}{(p+1)(p^2+1)(p^2+9)} + \frac{35p^3 - 48p^2 + 300p - 63}{(p^2+1)(p^2+9)} - \frac{30p}{(p^2+1)(p^2+4)(p^2+9)}$$

$$\Rightarrow \quad X = 30\cdot\frac{p}{p^2+1} - 45\cdot\frac{1}{p^2+9} + 3\cdot\frac{1}{p+1} + 2\cdot\frac{p}{p^2+4}$$

and $$Y = \frac{27p^3 - 55p^2 - 3p - 585}{(p^2+9)(p^2+1)} + \frac{30(p^2+3)}{(p^2+4)(p^2+9)(p^2+1)} + \frac{60p}{(p+1)(p^2+1)(p^2+9)}$$

$$\Rightarrow \quad Y = -\frac{3}{p+1} - 60\cdot\frac{1}{p^2+1} + \frac{2}{p^2+4} + 30\cdot\frac{p}{p^2+9}$$

Applying inverse Laplace transform on X and Y, we get

$$x = 3e^{-t} + 30 \cos t + 2 \cos 2t - 15 \sin 3t$$

$$y = -3e^{-t} - 60 \sin t + \sin 2t + 30 \cos 3t \quad \textit{Ans.}$$

Example 4 Solve $(D - 2)x - (D + 1)y = 6e^{3t}$

$(2D - 3)x + (D - 3)y = 6e^{3t}$. If $x(0) = 3$, $y(0) = 0$.

Solution: Given that

$$(D - 2)x - (D + 1)y = 6e^{3t}$$

$$(2D - 3)x + (D - 3)y = 6e^{3t}$$

or $$x' - 2x - y' - y = 6e^{3t} \quad \text{...(i)}$$

and $$2x' - 3x + y' - 3y = 6e^{3t} \quad \text{...(ii)}$$

Applying Laplace transform on equations (i) and (ii), we get

$$L(x') - 2L(x) - L(y') - L(y) = 6L(e^{3t}) \quad \text{...(iii)}$$

and $$2L(x') - 3L(x) + L(y') - 3L(y) = 6L(e^{3t}) \quad \text{...(iv)}$$

Assuming, $L(x) = X$ and $L(y) = Y$, in equations (iii) and (iv), we get

$$pX - x(0) - 2X - (pY - y(0)) - Y = \frac{6}{p-3}$$

$$2[pX - x(0)] - 3X + [pY - y(0)] - 3Y = \frac{6}{p-3}$$

or $$(p - 2)X - (p + 1)Y = \frac{6}{p-3} + 3 = \frac{3p-3}{p-3}$$

$$(2p - 3)X + (p - 3)Y = \frac{6}{p-3} + 6 = \frac{6p-12}{p-3}$$

On solving above equation for X and Y, we get

$$X = \frac{-1-6p+3p^2}{(p-1)^2\,(p-3)} = \frac{1}{p-1} + \frac{2}{p-3} + \frac{2}{(p-1)^2}$$

and $$Y = \frac{5-3p}{(p-1)^2\,(p-3)} = \frac{1}{p-1} - \frac{1}{p-3} - \frac{1}{(p-1)^2}$$

Applying inverse Laplace transform,

$$x = L^{-1}(X) = L^{-1}\left(\frac{1}{p-1}\right) + L^{-1}\frac{2}{(p-3)} + L^{-1}\frac{2}{(p-1)^2}$$

$$L^{-1}(Y) = y = L^{-1}\left(\frac{1}{p-1}\right) - L^{-1}\frac{1}{p-3} - L^{-1}\left(\frac{1}{(p-1)^2}\right)$$

$\Rightarrow$ $$x = e^t + 2e^{3t} + 2e^{3t}\cdot t$$

$\Rightarrow$ $$y = e^t - e^{3t} - te^t$$ ***Ans.***

Example 5 Solve $Dx + Dy = t$

$$D^2x - y = e^{-t}$$

Given that $x(0) = 3$, $x'(0) = -2$, $y(0) = 0$

Solution: Given differential equation

$$x' + y' = t \quad \text{...(i)}$$

$$x'' - y = e^{-t} \quad \text{...(ii)}$$

Applying Laplace transform on equations (i) and (ii), we get

$$L(x') + L(y') = L(t) \quad \text{...(iii)}$$

and $$L(x'') - L(y) = L(e^{-t}) \quad \text{...(iv)}$$

Let $L(x) = X$ and $L(y) = Y$; then

$pX - x(0) + pY - y(0) = 1/p^2$

and $p^2X - px(0) - x'(0) - Y = \dfrac{1}{p+1}$

Replacing given conditions;

$$pX + pY = 3 + 1/p^2$$

$$p^2X - Y = 3p - 2 + \frac{1}{p+1}$$

Solving for X and Y, we get

$$X = \frac{3p^2+1}{(1+p^2)p^3} + \frac{3p}{1+p^2} - \frac{2}{1+p^2} - \frac{1}{(1+p)(1+p^2)}$$

$$= \frac{1}{p^3}\left(1+2p^2\frac{2p^4}{1+p^4}\right) + \frac{3p}{1+p^2} - \frac{2}{1+p^2} + \frac{1}{2}\cdot\frac{1}{p+1} - \frac{1}{2}\cdot\frac{p}{1+p^2} + \frac{1}{2}\cdot\frac{p}{1+p^2}$$

$$= \frac{2}{p} + \frac{1}{p^3} + \frac{1}{2}\cdot\frac{1}{1+p} + \frac{1}{2}\cdot\frac{p}{1+p^2} - \frac{3}{2}\cdot\frac{2}{1+p^2}$$

and $$Y = \frac{1}{p(1+p)(1+p^2)} + \frac{2}{1+p^2}$$

$$= \frac{1}{p} - \frac{1}{2}\cdot\frac{p}{1+p^2} - \frac{1}{2}\cdot\frac{1}{1+p^2} - \frac{1}{2}\cdot\frac{1}{1+p^2} + \frac{2}{1+p^2}$$

$$= \frac{1}{p} - \frac{1}{2}\cdot\frac{1}{1+p} - \frac{1}{2}\cdot\frac{p}{1+p^2} + \frac{3}{2}\cdot\frac{1}{1+p^2}$$

Applying inverse Laplace transform on X, we get

$$L^{-1}(X) = x = 2L^{-1}\left(\frac{1}{p}\right) + L^{-1}\left(\frac{1}{p^3}\right) + \frac{1}{2}L^{-1}\left(\frac{1}{1+p}\right)$$

$$+ \frac{1}{2}L^{-1}\left(\frac{p}{1+p^2}\right) - \frac{3}{2}\cdot L^{-1}\left(\frac{1}{1+p^2}\right)$$

$$x = 2 + \frac{1}{2}t^2 + \frac{1}{2}e^{-t} + \frac{1}{2}\cos t - \frac{3}{2}\sin t$$

and, applying inverse Laplace transform on Y, we get

$$L^{-1}(Y) = y = L^{-1}(1/p) - \frac{1}{2}L^{-1}\left(\frac{1}{1+p}\right) - \frac{1}{2}L^{-1}\left(\frac{p}{1+p^2}\right) + \frac{3}{2}L^{-1}\left(\frac{1}{1+p^2}\right)$$

$$\Rightarrow \qquad y = 1 - \frac{1}{2}e^{-t} - \frac{1}{2}\cos t + \frac{3}{2}\sin t$$

$$\Rightarrow \qquad x = 2 + \frac{1}{2}t^2 + \frac{1}{2}e^{-t} + \frac{1}{2}\cos t - \frac{3}{2}\sin t \qquad \textbf{\textit{Ans.}}$$

EXERCISE

Solve the following system of differential equations:

1. $(D^2 + 2)x - Dy = 1$, $Dx + (D^2 + 2)y = 0$, $t > 0$, $x'(0) = y'(0) = 0 = x(0) = y(0)$.

$$\left[\textbf{\textit{Ans.}}\ x = \frac{1}{2} - \frac{1}{3}\cos t - \frac{1}{6}\cos 2t\right.$$

$$\left. y = -\frac{1}{3}\sin t + \frac{1}{6}\sin 2t\right]$$

2. $x'' + 4y = 4t$, $4x + y'' = 0$; $x'(0) = y'(0) = 0 = x(0) = y(0)$.

$$\left[\textbf{\textit{Ans.}}\ x = \frac{1}{4}(\sinh 2t - \sin 2t)\right.$$

$$\left. y = t - \frac{1}{4}(\sinh 2t + \sin 2t)\right]$$

3. $x' - 4x + y' - 6y = -2$, $2x' + x + 3y' + 2y = 0$; $x(0) = 7$, $y(0) = -4$.

[***Ans.*** $x = 2 + 5e^t$

$y = -1 - 3e^t$]

4. $x'' - y'' + y' - x = e^t - 2,\ 2x'' - y'' - 2x' + y = -t;\ x'(0) = y'(0) = 0 = x(0) = y(0).$

[**Ans.** $x = 1 - e^t + t\,e^t$

$y = -t + t\,e^t$]

5. $x' = 2x - 3y,\ y' = y - 2x;\ x(0) = 8,\ y(0) = 3.$

[**Ans.** $x = 5e^{-t} + 3e^{4t}$

$y = 5e^{-t} - 2e^{4t}$]

6. $x' + y' = 2z,\ y' + z' = zx,\ z' + x' = 2y.$ Given that $x(0) = a,\ y(0) = b,\ z(0) = c.$

$$\left[\textbf{Ans.}\quad x = \frac{1}{3}\,d\,e^t\left(a - \frac{1}{3}d\right)e^{-2t}\right.$$

$$y = \frac{1}{3}\,d\,e^t\left(b - \frac{1}{3}d\right)e^{-2t}$$

$$\left.z = \frac{1}{3}\,d\,e^t\left(c - \frac{1}{3}d\right)e^{-2t}\right]$$

5.5 PARTIAL DIFFERENTIAL EQUATIONS

To solve partial differential equations the Laplace transform technique is very useful. It can be applied to some specific boundary value problems. The Laplace transform gives rise to an ordinary differential equation for the Laplace transform. In next section, we are deriving some important results to solve partial differential equations.

Theorem 1. For a functions $y(x, t)$, we have

(i) $L\left[\dfrac{\partial y}{\partial t}\right] = p\,L(y) - y(x, 0)$

(ii) $L\left(\dfrac{\partial^2 y}{\partial t^2}\right) = p^2\,L(y) - p\,y(x, 0) - y_t(x, 0)$

(iii) $L\left(\dfrac{\partial y}{\partial x}\right) = \dfrac{d}{dx}(Y)$

(iv) $L\left(\dfrac{\partial^2 y}{\partial x^2}\right) = \dfrac{d^2}{dx^2}(Y)$, where $L(y) = Y$.

Proof: (i) We know that

$$L\left(\frac{\partial y}{\partial t}\right) = \int_0^\infty e^{-pt}\,\frac{\partial y}{\partial t}\,dt$$

$$= \lim_{k\to\infty}\int_0^k e^{-pt}\,\frac{\partial y}{\partial t}\,dt$$

$$= \lim_{k\to\infty}\left[(e^{-pt}\cdot y)_0^k + p\int_0^k e^{-pt}\, y\, dt\right]$$

since $\lim_{k\to\infty}(e^{-pt}\cdot y)_0^k = 0$

$$L\left(\frac{\partial y}{\partial t}\right) = p\int_0^\infty e^{-pt}\, y\, dt - y(x, 0)$$

$$= pY - y(x, 0)$$

(ii) Let us assume, $u = \dfrac{\partial y}{\partial t}$

Then $$L\left(\frac{\partial^2 y}{\partial t^2}\right) = L\left(\frac{\partial u}{\partial t}\right)$$

$$= p\, L(u) - u(x, 0)$$

$$= p[pL(y) - y(x, 0)] - y_t(x, 0)$$

$$= p^2 Y - py(x, 0) - y_t$$

(iii) We know that

$$L\left(\frac{\partial y}{\partial x}\right) = \int_0^\infty e^{-pt}\frac{\partial y}{\partial x}\, dt$$

$$= \frac{d}{dx}\int_0^\infty e^{-pt} y\, dt$$

$$= \frac{d}{dx}(y)$$

(iv) By using Leibnitz's rule for differentiation under the integral sign, we get

$$L\left(\frac{\partial^2 y}{\partial x^2}\right) = \int_0^\infty e^{-pt}\frac{\partial^2 y}{\partial x^2}\, dt$$

$$= \frac{d^2}{dx^2}\int_0^\infty e^{-pt}\, y(x, t)\, dt$$

$$\Rightarrow \quad L\left(\frac{\partial^2 y}{\partial x^2}\right) = \frac{d^2}{dx^2}(Y)$$

Here, $Y = L\, y(x, t)$

Example 1 Find the bounded solution of

$$\frac{\partial Y}{\partial x} = 2\frac{\partial Y}{\partial t} + Y \quad \text{if} \quad Y(x, 0) = 6e^{-3x}$$

Solution: Let us consider

$$L(Y(x, t)) = y(x, p) \qquad ...(i)$$

Applying Laplace transform to both sides of the given differential equation, we have

$$\frac{dy}{dx} = 2[py - Y(x, 0)] + y$$

or $$\frac{dy}{dx} - (2p + 1)y = -12e^{-3x} \qquad ...(ii)$$

which is linear in y, then, solution to the equation (ii)

$$y \cdot e^{-(2p + 1)x} = -12 \int e^{-3x} \cdot e^{-(2p + 1)x}\, dx + c$$

$$= -12 \int e^{-(2p + 4)x}\, dx + c$$

$$= \frac{6}{p+2}\, e^{-(2p + 4)x} + c$$

$\Rightarrow$ $$y = \frac{6e^{-3x}}{p+2} + c\, e^{-(2p + 1)x}$$

since, $Y(x, t)$ is bounded for all values of x and $y(x, p)$ is bounded for all x and $p > 0$. Hence, $c = 0$, for finite value of u as $x \to \infty$, we get

$$u = \frac{6e^{-3x}}{p+2}$$

$\Rightarrow$ $$Y(x, t) = 6e^{-3x}\, L^{-1}\, \frac{1}{p+2} = 6e^{-3x}\, e^{2t}$$

or $$Y(x, t) = 6e^{-(2t + 3x)}$$

which is the required solution to the given differential equation. ***Ans.***

Example 2 If $U(\pi/2, t) = 0$, $U(x, 0) = 30 \cos 5x$ and $\left(\frac{\partial U}{\partial x}\right)_{x=0} = 0$, then solve the equation

$$\frac{\partial U}{\partial t} = 3 \cdot \frac{\partial^2 U}{\partial x^2}$$

Solution: Given that

$$\frac{\partial U}{\partial t} = 3 \cdot \frac{\partial^2 U}{\partial x^2} \qquad ...(i)$$

and $$\left(\frac{\partial U}{\partial x}\right)_{x=0} = 0,\; U(\pi/2, t) = 0,\; U(x, 0) = 30 \cos 5x \qquad ...(ii)$$

Let $L[U(x, t)] = u(x, p)$, then applying Laplace transform to the given differential equation, we get

$$pu(x, p) - U(x, 0) = 3\frac{d^2 u}{dx^2}$$

$$\Rightarrow \quad 3\cdot\frac{d^2u}{dx^2} - p\cdot u(x, p) + U(x, 0) = 0$$

by equation (ii), we have

$$\frac{d^2u}{dx^2} - \frac{p}{3}u = -10\cos 5x \qquad \text{...(iii)}$$

which is linear in u, hence the solution of the equation (iii), we get

$$u(x, p) = c_1\, e^{\sqrt{(p/3)}\,x} + c_2\, e^{-\sqrt{(p/3)}\,x} + \frac{30}{75+p}\cos 5x \qquad \text{...(iv)}$$

Applying given conditions given equation (ii)

$$c_1\, e^{\pi/2\sqrt{(p/3)}} + c_2\, e^{-\pi/2\sqrt{(p/3)}} = 0$$

$$c_1\sqrt{p/3} - c_2\sqrt{p/3} = 0$$

On solving, we get $c_1 = 0$, $c_2 = 0$, then equation (iv) becomes

$$u(x, p) = \frac{30}{p+75}\cos 5x$$

Applying inverse Laplace, we get

$$U(x, t) = 30\, e^{-75t}\cos 5x$$ ***Ans.***

which is the required solution.

Example 3 Solve $\frac{\partial U}{\partial t} = \frac{\partial^2 U}{\partial x^2}$. If $u(x, 0) = 3\sin 2\pi x$,

$U(0, t) = 0$, $U(1, t) = 0$ and $0 < x < 1$, for $t > 0$

Solution: Given differential equation

$$\frac{\partial U}{\partial t} = \frac{\partial^2 U}{\partial x^2} \qquad \text{...(i)}$$

and conditions

$$U(x, 0) = 3\sin 2\pi x,\ U(0, t) = 0 = U(1, t)$$

$0 < x < 1$ for $t > 0$ ...(ii)

Let $L\, U(x, t) = u(x, p)$, applying Laplace transform on both sides of equation (i), we get

$$p\, u(x, p) - U(x, 0) = \frac{d^2u}{dx^2}$$

or $\frac{d^2u}{dx^2} - p\, u(x, p) + U(x, 0) = 0$

On applying conditions on given equation (ii), we get

$$\frac{d^2u}{dx^2} - pu = -\sin 2\pi x \qquad \text{...(iii)}$$

which is a linear equation of second order in u, hence, on solving equation (iii), we get

$$u(x, p) = c_1 e^{x\sqrt{p}} + c_2 e^{-x\sqrt{p}} + \frac{1}{4\pi^2 + p} 3 \sin 2\pi x \qquad \text{...(iv)}$$

Applying Laplace transform involving t in boundary conditions, we get

$$L\,U(0, t) = u(0, p) = 0 \qquad \text{...(v)}$$

$$L\,U(1, t) = u(1, p) = 0$$

replacing $x = 0$ and 1 in equations (iv) and (v), we get

$$c_1 + c_2 = 0$$

$$c_1 e^{\sqrt{p}} + c_2 e^{-\sqrt{p}} = 0$$

$$\Rightarrow \qquad c_1 = 0 = c_2$$

Putting the values of c_1 and c_2 in equation (iv), we have

$$u(x, p) = \frac{3}{4\pi^2 + p} \sin 2\pi x$$

Applying inverse Laplace, we get

$$U(x, t) = 3\, e^{-4\pi^2 t} \sin 2\pi x$$ ***Ans.***

Example 4 Solve $\dfrac{\partial^2 U}{\partial t^2} = 9 \dfrac{\partial^2 U}{\partial x^2}$

subject to conditions: $U(0, t) = 0$, $U(2, t) = 0$

$$U(x, 0) = 20 \sin 2\pi x - 10 \sin 5\pi x \text{ and } U_t(x, 0) = 0$$

Solution: Given partial differential equation

$$\frac{\partial^2 U}{\partial t^2} = 9 \frac{\partial^2 U}{\partial x^2} \qquad \text{...(i)}$$

and the conditions are

$$U(0, t) = 0 = U(2, t) = U_t(x, 0)$$

$$U(x, 0) = 2 \sin 2\pi x - 10 \sin 5\pi x \qquad \text{...(ii)}$$

Let $L[U(x, t)] = u(x, p)$, applying Laplace transform on equation (i), we get

$$p^2 u(x, p) - p\, U(x, 0) - U_t(x, 0) = 9 \frac{\partial^2 u}{\partial x^2}$$

Using equation (ii) for initial conditions, we get

$$9\frac{d^2u}{dx^2} - p^2u = -(20 \sin 2\pi x - 10 \sin 5\pi x)p$$

or $$\frac{d^2u}{dx^2} - \frac{p^2}{9}u = -(20 \sin 2\pi x - 10 \sin 5\pi x)\frac{p}{9} \quad \text{...(iii)}$$

which is a linear differential equation of second order, hence, solution

$$u(x, p) = c_1 e^{(p/3)x} + c_2 e^{-(p/3)x} - \frac{p}{9}\frac{1}{D^2 - p^2/9}(2 \sin 2\pi x - 10 \sin 5\pi x)$$

$$\Rightarrow \quad u(x, p) = c_1 e^{(p/3)x} + c_2 e^{-(p/3)x} + \left[\frac{20 \sin 2\pi x}{p^2 + 36\pi^2} - \frac{10 \sin 5\pi x}{p^2 + 225\pi^2}\right] \quad \text{...(iv)}$$

Applying Laplace transform on the conditions which involve t, we have

$$L[U(0, t)] = u(0, p) = 0 \quad \text{...(v)}$$

$$L[U(2, t)] = u(2, p) = 0$$

Replacing $x = 0$ and $x = 2$ in equations (iv) and (v), we get

$$c_1 + c_2 = 0$$

$$c_1 e^{2p/3} + c_2 e^{-2p/3} = 0$$

On solving, we get

$$c_1 = 0, c_2 = 0$$

Putting the values of c_1 and c_2 in equation (iv), we get

$$u(x, p) = 20 (\sin 2\pi x) \frac{p}{p^2 + 36\pi^2} - 10 (\sin 5\pi x) \frac{p}{p^2 + 225\pi^2}$$

Applying inverse Laplace Transform, we get

$$U(x, t) = 20 \sin 2\pi x \cos 6\pi t - 10 \sin 5\pi x \cos 15\pi t$$ ***Ans.***

EXERCISE

1. Find the solution of

$$\frac{d^2u}{dx^2} - \frac{\partial^2 U}{\partial x^2} = xt, \; 0 < x < 1, \; t > 0, \text{ satisfying the conditions}$$

$$U(x, 0) = 0 = U_t(x, 0) = U(0, t) \text{ and } U(1, t) = -\frac{1}{6}t^3.$$

[***Ans.*** $U(x, t) = \frac{xt^3}{6}$]

2. Find the solution of

$$\frac{\partial U}{\partial t} = \frac{2 \cdot \partial^2 U}{\partial x^2}, \text{ satisfying the conditions}$$

$$U(0, t) = 0 = U(5, t),\ U(x, 0) = 10 \sin 4\pi x.$$

[***Ans.*** $U(x, t) = 10\, e^{-32\pi^2 t} \sin 4\pi x$]

3. Solve

$$\frac{\partial^2 U}{\partial t^2} = 16 \frac{\partial^2 U}{\partial x^2}$$

Subject to conditions: $U(x, 0) = U(3, t) = U_x(0, t) = 0$ and

$$U_t(x, 0) = 12 \cos \pi x + 16 \cos 3\pi x - 8 \cos 5\pi x.$$

[***Ans.*** $U(x, t) = \frac{3}{\pi} \cos \pi x \sin 4\pi t + \frac{4}{3\pi}$

$\cos 3\pi x \sin 12\, \pi t - \frac{4}{3\pi} \cos 5\pi x \sin 20\, \pi t$]

4. Find the solution of $\frac{\partial U}{\partial t} = \frac{\partial^2 U}{\partial x^2} - 4U$ satisfying the conditions

$$U(0, t) = 0,\ U(\pi, t) = 0,\ U(x, 0) = 6 \sin x - 4 \sin 2x.$$

[***Ans.*** $U(x, t) = 6e^{-5t} \sin x - 4e^{-8t} \sin 2x$]

5. Find the bounded solution $U(x, t)$, $0 < x < 1$, $t > 0$ of the boundary value problem

$$\frac{\partial U}{\partial x} - \frac{\partial U}{\partial t} = 1 - e^{-t},\ U(x, 0) = x.$$

[***Ans.*** $U(x, t) = x + 1 - e^t$]

6. Solve the equation

$$\frac{\partial^2 U}{\partial t^2} = 16 \frac{\partial^2 U}{\partial x^2} - g,\ x > 0,\ t > 0.$$

Subject to conditions: $U(x, 0) = 0 = U_t(x, 0) = U(0, t)$

and $\lim_{x \to \infty} U_x(x, t) = 0$

[***Ans.*** $U(x, t) = \frac{1}{2} g\left(t - \frac{1}{4}x\right) H\left(t - \frac{1}{4}x\right) - \frac{1}{2} gt^2, H\left(t - \frac{2}{4}\right)$

is the Heaviside unit step function]

7. Find the bounded solution of $\frac{\partial U}{\partial t} = \frac{\partial^2 U}{\partial x^2}$, $x > 0$, $t > 0$ such that $U(0, t) = 1$, $U(x, 0) = 0$.

$\left[\textbf{\textit{Ans.}}\quad U(x, t) = erfc\left(\frac{x}{2\sqrt{t}}\right)\right]$

5.6 DIFFERENCE AND DIFFERENTIAL–DIFFERENCE EQUATIONS

The equation containing function and their differences is called difference equation. An equation containing differences and derivatives of function along with function is said to be differential-difference equation.

Example 1 Solve $3Y(x) - 4Y(x-1) + Y(x-2) = x$, $Y(x) = 0$, $x < 0$.

Solution: Given

$3Y(x) - 4Y(x-1) + Y(x-2) = x,$

applying Laplace transform on both sides of equation, we get

$$3LY(x-1) \; - 4LY(x-1) + LY(x-2) = L(x)$$

Now,

$$Ly(x-1) = \int_0^\infty e^{-px}\, y(x-1)dx$$

Assume $\quad x - 1 = u \;\Rightarrow\; x = 1 + u$

$$dx = du$$

Then, $\quad L\,y(x-1) = LY(u)$

$$= \int_{-1}^\infty e^{-p(u+1)}\, Y(u)\, du$$

$$= \int_{-1}^\infty e^{-p} \cdot e^{-pu}\, Y(u)\, du$$

$$= e^{-p} \int_{-1}^\infty e^{-pu}\, Y(u)\, du$$

Since, $\quad Y(u) = 0,\; u < 0.$

Then,

$$= e^{-p} \int_{-1}^\infty e^{-pu} \cdot Y(u)\, du + e^{-p} \int_{-1}^\infty e^{-pu}\, Y(u)\, du$$

$$= 0 + e^{-pu} \cdot y, \quad LY(x) = y$$

$$L\,Y(x-2) = \int_0^\infty e^{-px}\, Y(x-2)dy$$

Assume $\quad x - 2 = u$

$\Rightarrow \quad x = u + 2$

and $\quad dx = du$

Then, $\quad LY(x-2) = LY(u)$

$$= \int_{-2}^{\infty} e^{-p(u+2)} Y(u)\, du$$

$$= e^{-2p} \int_{-2}^{\infty} e^{-pu} Y(u) du$$

$$= e^{-2p}\left[\int_{-2}^{0} e^{-pu} Y(u)\, du + \int_{0}^{\infty} e^{-pu} Y(u)\, du\right]$$

Since $Y(u) = 0$, for $u < 0$, then

$$= 0 + e^{-2p} \int_{0}^{\infty} e^{-pu} Y(u) du$$

$$= e^{-2p}\, y$$

Using equation (i), we get

$$3y - 4e^{-p}\, y + e^{-2p}\, y = \frac{1}{p^2}, \quad L(x) = \frac{1}{p^2}$$

$$y[3 - 4e^{-p} + e^{-2p}] = \frac{1}{p^2}$$

$$\Rightarrow \qquad y = \frac{1}{p^2[3 - 4e^{-p} + e^{-2p}]} = \frac{1}{p^2(1 - e^{-p})(3 - e^{-p})}$$

$$= \frac{1}{2p^2}\left[\frac{1}{1 - e^{-p}} - \frac{1}{3 - e^{-p}}\right]$$

$$= \frac{1}{2p^2}\left[(1 + e^{-p} + e^{-2p} + e^{-3p} + \ldots) - \frac{1}{3}(1 - (e^{-p}/3)^{-1}\right]$$

$$= \frac{1}{2p^2}\left[(1 + e^{-p} + e^{-2p} + \ldots) - \frac{1}{3}\left(1 + \frac{e^{-p}}{3} + \frac{e^{-2p}}{3^2} + \ldots\right)\right]$$

$$= \frac{1}{3p^2} + \frac{1}{2}\sum_{n=1}^{\infty}\left(1 - \frac{1}{3^{n+1}}\right)\frac{e^{-np}}{p^2}$$

Applying inverse Laplace transform, we get

$$Y = \frac{x}{3} + \frac{1}{2}\sum_{n=1}^{[x]}\left(1 - \frac{1}{3^{n+1}}\right)(x - n)$$

$[x]$ is the greatest integer less than or equal to x.

Example 2 Solve the difference equation

$a_{n+2} - 5a_{n+1} + 6a_n = 4^n$, $a_0 = 0$, $a_1 = 1$.

Hence, find a_5.

Solution: Given that

$$a_{n+2} - 5a_{n+1} + 6a_n = 4^n \qquad ...(i)$$

Let us assume

$$a_n = Y(x),\ a_{n+1} = Y(x+1),\ a_{n+2} = Y(x+2) \text{ etc.}$$

and
$$4^n = f(x)$$

Then equation (i) becomes

$$Y(x+2) - 5Y(x+1) + 6Y(x) = f(x) \qquad ...(ii)$$

Now, applying Laplace transform on both sides of equation (ii), we get

$$e^{2p}y - \frac{(1-e^{-p})e^p}{p} - 5\,e^p y + 6y = \frac{1-e^{-p}}{(1-4\,e^{-p})p}$$

$$\Rightarrow \qquad y = \frac{(1-e^{-p})e^p}{(e^p-2)(e^p-3)p} + \frac{1-e^{-p}}{(1-4e^{-p})(e^p-2)(e^p-3)p}$$

$$= \left(\frac{1}{e^p-3} - \frac{1}{e^p-2}\right)\frac{(1-e^{-p})e^p}{p} + \frac{e^p-1}{(e^p-2)(e^p-3)(e^p-4)p}$$

$$\Rightarrow \qquad y = \left(\frac{1}{1-3e^{-p}} - \frac{1}{1-2e^{-p}}\right)\left(\frac{1-e^{-p}}{p}\right) + \left(\frac{1/2}{e^p-2} - \frac{1}{e^p-3} + \frac{1/2}{e^p-4}\right)\left(\frac{e^p-1}{p}\right) \qquad ...(iii)$$

Applying inverse Laplace transform on equation (iii), we get

$$Y(x) = 3^n - 2^n + \frac{1}{2}\cdot 2^n - 3^n + \frac{1}{2}\cdot 4^n$$

$$= \frac{1}{2}(4^n - 2^n)$$

On putting $Y(x) = a_n$ and $n = 5$, we have

$$a_n = \frac{1}{2}(4^5 - 2^5) = 496$$

Example 3 Prove that

$$Y(x) = \sum_{n=0}^{[x]} \frac{2^n (x-n)^{n+2}}{(n+2)!}$$

is the solution of the differential–difference equation.

$$Y'(x) = 2Y(x-1) + x, \quad \text{If } Y(x) = 0,\ x \le 0.$$

Solution: Given that

$$Y'(x) = 2Y(x-1) + x, \quad Y(x) = 0,\ x \le 0 \qquad \text{...(i)}$$

Applying Laplace transform on equation (i), we get

$$L[Y'(x)] = 2\,L[Y(x-1)] + L[x]$$

$$\Rightarrow \quad py - Y(0) = 2\,e^{-p}\,y + \frac{1}{p^2}$$

$$\Rightarrow \quad py - 0 = 2\,e^{-p}\,y + \frac{1}{p^2}$$

$$\Rightarrow \quad y = \frac{1}{p^2(p-2e^{-p})}$$

$$= \frac{1}{p^3}\cdot\frac{1}{[1-(2e^{p}/p)]}$$

$$= \frac{1}{p^3}\left(1-\frac{2e^{-p}}{p}\right)^{-1}$$

$$= \frac{1}{p^3}\left[1+\frac{2e^{-p}}{p}+\left(\frac{2e^{-p}}{p}\right)^2+\ldots\right]$$

$$= \frac{1}{p^3}+\frac{2e^{-p}}{p^4}+\frac{4e^{-2p}}{p^5}+\ldots$$

$$\Rightarrow \quad y = \sum_{n=0}^{\infty}\frac{2^n\,e^{-np}}{p^{n+3}}$$

Applying inverse Laplace transform, we get

$$L^{-1}\left[\frac{e^{-np}}{p^{n+3}}\right] = \begin{cases}\dfrac{(x-n)^{n+2}}{(n+2)!}, & x \ge n\\ 0, & \text{Otherwise}\end{cases}$$

$$\Rightarrow \quad Y(x) = \sum_{n=0}^{[x]}\frac{2^n\,(x-n)^{n+2}}{(n+2)!}$$

Hence proved.

EXERCISE

1. Solve the difference equations:

(i) $a_{n+2} - 2a_{n+1} + 2a_n = 0, a_0 = 0, a_1 = 1.$

$$\left[\textbf{Ans. } a_n = \frac{[(1+i)^n - (1-i)^n]}{2i}\right]$$

(ii) $a_{n+3} - 2a_{n+2} - a_{n+1} + 2a_n = 0, \quad a_0 = 0, a_1 = 1, a_2 = 1.$

$$\left[\textbf{Ans. } a_n = \frac{1}{3}[2^n - (-1)^n]\right]$$

(iii) $3a_{n+2} - 5a_{n+1} + 2a_n = 0, \quad a_0 = 1, a_1 = 0.$

$$\left[\textbf{Ans. } a_n = 3\left(\frac{2}{3}\right)^n - 2\right]$$

(iv) $a_{n+2} - 5a_{n+1} + 6a_n = 0, \quad a_0 = 0, a_1 = 1.$

$$\left[\textbf{Ans. } a_n = 3^n - 2^n\right]$$

(v) $a_{n+2} - 6a_{n+1} + 5a_n = 2^n, \quad a_0 = 0, a_1 = 0.$

$$\left[\textbf{Ans. } a_n = \frac{1}{4} + \frac{1}{12}\cdot 5^n \frac{1}{3}\cdot 2^n\right]$$

(vi) $a_{n+2} - 5a_{n+1} + 6a_n = 2n + 1 \quad a_0 = 0, a_1 = 1.$

$$\left[\textbf{Ans. } a_n = \frac{5}{2}\cdot 3^n - 5^n\cdot 2^n + n + \frac{5}{2}\right]$$

2. Solve the following equations

(i) $8Y(x) - 12Y(x-1) + 4Y(x-2) = f(x), \quad \text{if } Y(x) = 0 \ \forall\ x \le 0$

and $f(x) = \begin{matrix} 0, & x < 0 \\ e^{-x}, & x > 0 \end{matrix}$

$$\left[\textbf{Ans. } Y(x) = \frac{1}{8}e^{-x}\left[1 + \sum_{n=1}^{[x]}(2 - 2^{-n})e^n\right]\right]$$

(ii) $Y'(x) - Y(x-1) = x^2, \quad \text{if } Y(x) = 0 \ \forall\ x \le 0.$ Find $y(\pi)$

$$\left[\textbf{Ans. } Y(x) = 2\sum_{n=0}^{[x]}\frac{(x-n)^{n+3}}{(n+3)!}, y(\pi) = 12.12\right]$$

(iii) $Y''(x) - Y(x-1) = f(x), \quad \text{if } Y(x) = 0, Y'(x) = 0 \ \forall\ x \le 0$

and $f(x) = \begin{cases} 0, & x \le 0 \\ 2x, & x > 0 \end{cases}$

$$\left[\textbf{Ans. } Y(x) = 2\sum_{n=0}^{[x]}\frac{(x-n)^{zn+3}}{(2n+3)!}\right]$$

5.7 SOLUTION OF INTEGRAL EQUATIONS BY USING LAPLACE TRANSFORMS

Integral Equation: The equation

$$F(t) = y(t) + \int_a^b k(u, t)\, F(u)\, du$$

is said to be an integral equation, where $y(t)$ and $k(u, t)$ are known while a and b are functions of t or constants.

Integral equation of convolution type: If an equation can be put in the form as given below.

$$F(t) = y(t) + \int_0^t K(t-u)\, F(u) du$$

is said to be an integral equation of convolution type which can be expressed as:

$$F(t) = y(t) + K(t) * F(t)$$

Abel's integral equation: If an equation can be put on

$$G(t) = \int_0^t \frac{F(u) du}{(t-u)^n}$$

is said to be Abel's integral equation. $F(u)$ is unknown and $G(t)$ is known, $0 < n < 1$.

integro-differential equation: An equation of the following form:

$$F'(t) = F(t) + y(t) + \int_0^t F(u) \sin(t-u)\, du$$

is said to be Integro-differential equation.

Example 1 Solve the integral equation

$$F(t) = 1 + \int_0^t F(u) \sin(t-u)\, du$$

and verify your solution.

Solution: The given integral equation

$$F(t) = 1 + \int_0^t F(u) \sin(t-u)\, du \qquad \text{...(i)}$$

Equation (i) can be written as (by using convolution theorem)

$$F(t) = 1 + F(t) * \sin t \qquad \text{...(ii)}$$

Applying Laplace transform,

$$f(p) = \frac{1}{p} + \frac{1}{1+p^2} f(p)$$

$$\Rightarrow \quad \left(1 - \frac{1}{1+p^2}\right) f(p) = \frac{1}{p}$$

$$f(p) = \frac{1+p^2}{p^3} = \frac{1}{p^3} + \frac{1}{p}$$

Applying inverse Laplace transform, we get

$$F(t) = 1 + \frac{1}{2}t^2$$

Verification:

$$= 1 + \int_0^t \left(1 + \frac{1}{2}u^2\right)\sin(u-t)du$$

$$= 1 - \int_0^t \sin(u-t)du - \frac{1}{2}\int_0^t u^2 \sin(u-t)du$$

$$= 1 + [\cos(u-t)]_0^t - \frac{1}{2}\left[-u^2\cos(u-t) + \int 2u\cos(u-t)du\right]_0^t$$

$$= 1 + (1 - \cos t) + \frac{t^2}{2} - \int_0^t u\cos(u-t)du$$

$$= 2 - \cos t + \frac{t^2}{2} - [u\sin(u-t) + \cos(u-t)]_0^t$$

$$= 2 - \cos t + \frac{t^2}{2} - [0 + (1 - \cos t)]$$

$$= 1 + \frac{1}{2}t^2$$

Hence, proved

Example 2 Solve the integral equation

$$F(t) = t + 2\int_0^t \cos(t-u)F(u)du$$

Solution: Given integral equation

$$F(t) = t + 2\int_0^t \cos(t-u)F(u)du \quad \text{...(i)}$$

Equation (i) can be expressed as

$$F(t) = t + 2\cos t * F(t) \quad \text{...(ii)}$$

Applying Laplace transform on equation (ii), we get

$$L[F(t)] = L[t] + 2L[\cos t * F(t)]$$

$$\Rightarrow \quad f(p) = \frac{1}{p^2} + 2\frac{p}{1+p^2}f(p)$$

or $$\left(1-\frac{2p}{1+p^2}\right)f(p) = \frac{1}{p^2}$$

$\Rightarrow$ $$f(p) = \frac{1+p^2}{(p-1)^2 p^2}$$

or $$f(p) = \frac{1}{(p-1)^2}+\frac{1}{(p-1)^2 p^2} \quad \text{...(iii)}$$

Applying inverse Laplace transform on equation (iii), we get

$$F(t) = L^{-1}\left[\frac{1}{(p-1)^2}\right]+L^{-1}\left[\frac{1}{(p-1)^2 p^2}\right]$$

$$= t\cdot e^t + (t-2)\, e^t + t + 2$$

$\Rightarrow$ $$F(t) = 2(t-2)\, e^t + t + 2$$ ***Ans.***

Example 3 Solve the integral equation

$$F'(t) = t+\int_0^t F(t-u)\cos u\, du,\ F(0) = 4.$$

Solution: Given integral equation

$$F'(t) = t+\int_0^t F(t-u)\cos u\, du \quad \text{...(i)}$$

Equation (i) can be written as

$$F'(t) = t + F(t) * \cos t \quad \text{...(ii)}$$

Applying Laplace transform on equation (ii), we get

$$L[F'(t)] = L[t] + L[F(t) * \cos t]$$

$\Rightarrow$ $$p\, f(p) - F(0) = \frac{1}{p^2}+\frac{p}{1+p^2}f(p)$$

$\Rightarrow$ $$\left(p-\frac{p}{1+p^2}\right)f(p) = \frac{1}{p^2}+4$$

or $$f(p) = \frac{p^2+1}{p^5}+\frac{4(p^2+1)}{p^3}$$

$\Rightarrow$ $$f(p) = \frac{4}{p}+\frac{5}{p^3}+\frac{1}{p^5} \quad \text{...(iii)}$$

Applying inverse Laplace transform on equation (iii), we get

$$F(t) = 4+\frac{5t^2}{2}+\frac{t^4}{4!}$$

or $$F(t) = 4+\frac{5t^2}{2}+\frac{t^4}{24}$$

Example 4 Solve the given Abel's equation

$$1 + t + t^2 = \int_0^t \frac{F(u)du}{(t-u)^{1/2}}.$$

Solution: Solve the given Abel's equation

$$t^2 + t + 1 = \int_0^t \frac{F(u)du}{(t-u)^{1/2}} \qquad \text{...(i)}$$

Equation (i) can be written as

$$t^2 + t + 1 = F(t) * t^{1/2} \qquad \text{...(ii)}$$

Applying Laplace transform on equation (ii), we get

$$\frac{2!}{p^3}+\frac{1}{p^2}+\frac{1}{p} = \frac{\Gamma(1/p)}{p^{1/2}}\cdot f(p)$$

$$\Rightarrow \quad \frac{1}{p}+\frac{1}{p^2}+\frac{2}{p^3} = \frac{\sqrt{\pi}}{p^{1/2}}\cdot f(p)$$

or $$\sqrt{\pi}\, f(p) = \frac{1}{p^{1/2}}+\frac{1}{p^{3/2}}+\frac{2}{p^{5/2}} \qquad \text{...(iii)}$$

Applying inverse Laplace to equation (iii), we get

$$\sqrt{\pi}\, L^{-1}[f(p)] = L^{-1}\left[\frac{1}{p^{1/2}}\right]+L^{-1}\left[\frac{1}{p^{3/2}}\right]+2L^{-1}\left[\frac{1}{p^{5/2}}\right]$$

$$\Rightarrow \quad \sqrt{\pi}\, F(t) = \frac{t^{-1/2}}{\Gamma(1/2)}+\frac{t^{1/2}}{\Gamma(3/2)}+\frac{2t^{3/2}}{\Gamma(5/2)} \qquad \left[As\ L[t^n]=\frac{\Gamma(n+2)}{p^{n+1}}\right]$$

or $$F(t) = \frac{1}{\pi}\left[t^{-1/2}+2t^{1/2}+\frac{8}{3}t^{3/2}\right].$$

Example 5 Solve

$$F'(t) = \sin t+\int_0^t F(t-u)\cos u\, du\ ,\ F(0) = 0.$$

Solution: Given equation

$$F'(t) = \sin t + \int_0^t F(t-u)\cos u \, du \qquad \text{...(i)}$$

Equation (i) can be written as

$$F'(t) = \sin t + F(t) * \cos t \qquad \text{...(ii)}$$

Applying Laplace transform on equation (ii), we get

$$p\,f(p) - F(0) = \frac{1}{1+p^2} + \frac{p}{1+p^2} \cdot f(p)$$

$$\Rightarrow \quad \left(p - \frac{p}{1+p^2}\right) f(p) = \frac{1}{1+p^2}$$

or

$$\frac{p^3}{1+p^2} f(p) = \frac{1}{1+p^2}$$

$$\Rightarrow \quad f(p) = \frac{1}{p^3} \qquad \text{...(iii)}$$

Applying inverse Laplace transform on equation (iii), we get

$$F(t) = \frac{1}{2}t^2$$

Example 6 Convert the given differential equations as an integral equation

$2F''(t) - 3F'(t) - 2F(t) = 4e^{-t} + 2\cos t$, if $F(0) = 4$, $F'(0) = -1$.

Solution: Given that

$$2F''(t) - 3F'(t) - 2F(t) = 4e^{-t} + 2\cos t \qquad \text{...(i)}$$

and $F(0) = 4, \; F'(0) = -1$

Process 1. Integrating equation (i) under the limit 0 to t, we get

$$2[F'(u)]_0^t - 3[F(u)]_0^t - 2\int_0^t F(u)du = -4[1 - e^{-t}]_0^t + 2\,[\sin u]_0^t \sin t$$

$$\Rightarrow \quad 2F'(t) - 2F'(0) - 3F(t) + 3F(0) - 2\int_0^t F(u)du = 4(1 - e^{-t}) + 2\sin t$$

On applying given conditions, we get

$$2F'(t) - 3F(t) - 2\int_0^t F(u)du = -4e^{-t} + 2\sin t - 10 \qquad \text{...(ii)}$$

Now, integrating equation (ii) under the limits 0 to t, we have

$$2[F(u)]_0^t - 3\int_0^t F(u)du - 2\int_0^t (t-u)F(u)du = 4\left[e^{-u}\right]_0^t - 2[\cos u]_0^t - 10t$$

$$\Rightarrow \quad 2F(t) + \int_0^t (-2t + 2u - 3)\, F(u)du = 4e^{-t} - 2\cos t - 10t + 6$$

Example 7 Convert the given integral equation

$$F(t) - \int_0^t (t-u)\sec t \cdot F(u)du = t$$

into differential equation and associated conditions.

Solution: Given integral equation

$$F'(t) - \int_0^t (t-u)\sec t \cdot F(u)du = t \qquad \text{...(i)}$$

We have

$$\frac{d}{dt}\int_a^b k(u,t)du = \int_a^b \frac{\partial k}{\partial t}du + k(b,t)\frac{db}{dt} - k(a,t)\frac{da}{dt} \qquad \text{...(ii)}$$

Differentiating equation (i) and using equation (ii), we have

$$F'(t) - \int_0^t [\sec t + (t-u)\sec t \tan t]F(u)du = 1$$

$$\Rightarrow \quad F'(t) - \int_0^t \sec t\, F(u)du - \tan t \int_0^t (t-u)\sec t\, F(u)du = 1$$

or
$$F'(t) - \int_0^t \sec t\, F(u)du + \tan t \cdot [t - F(t)] = 1 \qquad \text{...(iii)}$$

Now, differentiating equation (iii), we get

$$F''(t) + \sec^{2t}[t - F(t)] + \tan t[1 - F'(t)] - \sec t \tan t \int_0^t F(u)du - \sec t\, F(t) = 0$$

$$\Rightarrow \quad F''(t) + \sec^{2t}[t - F(t)] + \tan t[1 - F'(t)] - \sec t\, F(t)$$
$$+ \tan t\, [1 - F'(t) - t\tan t + \tan t\, F(t)] = 0$$

$$F''(t) - (\tan t + \tan t)\, F'(t) + [-\sec^2 t + \tan^2 t - \sec t]\, F(t)$$
$$+ [\tan t - t\tan^2 t + \tan t + t\sec^2 t] = 0$$

$$\Rightarrow \quad F''(t) - 2\tan t\, F'(t) - (1 + \sec t)F(t) + (t + 2\tan t) = 0 \qquad \text{...(iv)}$$

On replacing $t = 0$ in equation (i) and equation (iii), we get

$$F(0) = 0$$

$$F'(0) = 1. \qquad \text{...(v)}$$

Equation (iv) is required differential equation and equation (v) represents boundary conditions.

EXERCISE

1. Prove that solution of the given integral equation

$$F(t) = 4t - 3\int_0^t F(u)\sin(t-u)du$$

is $$F(t) = t + 3\sin t\cos t$$

2. Prove that $F(t) = \dfrac{3\sqrt{3}}{4\pi}t^{1/3}(2+3t)$ is the solution of the integral equation

$$\int_0^t \frac{F(u)du}{(t-u)^{1/3}} = t(1+t)$$

3. Solve the following integral equations:

(i) $F(t) = 1 + 2\int_0^t e^{-2u} F(t-u)du.$

[***Ans.*** $F(t) = 1 + 2t$]

(ii) $2F(t) = 2 - t\int_0^t F(t-u)F(u)du.$

[***Ans.*** $F(t) = 1, -1$]

(iii) $F(t) = 1 + 2\int_0^t F(t-u)\cos u du\cdot$

[***Ans.*** $F(t) = 1 + 2t\cdot e^t$]

(iv) $y(t) = t + \frac{1}{6}\int_0^t (t-u)^3 y(u)du.$

[***Ans.*** $y(t) = 1/2(\sinh t + \sin t)$]

5.8 APPLICATIONS TO DYNAMICS

Example 1 A mass m moves along the x-axis under the influence of a force which is proportional to its instantaneous speed and in a direction opposite to the direction of motion. Assuming that at $t = 0$, location of particle is $x = a$ and moving to the right with speed u. Find the position where the mass comes to rest.

Solution: Particle's equation of motion

$$m\cdot\frac{d^2x}{dt^2} = -k\cdot\frac{dx}{dt}$$

or $$mx'' = -kx' \qquad ...(i)$$

Given boundary conditions are

$$x(0) = a\ ,\ x'(0) = u \qquad ...(ii)$$

Then by applying Laplace transform on equation (i), we get

$$m\,L[x''] = -\,k[x']$$

$$\Rightarrow \quad m[p^2X - px(0) - x'(0)] + k[pX - x(0)] = 0$$

$$\Rightarrow \quad m[p^2X - ap - u] + k[pX - a] = 0$$

Then

$$X = \frac{mu + ka + mpa}{mp[p + k/m]}$$

$$= \frac{mu + ka}{mp[p + k/m]} + \frac{a}{p + k/m}$$

$$= \frac{mu + ka}{m}\left[\frac{m}{kp} - \frac{1}{p + k/m}\right] + \frac{a}{p + k/m} \qquad ...(iii)$$

Applying inverse Laplace transform on equation (iii), we get

$$L^{-1}[X] = x = \left[\frac{mu + ka}{m}\right]\left[\frac{m}{k} - e^{-kt/m}\right] + a\,e^{-kt/m} \qquad ...(iv)$$

$$\Rightarrow \quad \frac{dx}{dt} = \left[\frac{(mu + ka)k}{m^2} - \frac{ak}{m}\right]e^{-kt/m}$$

Given

$$\frac{dx}{dt} = 0$$

$$\Rightarrow \quad \left[\frac{(mu + ka)k}{m^2} - \frac{ak}{m}\right]e^{-kt/m} = 0$$

Then, $\dfrac{mu + ak}{m} - a = 0$ as $e^{-kt/m} \neq 0$

Using equation (iv), we get

$$x = a\left[\frac{m}{k} - e^{-kt/m}\right] + a\,e^{-kt/m}$$

$$= \frac{am}{k}$$

$$= \frac{m}{k}\left[\frac{mu + ak}{m}\right]$$

$$= \frac{mu + ak}{k}$$

Then, if $\frac{dx}{dt}=0$, $x=\frac{mu}{k}+a$

Therefore, mass comes to rest at a distance $\frac{mu}{k}+a$ from the centre.

Example 2 A mass of 2 grams moves on the x-axis and is attracted towards origin O with a force numerically equal to $8x$. If it is initially at rest at $x = 10$.

Find its position at any time t. Also a damping force numerically equal to eight times the instantaneously velocity acts.

Solution: Given that

$$m = 2$$

Resistance = $8x$, $x(0) = 10$, $x'(0) = 0$

$$\text{Damping force} = 8\cdot\frac{dx}{dt}$$

Now, equation of motion

$$\Rightarrow \quad 2\cdot\frac{d^2x}{dt^2} = -8x - 8\cdot\frac{dx}{dt}$$

$$\Rightarrow \quad \frac{d^2x}{dt^2}+4\cdot\frac{dx}{dt}+4x = 0$$

Applying Laplace transform, we get

$$p^2X - p(x(0)) - x'(0) + 4X + [pX - x(0)] = 0$$

Since, $x'(0) = 0$ and $x(0) = 10$, then

$$p^2X - 10p + 4X + 4pX - 40 = 0$$

$$\Rightarrow \quad (p + 2)^2 X = 40 + 10p$$

$$X = \frac{10p+40}{(p+2)^2} = \frac{40}{(p+2)^2}+\frac{10(p+2)-20}{(p+2)^2}$$

$$\Rightarrow \quad x = L^{-1}\left[\frac{40}{(p+2)^2}+\frac{10}{(p+2)}-\frac{20}{(p+2)^2}\right]$$

$$= L^{-1}\frac{20}{(p+2)^2+10/p+2}$$

$$= 20\,e^{-2t}L^{-1}\frac{1}{p^2}+10\,e^{-2t}$$

$$= (20t + 10)e^{-2t}$$

Application to Beams

Example 3 A beam which is clamped at its ends $x = 0$, $x = 1$, carries a uniform load w_0 per unit length, show that the deflection at any point is

$$y(x) = \frac{w_0(x^2(1-x)^2)}{24\,EI}$$

Solution: We know that the deflection $y(x)$ can be represented by the differential equation as given below:

$$\frac{d^4y}{dx^4} = w_0\frac{1}{EI}$$

Let $\quad w_0 \cdot \frac{1}{EI} = k$

Then, $\quad \frac{d^4y}{dx^4} - k = 0 \qquad \text{...(i)}$

Applying Laplace transform on equation (i), we get

$p^4Y - p^3y(0) - p^2y'(0) - py''(0) - y'''(0) - k/p = 0 \qquad \text{...(ii)}$

Assuming $\quad L(y) = Y$

Boundary conditions are given as

$$y(0) = y'(0) = 0$$

$$y(l) = y'(l) = 0 \text{ and assume } y''(0) = k_1,\ y'''(0) = k_2$$

By equation (ii), we get

$p^4Y - py''(0) - y'''(0) - k/p = 0$

$$\Rightarrow \qquad p^4Y = k_1p + k_2 + k/p$$

$$\Rightarrow \qquad Y = \frac{k_1}{p^3} + \frac{k_2}{p^4} + \frac{k}{p^5} \qquad \text{...(iii)}$$

Applying inverse Laplace transform on equation (iii), we have

$$y = k_1\frac{x^2}{2!} + k_2\frac{x^3}{3!} + k\frac{x^4}{4!}$$

or

$$y = \frac{k_1x^2}{2} + \frac{k_2x^3}{6} + \frac{kx^4}{24}$$

Using $y(l) = y'(l) = 0$

$$0 = \frac{k_1 l^2}{2} + \frac{k_2 l^3}{6} + \frac{kl^4}{24} \quad \text{...(iv)}$$

$$0 = k_1 l + k_2 \frac{l^2}{2} + k\frac{l^3}{6} \quad \text{...(v)}$$

Solving equations (iv) and (v), we get

$$k_1 = \frac{al^2}{12}, \quad k_2 = -\frac{al}{2}$$

Then, $$y = \frac{kx^2}{24}(x^2 - 2lx + l^2)$$

$$= \frac{kx^2}{24}(l - x)^2$$

$$= \frac{w_0 x^2}{24\,EI}(l - x)^2$$

Example 4 An infinitely long string having one end at $x = 0$ is initially at rest on the x-axis. The end $x = 0$ undergoes a periodic transverse displacement given by $a_0 \sin nt$, $t > 0$. Find the displacement of any point on the string at $t > 0$.

Solution: The displacement $y(x, t)$ of any point of the string can be given by the equation

$$\frac{\partial^2 y}{\partial t^2} = a^2 \frac{\partial^2 y}{\partial x^2}, \; x, t > 0 \quad \text{...(i)}$$

Subject to boundary condition

$$y(0, t) = a_0 \sin nt$$

$$y(x, 0) = 0$$

$$\frac{dy}{dt} = 0 \text{ at } t = 0$$

and y is finite for every x.

Applying Laplace transform of the equation (i)

$$p^2 Y - py(x,0) - y'(x,0) = a^2 \frac{d^2 Y}{dx^2}$$

$$\Rightarrow \quad p^2 Y = a^2 \frac{d^2 Y}{dx^2}$$

or $$\frac{d^2 Y}{dx^2} - \frac{p^2}{a^2} Y = 0$$

$$\Rightarrow \quad Y(x, p) = k_1 e^{xp/a} + k_2 e^{-xp/a} \quad \text{...(ii)}$$

since y is finite.

Therefore, $Y(x, p) = k_2 e^{-xp/a}$ ...(iii)

Now,

$$y(0, t) = a_0 \sin t$$

Applying Laplace transform, we get

$$Y(0, p) = a_0 \frac{n}{n^2 + p^2}$$

$$\Rightarrow \quad k_2 = a_0 \frac{n}{n^2 + p^2}$$

$$\Rightarrow \quad Y = \frac{n\, a_0 e^{-xp/a}}{n^2 + p^2} \qquad \text{...(iv)}$$

Applying inverse Laplace transform, on equation (iv), we get

$$y(x, t) = a_0\, n\, L^{-1}\left[\frac{1}{p^2 + n^2} e^{-xp/a}\right]$$

$$= a_0 \sin n\left(t - \frac{x}{a}\right) H\left(t - \frac{x}{a}\right)$$

here, $H(t - x/a)$ is Heaviside unit step function.

Example 5 A beam which is hinged at its ends $x = 0$ and $x = l$ carries a uniform load (w_0) per unit length. Find the deflection at any point.

Solution: The deflection can be given as

$$\frac{d^4 y}{dx^4} = \frac{w_0}{EI},\ 0 < x < l. \qquad \text{...(i)}$$

Assuming $k = w_0/EI$, we have

$$\frac{d^4 y}{dx^4} = k \qquad \text{...(ii)}$$

With boundary conditions

$$y(0) = y''(0) = 0 \qquad \text{...(iii)}$$

and

$$y(l) = y''(l) = 0 \qquad \text{...(iv)}$$

Applying Laplace transform on equation (iv), we get

$$p^4 Y - p^3 y(0) - p^2 y'(0) - p y''(0) - y'''(0) - k/p = 0$$

$$\Rightarrow \quad p^4 Y - p^2 y'(0) - y'''(0) = k/p$$

Assume $y' = k_1$ and $y'''(0) = k_2$, then

$$p^4 Y = k_1 p^2 + k_2 + k/p$$

$$Y = \frac{k_1}{p^2} + \frac{k_2}{p^4} + \frac{k}{p^5}$$

Applying inverse Laplace transform, we get

$$y = k_1 x + k_2 \frac{x^3}{3!} + k \frac{x^4}{4!}$$

$$y = k_1 x + \frac{k_2 x^3}{6} + \frac{kx^4}{24}$$

Now, applying condition $y(l) = 0 = y''(l)$, we get

$$k_1 = kl^3/24$$

$$k_2 = -kl/2$$

Replacing the values, we get

$$y = \frac{kx}{24}(l-x)(l^2 + lx - x^2)$$

or

$$y = \frac{w_0 x}{24\, EI}(l-x)(l^2 + lx - x^2)$$

Example 6 An inductor of 4 henrys is in series with a resistance of 40 ohms and an emf of 160 volts. Assuming that at $t = 0$, the current in zero. Find the current at time $t > 0$.

Solution: Given that

$$L = 4H,\ R = 40\Omega,\ E = 160 \text{ V}$$

$$i = 0 \text{ at time } t = 0.$$

In the absence of conductor, we know that

$$L\frac{d^2Q}{dt^2} + R\frac{dQ}{dt} = E \qquad \text{...(i)}$$

Substitute the given values, we get

$$4Q'' + 40Q' = 160 \qquad \text{...(ii)}$$

Applying Laplace transform, we get by equation (ii)

$$4L[Q''] + 40L[Q]' = 160L[1]$$

$$\Rightarrow\ 4[p^2 q - pQ(0) - Q'(0)] - 40[pq - Q(0)] = 160 \cdot \frac{1}{p}, \text{ assuming } L[Q] = q \qquad \text{...(iii)}$$

$$i = \frac{dQ}{dt}$$

$\Rightarrow \qquad Q' = 0 \text{ at } t = 0$

and also $\qquad Q = 0 \text{ at } t = 0$

then by equation (iii), we have

$$4[p^2q] + 40\,pq = 160\cdot\frac{1}{p}$$

$$q = \frac{40}{p^2(p+10)} \qquad \text{...(iv)}$$

Applying inverse Laplace transform, we get

$$L^{-1}[q] = Q = 40\,L^{-1}\left[\frac{1}{p^2(p+10)}\right] \qquad \text{...(v)}$$

We know $L^{-1}\left[\dfrac{1}{p+10}\right] = e^{-10t}$

Then, $L^{-1}\left[\dfrac{1}{p(p+10)}\right] = \displaystyle\int_0^t e^{-10u}du = \frac{1}{10}(1-e^{-10t})$

Then, $L^{-1}\left[\dfrac{1}{p^2(p+10)}\right] = \displaystyle\int_0^t \frac{1}{10}(1-e^{-10u})du$

$$= \frac{1}{10}\left[t+\frac{e^{-10t}-1}{10}\right]$$

Therefore, $\qquad Q = 40\left(t+\dfrac{e^{-10t}-1}{10}\right)$

and current $\qquad i = \dfrac{dQ}{dt}$

$$= 40\left[1+\frac{-10e^{-10t}-0}{10}\right]$$

$\Rightarrow \qquad i = 4(1 - e^{-10t}).$

Example 7 An alternating emf, E sin wt is applied to an inductance and a capacitance C in series. Show that the current in the circuit is $\dfrac{wE}{L(n^2 - w^2)}(\cos wt - \cos nt)$, where $n^2 = I/LC$.

Solution: Differential equation to determine current I, can be given as

$$L\frac{dI}{dt}+\frac{Q}{C} = E\sin wt \qquad \text{...(i)}$$

at $$t = 0, I = Q = 0 \qquad \text{...(ii)}$$

Applying Laplace transform on equation (i), we get

$$L[pi - I(0)] + \frac{1}{C}q = \frac{Ew}{p^2 + w^2}$$

when $L[I] = i$ $L(Q) = q$

Then, $$L\,pi + \frac{1}{C}q = \frac{Ew}{p^2 + w^2} \qquad \text{...(iii)}$$

We know that

$$L[I] = L\left[\frac{dQ}{dt}\right]$$

$\Rightarrow$ $$i = pq - Q(0) \text{ given that } Q(0) = 0$$

Then, $$i = pq$$

By equation (iii), we have

$$L\,pi + \frac{1}{C}\cdot\frac{i}{p} = \frac{Ew}{p^2 + w^2}$$

$\Rightarrow$ $$i = \frac{Ew}{\left(Lp + \frac{1}{Cp}\right)\left(p^2 + w^2\right)}$$

$\Rightarrow$ $$Lp + \frac{1}{Cp} = \frac{1}{Cp}\left(\frac{p^2}{n^2} + 1\right)$$

$$= \frac{(p^2 + n^2)}{n^2 pC}$$

$\Rightarrow$ $$i = \frac{Ewp}{(p^2 + w^2)(p^2 + n^2)L}$$

$$i = \frac{Ew}{(n^2 + w^2)L}\left[\frac{p}{p^2 + w^2} - \frac{p}{p^2 + n^2}\right]$$

Applying inverse Laplace transform, we get

$$I = \frac{Ew}{L(n^2 + w^2)}[\cos wt - \cos nt]$$

Example 8 An inductor of 2*H*, a resistance of 16 ohms and a capacitor of 0.02 farads are connected is series with an emf of 100 sin 3*t* volts. At $t = 0$ the charge on the capacitor and current in the circuit are zero. Find the charge and current at any time $t > 0$.

Solution: We know that

$$L\frac{d^2Q}{dt^2} + R\cdot\frac{dQ}{dt} + \frac{Q}{C} = E \quad ...(i)$$

Given that

$$L = 2,\ R = 16,\ C = 0.02 \text{ and } E = 100 \sin 3t$$

Then by equation (i), we get

$$2\frac{d^2Q}{dt^2} + 16\cdot\frac{dQ}{dt} + \frac{Q}{0.02} = 100 \sin 3t$$

$$\Rightarrow \quad Q'' + 8Q' + 25Q = 50 \sin 3t \quad ...(ii)$$

Applying Laplace transform on equation (ii), we get

$$L[Q''] + 8L[Q'] + 25L[Q] = 50\ L[\sin 3t]$$

$$\Rightarrow \quad [p^2q - pQ(0) - Q'(0)] + 8[pq - Q(0)] + 259 = \frac{50(3)}{p^2+9}$$

Replacing $Q(0) = 0 = Q'(0)$

$$\text{Then, } p^2q + 8pq + 259 = \frac{150}{p^2+9}$$

$$q = \frac{150}{(p^2+8p+25)(p^2+9)}$$

$$= \frac{75}{52}\left[\frac{2-p}{p^2+9} + \frac{p+6}{p^2+8p+25}\right]$$

$$= \frac{75}{52}\left[\frac{2}{p^2+9} + \frac{-p}{p^2+9} + \frac{p+4}{(p+4)^2+9} + \frac{2}{(p+4)^2+9}\right]$$

Applying inverse Laplace transform, we have

$$L^{-1}[q] = Q = \frac{75}{52}\left[\frac{2}{3}\sin 3t - \cos 3t + e^{-4t}\cos 3t + \frac{2}{3}e^{-4t}\sin 3t\right]$$

$$= \frac{75}{156}\left[(2\sin 3t - 3\cos 3t) + e^{-4t}(2\cos 3t + 3\sin 3t)\right]$$

$$\text{Now,} \quad I = \frac{dQ}{dt}$$

$$= \frac{25}{52}\left[3(2\cos 3t + 3\sin 3t) + e^{-4t}\left(-4(2\sin 3t + 3\cos 3t)\right) + 3(2\cos 3t - 3\sin 3t)\right]$$

$$= \frac{25}{52}\left[3(2\cos 3t + 3\sin 3t) - e^{-4t}(17\sin 3t + 6\cos 3t)\right]$$

EXERCISE

1. Solve the boundary value problem

$$\frac{\partial^2 y}{\partial t^2} = a^2 \frac{\partial^2 y}{\partial x^2} - g \ (x > 0,\ t > 0)$$

with the boundary conditions:

$$y(x,0) = 0,\ y'_t(x,0) = 0,\ x \geq 0$$

$$y(0,t) = 0,\ \lim_{x\to\infty} y_x(x,t) = 0,\ t \geq 0$$

$$\left[\textbf{Ans. } y(x,t) = \frac{-gt^2}{2} + \frac{g}{2}\left(t - \frac{x}{a}\right)^2 H\left(t - \frac{x}{a}\right)\right]$$

2. Solve the boundary value problem

$$\frac{\partial^2 y}{\partial t^2} = a^2 \frac{\partial^2 y}{\partial x^2} \ (x > 0,\ t > 0)$$

with the boundary conditions

$$y(x,0) = 0,\ x \geq 0$$

$$y'_t(x,0) = 0,\ x > 0$$

$$y(0,t) = t$$

and $\lim_{x\to\infty} y(x,t) = 0,\ t \geq 0$

$$\left[\textbf{Ans. } y(x,t) = \left(t - \frac{x}{a}\right) H\left(t - \frac{x}{a}\right)\right]$$

Chapter 6

Z-Transform

6.1 INTRODUCTION

The Z-transform plays an important role to study various branches of science and technoogy. The Z-transform in the discrete form was studied for the first time by W. Hurowicz in 1947.

Sequence: Representation of real or complex numbers in ordered list form is said to be sequence

$$\{f(k)\} = \{a_0, a_1, a_2,..., a_k,...\}$$

The symbol ↑ denotes the zero or initial term position. If the symbol ↑ is absent in the given sequence then the left most element can be regarded as the initial or zero position term.

Example

(i) $f_1(k) = \{1, 2, 3, ..., 20\}$

In this sequence symbol ↑ is absent and 1 can be assumed as initial or zero position element.

(ii) $f_2(k) = \{-3, -2, -1, \underset{\uparrow}{0}, 1, 2, 3\}$ In this sequence '0' is the initial term or zero position term.

A sequence can be represented by its general term as given below:

$$f_3(k) = 4^k \ \forall \ k \in I$$

$$\Rightarrow \quad f_3(k) = (... \ 4^{-5}, 4^{-4}, 4^{-3}, 4^{-2}, 4^{-2}, 4^{-1}, 4^0, 4^1, 4^2, 4^3,...)$$

$$f_4(k) = 3^k: -2 \leq k \leq 3$$

Then, $f_4(k) = (3^{-2}, 3^{-1}, 3^0, 3^1, 3^2, 3^3)$

6.2 Z-TRANSFORM

For a given sequence, $\{f(k)\}$, of real numbers, Z-transform defined as

$$Z\{f(k)\} = F(z) = \sum_{k=-\infty}^{\infty} f(k)z^{-k} \tag{i}$$

For any complex variable (z) in the region of absolute convergence of infinite sum, where Z denotes operator of Z-transform and equation (i) is known as bilateral Z-transform.

For the limit $k = 0$ to ∞,

$$Z\{f(k)\} = F(z) = \sum_{k=0}^{\infty} f(k)z^{-k}$$

is known as unilateral Z-transform.

6.2.1 Region of Convergence

The region of convergence is the set of points in the complex plane for which the Z-transform summation converges.

$$\Rightarrow \qquad \text{Region of convergence} = \left\{ z : \left| \sum_{k=0}^{\infty} f(k)z^{-k} \right| \right\}$$

Let us explain by taking an example:

$$f(k) = \begin{cases} 0 & , \quad k < 0 \\ 3^k & , \quad k \geq 0 \end{cases}$$

Then, $\qquad \{f(k)\} = 3^0, 3^1, 3^2, 3^3, \ldots$

Thus, Z-transform of the given sequence $f(k)$ can be defined as

$$F(z) = Z\{f(k)\} = \sum_{k=0}^{\infty} 3^k z^{-k}$$

$$= \sum_{k=0}^{\infty} \left(\frac{3}{z}\right)^k$$

$$= 1 + \frac{3}{z} + \frac{9}{z^2} + \frac{27}{z^3} + \frac{81}{z^4} + \ldots$$

$$= \frac{1}{1 - \frac{3}{z}} = \frac{z}{z-3}$$

$$\Rightarrow \qquad \left|\frac{3}{z}\right| < 1$$

or $\qquad |z| > 3$

$\Rightarrow$ Region of convergence is the exterior points of the circular region $|z| = 3$.

Illustrative Examples

Example 1 Find Z-transform of the following sequences:

(i) $\{f(k)\} = \{3, \underset{\uparrow}{2}, 5, 8, 7, 9\}$

(ii) $\{f(k)\} = \{8, 6, 3, -1, 0, 1, 4, 5\}$

Solution:

(i) Given that

$$\{f(k)\} = \{3, \underset{\uparrow}{2}, 5, 8, 7, 9\}$$

Since zero position is second element in the sequence, therefore, summation will be from –1 to 4.

Then
$$F(z) = \sum_{-1}^{4} f(k)z^{-k}$$
$$= f(-1)z^1 + f(0)z^0 + f(1)\,z^{-1} + f(2)\,z^{-2} + f(3)z^{-3} + f(4)\,z^{-4}$$
$$= 3z + 3z^0 + 5z^{-1} + 8z^{-2} + 7z^{-3} + 9z^{-4}$$
$$= 3z + 2 + \frac{5}{z} + \frac{8}{z^2} + \frac{7}{z^3} + \frac{9}{z^4}$$

(ii) Given that

$$f(k) = \{8, 6, 3, -1, 0, 1, 4, 5\}$$

Since, no zero is position given in the sequence therefore, element '8' is considered zero position element or element corresponding to $k = 0$. Then,

$$f(z) = Z\{f(k)\} = \sum_{k=0}^{7} f(k)z^{-k}$$
$$= f(0)z^0 + f(1)z^{-1} + f(2)z^{-2} + f(3)z^{-3} + f(4)z^{-4} + f(5)z^{-5} + f(6)z^{-6} + f(7)z^{-7} + \ldots$$
$$= 8\cdot z^0 + 6\cdot z^{-1} + 3\cdot z^{-2} + (-1)\cdot z^{-3} + 0\cdot z^{-4} + 1\cdot z^{-5} + 4\cdot z^{-6} + 5\cdot z^{-7}$$
$$\Rightarrow \quad F(t) = 8 + \frac{6}{z} + \frac{3}{z^2} - \frac{1}{z^3} + \frac{1}{z^5} + \frac{4}{z^6} + \frac{5}{z^7}$$

6.3 Z-TRANSFORM OF SOME STANDARD SEQUENCES

(i)
$$Z(a^k) = \frac{z}{z-a}, k \geq 0$$
$$Z\{f(k)\} = \sum_{k=0}^{\infty} f(z)z^{-k}, k \geq 0$$
$$= \sum_{k=0}^{\infty} a^k\, z^{-k} = \sum_{k=0}^{\infty} \left(\frac{a}{z}\right)^k$$

$$= 1 + \frac{a}{z} + \frac{a^2}{z^2} + \cdots = \frac{1}{1 - \frac{a}{z}} = \frac{z}{z - a}$$

(ii) $$Z\{f(k)\} = \frac{z}{z-1}$$

where $f(k)$ is a unit step squence defined as

$$f(k) = \begin{cases} 0 & , \quad k < 0 \\ 1 & , \quad k \geq 0 \end{cases}$$

Then,

$$Z\{f(k)\} = \sum_{k=-\infty}^{\infty} f(k) z^{-k}$$

$$= \sum_{k=-\infty}^{0} f(k) z^{-k} + \sum_{k=0}^{\infty} f(k) z^{-k}$$

$$= \sum_{k=0}^{\infty} 1 \cdot z^{-k} = 1 + \frac{1}{z} + \frac{1}{z^2} + \cdots$$

$$= \frac{1}{1 - \frac{1}{z}} = \frac{z}{z - 1}$$

(iii) $$Z\{\delta(k)\} = 1$$

where, $\delta(k)$ is the unit impulse sequence defined as

$$\delta(k)\} = \begin{cases} 1 & , \quad n = 0 \\ 0 & , \quad n \neq 0 \end{cases}$$

We know that $$Z\{\delta(k)\} = \sum_{k=-\infty}^{\infty} \delta(k)\, z^{-k}$$

$$= 0 + 1 + 0 + \ldots = 1$$

(iv) $$Z(\cos k\,\theta) = \frac{z(z - \cos\theta)}{z^2 - 2z\cos\theta + 1}, k \geq 0$$

We know that $$e^{-ik\theta} = \cos k\theta - i \sin k\theta$$

or $$\cos k\theta = \text{Re}\,\{e^{-ik\theta}\}$$

Now, $$Z\{e^{-ik\theta}\} = Z\{e^{-i\theta}\}^k$$

$$= \frac{z}{z - e^{-i\theta}}, \qquad \left[Z(a^k) = \frac{z}{z-a} \right]$$

$$= \frac{z(z - e^{i\theta})}{(z - e^{i\theta})(z - e^{-i\theta})}$$

$$= \frac{z[z - (\cos\theta + i\sin\theta)]}{z^2 - z(e^{i\theta} + e^{-i\theta}) + 1}$$

$$= \frac{z[z - \cos\theta] - (iz\sin\theta)}{z^2 - 2z\cos\theta + 1}$$

Comparing real and imaginary parts, we have

$$Z(\cos k\theta) = \text{Re}\,[Z(e^{-ik\theta})]$$

$$= \frac{z(z - \cos\theta)}{z^2 - 2z\cos\theta + 1}, \; k \geq 0$$

and $$Z(\sin k\theta) = \frac{z\sin\theta}{z^2 - 2z\cos\theta + 1}, \; k \geq 0$$

Example 2 Find Z-transform of sequence $a^{|k|}$.

Solution: We know that

$$Z(a^{|k|}) = \sum_{k=-\infty}^{\infty} a^{|k|} z^{-k}$$

$$= \sum_{k=-\infty}^{-1} a^{-k} \cdot z^{-k} + \sum_{k=0}^{\infty} a^k z^{-k}$$

$$= \sum_{k=-\infty}^{-1} (az)^{-k} + \sum_{k=0}^{\infty} \left(\frac{a}{z}\right)^k$$

$$= [\ldots\ldots + a^3z^3 + a^2z^2 + az] + \left[1 + \frac{a}{z} + \frac{a^2}{z^2} + \ldots\right]$$

$$= \frac{az}{1 - az} + \frac{1}{1 - \frac{a}{z}}$$

$$= z\left[\frac{a}{1 - az} + \frac{1}{z - a}\right]$$

$$= \frac{z(1-a^2)}{(1-az)(z-a)}$$

Example 3 Find the Z-transform of $f(k)$ if

$$f(k) = \begin{cases} 4^k & , \quad k < 0 \\ 3^k & , \quad k \geq 0 \end{cases}$$

Solution: We know that

$$Z\{f(k)\} = \sum_{k=-\infty}^{\infty} f(k)\, z^{-k}$$

$$= \sum_{k=-\infty}^{1} f(k) z^{-k} + \sum_{k=0}^{\infty} f(k) z^{-k}$$

$$= \sum_{k=-\infty}^{-1} 4^k z^{-k} + \sum_{k=0}^{\infty} 3^k z^{-k}$$

$$= \sum_{k=-\infty}^{-1} \left(\frac{z}{4}\right)^{-k} + \sum_{k=0}^{\infty} \left(\frac{3}{z}\right)^{k}$$

$$= \left[\ldots + \left(\frac{z}{4}\right)^4 + \left(\frac{z}{4}\right)^3 + \left(\frac{z}{4}\right)^2 + \left(\frac{z}{4}\right)^1\right] + \left[1 + \frac{3}{z} + \left(\frac{3}{z}\right)^2 + \ldots\right]$$

$$= \frac{z/4}{1 - z/4} + \frac{1}{1 - 3/z}$$

$$= \frac{z}{4-z} + \frac{z}{z-3}$$

$$= \frac{-z}{z^2 - 7z + 12}$$

Since, both the series are convergent in the region $|z| < 4$ and $|z| > 3$. Then ROC of the z-transform is $3 < |z| < 4$.

6.4 PROPERTIES OF Z-TRANSFORM

(i) Linear Property

For any two sequences $f_1(k)$ and $f_2(k)$

$$Z\{c_1 f_1(k) + c_2 f_2(k)\} = c_1 Z\{f_1(k)\} + c_2 Z\{f_2(k)\}$$

We know that $$Z[f(k)] = F\{z\} = \sum_{-\infty}^{\infty} f(k) z^{-k}, k \in I$$

Then, $Z\{c_1 f_1(k) + c_2 f_2(k)\}$

$$= \sum_{-\infty}^{\infty} [c_1 f_1(k) + c_2 f_2(k)] z^{-k}$$

$$= \sum_{-\infty}^{\infty} [c_1 f_1(k) z^{-k} + c_2 f_2(k) z^{-k}]$$

$$= \sum_{-\infty}^{\infty} c_1 f_1(k) z^{-k} + \sum_{-\infty}^{\infty} c_2 f_2(k) z^{-k}]$$

$$= c_1 \sum_{-\infty}^{\infty} f_1(k) z^{-k} + c_2 \sum_{-\infty}^{\infty} c_2 f_2(k) z^{-k}$$

$$= c_1 Z\{f_1(k)\} + c_2 Z\{f_2(k)\}$$

$\Rightarrow \quad Z\{c_1 f_1(k) + c_2 f_2(k)]\}$

$$= c_1 Z\{f_1(k)\} + c_2 Z\{f_2(k)\}$$

(ii) Scaling in z-domain

If $\qquad Z\{f(k)\} = F(z)$

Then, $\qquad Z\{a^k f(k)\} = F(z/a)$

We have $\quad Z\{a^k f(k)\} = \sum_{-\infty}^{\infty} a^k f(k) \cdot z^{-k}$

$$= \sum_{-\infty}^{\infty} f(k) \cdot \left(\frac{a}{z}\right)^k$$

$$= \sum_{-\infty}^{\infty} f(k) \cdot \left(\frac{z}{a}\right)^{-k} = F\left(\frac{z}{a}\right)$$

$\Rightarrow \qquad Z\{a^k f(k)\} = F(z/a)$

(iii) Shifting Property

If $\qquad Z\{f(k)\} = F(z)$

Then,

(a) $Z\{f(k-n)\} = z^{-n} F(z)$

(b) $Z\{f(k+n)\} = z^{n} F(z)$

(a) We have, $Z\{f(k)\} = \sum_{k=-\infty}^{\infty} f(k) z^{-k}$

Then, $$Z\{f(k-n)\} = \sum_{k=-\infty}^{\infty} f(k-n)\cdot z^{-k}$$

$$= \sum_{k=-\infty}^{\infty} f(k-n)\, z^{-(k-n)}\cdot z^{-n}$$

$$= z^{-n} \sum_{k=-\infty}^{\infty} f(k-n)\cdot z^{-(k-n)}$$

$$= z^{-n}\cdot F(z)$$

Similarly, we can find,

$$Z\{f(k+n)\} = z^{n}\, F(z)$$

(iv) Left Shift Property for Unilateral Transform

$$Z\{f(k)\} = F(z),\ k \geq 0$$

Then, $$Z\{f(k+n)\} = z^{n}\, F(z) - \sum_{r=0}^{n-1} F(r) z^{n-r}$$

We know that $$Z\{f(k)\} = \sum_{k=-\infty}^{\infty} f(k) z^{-k}$$

Then, $$Z\{f(k+n)\} = \sum_{k=0}^{\infty} f(k+n)\, z^{-k},\ k \geq 0$$

$$= z^{+n} \sum_{k=0}^{\infty} f(k+n) z^{-(k+n)}$$

$$= z^{n} \sum_{r=n}^{\infty} f(r) z^{-r}$$

$$= z^{n}\left[\sum_{r=0}^{\infty} f(r) z^{-r} - \sum_{r=0}^{n-1} F(r)\cdot z^{-r}\right]$$

$$\Rightarrow \quad Z(f(k+n)) = z^{n}\left[F(z) - \sum_{r=0}^{n-1} F(r) z^{-r}\right]$$

Note: Special case

Let $k = 1,\ Z\{f(k+1)\} = z\,[F(z) - f(0)]$

$k = 2,\ Z\{f(k+2)\} = z^{2}\,[F(z) - F(0) - z^{-1} f(1)]$

(v) Unilateral Transform (Right Shift Property)

If $$Z\{f(k)\} = F(z)$$

Then, $$Z\{f(k-n)\} = z^{-n} F(z)$$

We know that

$$Z\{f(k-n)\} = \sum_{k=0}^{\infty} f(k-n)\, z^{-k}$$

$$= \sum_{k=0}^{\infty} f(k-n) z^{-(k+n)} \cdot z^{-n}$$

$$= z^{-n} \sum_{r=-n}^{\infty} F(r) z^{-r}$$

$$= z^{-n} \left[\sum_{r=0}^{\infty} f(r) z^{-n} + \sum_{r=-n}^{-1} f(r) z^{-r} \right]$$

$$= z^{-n} \left[F(z) + \sum_{r=-n}^{\infty} f(r) z^{-r} \right]$$

Since, $F(-1) = F(-2) = \ldots F(-n) = 0$

$\Rightarrow$ $Z\{f(k-n)\} = z^{-n} F(z)$

For $k = 1$, $Z\{f(k-1)\} = \dfrac{F(z)}{z}$

$k = 2$, $Z\{f(k-2)\} = \dfrac{F(z)}{z^2}$

(vi) Multiplication by *k*

If $Z\{f(k)\} = F(z)$

Then, $Z\{kf(k)\} = -z \cdot \dfrac{d}{dz} F(z)$

$Z\{kf(k)\}$

$$= \sum_{k=-\infty}^{\infty} k\, f(k) z^{-k}$$

$$= -z \sum_{k=-\infty}^{\infty} f(k)(-k\, z^{-k-1})$$

$$= -z \sum_{k=-\infty}^{\infty} f(k) \cdot \frac{d}{dz}(z^{-k})$$

$$= -z \sum_{k=-\infty}^{\infty} \frac{d}{dz} [f(k) \cdot z^{-k}]$$

$$= -z \frac{d}{dz} \sum_{k=-\infty}^{\infty} f(k) z^{-k}]$$

$$= -z \cdot \frac{d}{dz} F(z)$$

(vii) Division by *k*

If $\quad Z\{f(k)\} = F(z)$

Then, $\quad Z\left[\frac{f(k)}{k}\right] = -\int_0^z \frac{F(z)}{z} dz$

We know that

$$Z\left[\frac{f(k)}{k}\right] = \sum_{k=-\infty}^{\infty} \frac{f(k)}{k} z^{-k}$$

$$= \sum_{k=-\infty}^{\infty} f(k) \cdot \frac{z^{-k}}{k}$$

$$= -\sum_{k=-\infty}^{\infty} f(k) \cdot \int_0^z z^{-k-1} dz$$

$$= -\int_0^z \sum_{k=-\infty}^{\infty} f(k) z^{-k-1} dz$$

$$= -\int_0^z \sum_{k=-\infty}^{\infty} f(k) z^{-k-1} dz$$

$$= -\int_0^z z^{-1} \sum_{k=-\infty}^{\infty} f(k) z^{-k} dz$$

$$= -\int_0^z z^{-1} F(z) dz$$

$$= -\int_0^z \frac{F(z)}{z} dz$$

Example 4 Find the Z-transform of the following sequence for $k \geq 0$

(i) $\{k\, a^k\}$ (ii) $\{k^2\, a^k\}$.

Solution:

(i) Given that $f(k) = ka^k$

and, we know that $Z\{a^k\} = \dfrac{z}{z-a}$, $k \geq 0$

Then, $\quad Z\{ka^k\} = -z \dfrac{d}{dz} [Z\{f(k)\}]$

$$= -z \cdot \frac{d}{dz} \cdot \left(\frac{z}{z-a}\right)$$

$$= -z \cdot \left[\frac{z-a-z}{(z-a)^2}\right]$$

$$= \frac{az}{(z-a)^2}$$

$$\Rightarrow \qquad Z\{ka^k\} = \frac{az}{(z-a)^2}$$

(ii) $$f(k) = \{k^2 a^k\}$$

$$Z\{k^2 a^k\} = Z[k\{ka^k\}]$$

$$= -z \cdot \frac{d}{dz}\left[\frac{az}{(z-a)^2}\right]$$

$$= -za\left[\frac{(z-a)^2 - 2z(z-a)}{(z-a)^4}\right]$$

$$= \frac{az\,(z+a)}{(z-a)^3}$$

$$\Rightarrow \qquad Z\{k^2 a^k\} = \frac{az(z+a)}{(z-a)^3}$$

Example 5 Find Z-transform of the following sequences for $k \geq 0$.

$f(k) =$ (i) $ak/k!$

(ii) $\frac{a^k}{k!}e^{-a}$

(iii) $\frac{1}{k}, k > 0$

Solution: We know that $Z\{f(k)\} = \sum_{k=0}^{\infty} f(k)z^{-k}$ ...(i)

(i) $$Z\left\{\frac{a^k}{k!}\right\} = \sum_{k=0}^{\infty} \frac{a^k}{k!} z^{-k} = \sum_{k=0}^{\infty} \frac{1}{k!}\left(\frac{a}{z}\right)^k$$

$$= 1 + \frac{a}{z} + \frac{1}{2!}\left(\frac{a}{z}\right)^2 + \frac{1}{3!}\left(\frac{a}{z}\right)^3 + \ldots = e^{a/z} \quad \Rightarrow \quad Z\left\{\frac{a^k}{k!}\right\} = e^{a/z}$$

(ii) $$f(k) = \frac{a^k}{k!}e^{-a}$$

Then, $$Z\left\{\frac{a^k}{k!}e^{-a}\right\} = e^{-a}Z\left\{\frac{a^k}{k!}\right\} = e^{\frac{a}{z}-a} = e^{a(1-z)/z}$$

(iii) $$f(k) = \frac{1}{k}, \; k > 0$$

$$\Rightarrow \quad Z\{f(k)\} = \sum_{k=1}^{\infty} \frac{1}{k}z^{-k}$$

$$= z^{-1} + \frac{z^{-2}}{2} + \frac{z^{-3}}{3} + \frac{z^{-4}}{4} + \ldots$$

$$= -\left[-z^{-1} - \frac{z^{-2}}{2} - \frac{z^{-3}}{3} - \frac{z^{-4}}{4} + \ldots\right]$$

$$= -\log(1 - z^{-1})$$

$$= -\log\left(1 - \frac{1}{z}\right) = -\log\frac{z-1}{z}$$

$$= \log\frac{z}{z-1}$$

$$\Rightarrow \quad Z\left\{\frac{1}{(k+1)!}\right\} = \log\frac{z}{z-1}$$

Example 6 If $Z\left\{\frac{1}{(k+1)!}\right\} = e^{1/z}$, $k \geq 0$, then,

Find

(i) $Z\left\{\frac{1}{(k+1)!}\right\} = e^{1/z}$ (ii) $Z\left\{\frac{1}{(k-1)!}\right\}$.

Solution: Given that

(i) $Z\left\{\frac{1}{(k+1)!}\right\} = e^{1/z}$

We know that

$$Z\{f(k+n)\} = z^n F(z) - \sum_{r=0}^{n-1} f(r) z^{n-r}$$

$$\Rightarrow \quad Z\left\{\frac{1}{(k-1)!}\right\} = z^1 \cdot e^{1/z} - \sum_{r=0}^{n-1} \frac{1}{r!} z^{1-r}$$

(ii) $$= e^{1/z} - z$$
$$= z\,(e^{1/z} - 1)$$

(iii) Now $Z\,f(k-n) = z^{-n}\,F(z)$

$$\Rightarrow \quad Z\left\{\frac{1}{(k-1)!}\right\} = z^{-1}\cdot e^{1/z}$$
$$= z^{-1}\cdot e^{1/z}$$
$$\Rightarrow \quad Z\left\{\frac{1}{(k-1)!}\right\} = \frac{e^{1/z}}{z}$$

6.5 INITIAL VALUE THEOREM

If $$Z\{f(k)\} = F(z),\ k \geq 0$$

Then, $$f(0) = \lim_{k\to 0} f(z)$$
$$= \lim_{z\to\infty} F(z)$$

Proof: Given sequence is defined for $k \geq 0$. Then,

$$F(z) = Z\,\{f(k)\}$$
$$= \sum_{k=0}^{\infty} f(k)\,z^{-k}$$
$$\Rightarrow \quad F(z) = f(0) + \frac{f(1)}{z} + \frac{f(2)}{z^2} + \frac{f(3)}{z^3} + \ldots$$

Applying limit as $z \to \infty$, we get

$$\lim_{z\to\infty} F(z) = f(0)$$
$$\Rightarrow \quad F(0) = \lim_{k\to 0} F(k) = \lim_{z\to\infty} F(z)$$

Note: In some applications, we often need the value of $f(k)$ for $k = 0$ or as $k \to \infty$ without requiring complete process of $f(k)$. We can find these values by using initial and final value theorems.

6.6 FINAL VALUE THEOREM

If $$Z\{f(k)\} = F(z),\ k \geq 0$$

Then, $$\lim_{k\to 0} f(k) = \lim_{z\to 1}\,(z-1)\,F(z)$$

Proof: Using shifting property, we can write $(k \geq 0)$

$$Z\,\{f(k+1) - f(k)\}$$

$$= Z\{f(k+1)\} - Z\{f(k)\}$$
$$= [z\,F(z) - z\,f(0) - F(z)]$$
$$= (z-1)\,F(z) - z\,f(0) \qquad \text{...(i)}$$

Rewriting equation (i), we get

$$(z-1)\,F(z) - zf(0)$$
$$= Z\,[f(k+1) - f(k)]$$
$$= \sum_{k=0}^{\infty} [f(k+1) - f(k)]z^{-k}$$

Applying $z \to 1$, we get

$$\lim_{z\to 1} [(z-1)\,F(z)] - f(0) = \lim_{z\to 1} \sum_{k=0}^{\infty} [f(k+1) - f(k)]z^{-k}$$
$$= \sum_{k=0}^{\infty} (f(k+1) - f(k))$$
$$= \lim_{k\to\infty} [(f(1) - f(0)) + (f(2) - f(1))] + \ldots + (f(k+1) - f(k))$$
$$= \lim_{k\to\infty} [(f(k+1) - f(0)]$$

$$\Rightarrow \qquad \lim_{k\to\infty} f(k+1) = \lim_{z\to 1} [(z-1)\,F(z)] - f(0) + f(0)$$

or
$$\lim_{k\to\infty} f(k) = \lim_{z\to 1} (z-1)\,F(z) \qquad [\text{For, } k \to \infty, \text{ we assume } k+1 \simeq k]$$

$$\Rightarrow \qquad \lim_{k\to\infty} f(k) = \lim_{z\to 1} (z-1)\,F(z)$$

Partial Sum: If $Z\,[f(k)] = F(z)$

Then, prove that

$$Z\left[\sum_{n=-\infty}^{k} f(n)\right] = \frac{F(z)}{1-z^{-1}}$$
$$= \frac{z\,F(z)}{z-1}$$

Proof: Let us assume a function $h(k)$ such that

$$h(k) = \sum_{n=-\infty}^{k} f(n)$$

$$\Rightarrow \qquad Z\,[h(k)] = H(z)$$

Now, we have,

$$h(k) - h(k-1) = \sum_{n=-\infty}^{k} f(n) - \sum_{n=-\infty}^{k-1} f(n)$$
$$= f(k)$$

Applying Z-transform, we have

$$Z\{h(k) - h(k-1)\} = Z\{f(k)\}$$

or $$H(z) - z^{-1} H(z) = F(z)$$

$\Rightarrow$ $$(1 - z^{-1}) H(z) = F(z)$$

or $$H(z) = \frac{F(z)}{1-z^{-1}} = \frac{z\,F(z)}{z-1}$$

6.7 CONVOLUTION THEOREM

Convolution of two discrete sequences $f(k)$ and $g(k)$ can be defined as

$$f(k) * g(k) = \sum_{m=0}^{n} f(m)\, g(k-m) \quad \text{(* denotes the convolution operation)}$$

Z-transform of the convolution of two sequences is equal to the multiplication of its Z-transform

Let $$Z\{f(k)\} = F(z)$$

$$Z\{g(k)\} = G(z)$$

Then, $$Z\{f(k) * g(k)\}g = F(z)\, G(z)$$

Proof: Using definition of the Z-transform, we have

$$F(z)\, G(z) = \left(\sum_{k=0}^{\infty} f(k) z^{-k}\right)\left(\sum_{k=0}^{\infty} g(k) z^{-k}\right)$$

$$= \left[f(0) + \frac{f(1)}{z} + \frac{f(2)}{z^2} + \ldots\right]\left[g(0) + \frac{g(1)}{z} + \frac{f(2)}{z^2} + \ldots\right]$$

$\Rightarrow$ $$F(z)\, G(z) = \sum_{k=0}^{\infty} [f(0)\, f(k) + f(1)\, g(k-1) + f(2)\, g(k-2) + \ldots + f(k)\, g(0)] z^{-k}$$

$$= Z\,[f(0)\, f(k) + f(1)\, g(k-1) + \ldots + f(k)\, g(0)]$$

$$= Z\left[\sum_{m=0}^{\infty} f(k)\, g(k-m)\right]$$

$\Rightarrow$ $$F(z)\, G(z) = Z\,(f(k) * g(k))$$

Example 7 Find Z-transform of a^k, $k \geq 0$.

Solution: We have

$$Z\{1\} = \frac{z}{z-1}$$

$$\Rightarrow \qquad Z\{a^k, 1\} = \frac{z/a}{z/a-1}$$

$$= \frac{z}{z-a}$$

6.8 INVERSE Z-TRANSFORM

If $\qquad Z\{f(k)\} = F(z)$

Then the inverse of the Z transform, denoted by Z^{-1}, can be written as

$$Z^{-1}\{F(z)\} = f(k)$$

(i) ***Partial function method:*** In this method we decompose the function into simple fractions and then solve the problem.

(ii) ***Power series method:*** In this method, we find the inverse Z-transform by expanding $F(z)$ in power series, etc.

Example 8 Find the inverse Z-transform of

$$F(z) = \frac{z}{(2-z)(2z-1)}.$$

Solution: Given that

$$F(z) = \frac{z}{(2-z)(2z-1)} \qquad \text{(i)}$$

$$\Rightarrow \qquad \frac{F(z)}{z} = \frac{1}{(2-z)(2z-1)}$$

$$= -\frac{1}{3}\cdot\frac{1}{(z-2)} + \frac{2}{3}\cdot\frac{1}{(2z-1)}$$

$$\Rightarrow \qquad F(z) = \frac{1}{3}\cdot\frac{z}{z-1/2} - \frac{1}{3}\cdot\frac{z}{(z-2)} \qquad \text{...(ii)}$$

Applying inverse Z-transform, we get

$$Z^{-1}\{F(z)\} = \frac{1}{3}Z^{-1}\left\{\frac{z}{z-1/2}\right\} - \frac{1}{3}Z^{-1}\left\{\frac{z}{z-2}\right\} \qquad \text{...(iii)}$$

We know that

$$Z(a^k) = \frac{z}{z-a}$$

$$\Rightarrow \qquad Z^{-1}\left\{\frac{z}{z-a}\right\} = a^k$$

Then by equation (iii), we get

$$f(k) = \frac{1}{3}\left(\frac{1}{2}\right)^k - \frac{1}{3}(2)^k$$

$$\Rightarrow \quad Z^{-1}\left\{\frac{z}{(2-z)(2z-1)}\right\} = \frac{1}{3}\left[\left(\frac{1}{2}\right)^k - (2)^k\right]$$

Example 9 Find the inverse *Z*-transform of the function

$$F(z) = \frac{2z^2+3z}{(z+2)(z-4)}.$$

Solution: Given that

$$F(z) = \frac{2z^2+3z}{(z+2)(z-4)} \qquad \text{...(i)}$$

or $$\frac{F(z)}{z} = \frac{2z+3}{(z+2)(z-4)}$$

Now, $$\frac{2z+3}{(z+2)(z-4)} = \frac{A}{z+2}+\frac{B}{z-4}$$

or $$2z + 3 = A(z - 4) + B(z + 2)$$

$$\Rightarrow \quad A = \frac{1}{6}, B=\frac{11}{6}$$

Hence, $$\frac{2z+3}{(z+2)(z-4)} = \frac{1}{6}\cdot\left(\frac{z}{z+2}\right)+\frac{11}{6}\cdot\left(\frac{z}{z-4}\right)$$

or $$Z^{-1}\left[\frac{2z^2+3z}{(z+2)\,(z-4)}\right]=\frac{1}{6}Z^{-1}\left(\frac{z}{z+2}\right)+\frac{11}{6}Z^{-1}\left(\frac{z}{z+4}\right) \qquad \text{...(ii)}$$

Since, $$Z\{a^k\} = \frac{z}{z-a}$$

$$\Rightarrow \quad Z^{-1}\left(\frac{z}{z-a}\right) = a^k$$

Then by equation (ii), we have

$$Z^{-1}\left\{\frac{2z^2+3z}{(z+2)(z-4)}\right\} = \frac{1}{6}(-2)^k+\frac{11}{6}(4)^k$$

Example 10 Find the inverse Z-transform of

$$F(z) = \frac{z^3 - 20z}{(z-2)^3(z-4)}.$$

Solution: Given that

$$F(z) = \frac{z^3 - 20z}{(z-2)^3(z-4)} \quad ...(i)$$

or

$$\frac{F(z)}{z} = \frac{z^2 - 20}{(z-2)^3(z-4)}$$

For resolving in partial fraction, we have

$$\frac{z^2 - 20}{(z-2)^3(z-4)} = \frac{a + bz + cz^2}{(z-2)^3} + \frac{d}{(z-4)}$$

$$\Rightarrow \quad (a + bz + cz^2)(z - 4) + d(z - 2)^3 = z^2 - 20$$

On equating coefficients of z^0, z^1, z^2, z^3, we get

$$a = 6$$
$$b = 0$$
$$c = 1/2$$
$$d = -1/2$$

Then,

$$F(z) = \frac{1}{2}\cdot\frac{12z + z^3}{(z-2)^3} - \frac{1}{2}\frac{z}{(z-4)}$$

$$= \frac{1}{2}\left[\frac{z(z-2)^2 + 4z^2 + 8z}{(z-2)^3} - \frac{z}{z-4}\right]$$

$$= \frac{1}{2}\left[\frac{z}{z-2} + 2\cdot\frac{2z^2 + 4z}{(z-2)^3} - \cdot\frac{z}{z-4}\right]$$

Applying inverse Z-transform, we get

$$f(k) = \frac{1}{2}[2^k + 2 \cdot k^2 2^k - 4^k]$$

or

$$f(k) = 2^{k-1} + k^2\, 2^k - 2^{-1}\, 4^k$$

Example 11 Find the inverse Z-transform of the function

$$F(z) = \frac{3z^2 + z}{(5z-1)(5z+2)}.$$

Solution: Given that

$$F(z) = \frac{3z^2 + z}{(5z-1)(5z+2)} \quad ...(i)$$

or
$$\frac{F(z)}{z} = \frac{3z+1}{(5z-1)(5z+2)}$$

$$3z+1 = a(5z+2) + b(5z-1)$$
$$a = \frac{8}{15}$$
$$b = \frac{1}{15}$$

Then,
$$\frac{3z+1}{(5z-1)(5z+2)} = \frac{a}{5z-1} + \frac{b}{5z+2}$$

$$\Rightarrow \quad \frac{3z^2+z}{(5z-1)(5z+2)} = \frac{8}{15}\cdot\frac{z}{5z-1} + \frac{1}{15}\cdot\frac{z}{5z+2}$$

$$\Rightarrow \quad Z^{-1}\left[\frac{3z^2+z}{(5z-1)(5z+2)}\right] = \frac{8}{75}Z^{-1}\left(\frac{z}{z-1/5}\right) + \frac{1}{75}Z^{-1}\left(\frac{z}{z+2/5}\right)$$

or
$$Z^{-1}\,[F(z)] = \frac{8}{75}\left(\frac{1}{5}\right)^k + \frac{1}{75}\left(-\frac{2}{5}\right)^k$$

Example 12 Find the inverse Z-transform of the function

$$F(z) = \frac{8z-z^3}{(4-z)^3}.$$

Solution: Given that

$$F(z) = \frac{8z-z^3}{(4-z)^3}$$

or
$$F(z) = \frac{z(z^2-8)}{(z-4)^3}$$

We know that
$$Z\{4^k\} = \frac{z}{z-4}$$

$$\Rightarrow \quad Z\{k\,4^k\} = -z\cdot\frac{d}{dz}\left(\frac{z}{z-4}\right)$$
$$= \frac{4z}{(z-4)^2}$$

and
$$Z\{k^2\,4^k\} = -z\cdot\frac{d}{dz}\left[\frac{4z}{(z-4)^2}\right]$$
$$= \frac{4z^2+16z}{(z-4)^3}$$

Now, $$F(z) = z\left[\frac{a}{z-4} + \frac{4b}{(z-4)^2} + \frac{(4z+16)c}{(z-4)^3}\right]$$

Then, solving partial fractions, we get

$$F(z) = \frac{z}{z-4} + \frac{7}{4}\frac{4z}{(z-4)^2} + \frac{1}{4}\cdot\frac{4z^2+16}{(z-4)^3}$$

Applying inverse Z-transform and using derived results, we get

$$Z^{-1}\{F(z)\} = Z^{-1}\left(\frac{z}{z-4}\right) + \frac{7}{4}Z^{-1}\frac{4z}{(z-4)^2} + \frac{1}{4}Z^{-1}\left(\frac{4z^2+16}{(z-4)^3}\right)$$

$$\Rightarrow \qquad f(k) = 4^k + \frac{7}{4}k\cdot 4^k + \frac{1}{4}k^2 4^k$$

Example 13 Find the inverse Z-transform of the following function by power series method

$$F(z) = \log\frac{z}{z+1}.$$

Solution: Given that

$$F(z) = \log\frac{z}{z+1} \qquad \text{...(i)}$$

Let us assume $z = 1/u$, we get

$$\begin{aligned} F(z) &= \log\left(\frac{1/u}{1/u+1}\right) \\ &= -\log(1+u) \\ &= -u + \frac{1}{2}u^2 - \frac{1}{3}u^3 + \frac{1}{4}u^4 - \ldots \\ &= -z^{-1} + \frac{1}{2}z^{-2} - \frac{1}{3}z^{-3} + \ldots \\ &= \sum_{k=1}^{\infty}\frac{(-1)^k}{k}z^{-k} \qquad \text{...(ii)} \end{aligned}$$

We know that $$Z\{f(k)\} = \sum_{k=-\infty}^{\infty} f(k)z^{-k} \qquad \text{...(iii)}$$

On comparing equations (ii) and (iii), we get

$$f(k) = \begin{cases} (-1)^k/k, & k > 0 \\ 0, & k \le 0 \end{cases}$$

Example 14 Using convolution theorem, find inverse Z-transform of the function

$$E(z) = \frac{z^2}{(z-1)(z-2)}.$$

Solution: Given that

$$E(z) = \frac{z^2}{(z-1)(z-2)} \qquad \text{...(i)}$$

Let us assume $F(z) = \frac{z}{z-1}, |z| > 1$ ...(ii)

and $G(z) = \frac{z}{z-2}, |z| > 2$ ...(iii)

Here $|z| > 2$ is the intersection of two regions of convergence.

Now, from equations (ii) and (iii), we have

$$f(k) = (1)^k$$
$$g(k) = 2^k$$

We know that convolution of two discrete sequences $f(k)$ and $g(k)$ is defined as

$$f(k) * g(k) = \sum_{m=0}^{k} f(m)\, g(k-m)$$

Now, $Z^{-1}[F(z)\, G(z)] = f(k) * g(k)$

$$\Rightarrow \quad Z^{-1}\left[\frac{z^2}{(z-1)(z-2)}\right] = 1^k \cdot (2)^k$$

$$= \sum_{m=0}^{k} (1)^m (2)^{k-m}$$

$$= 2^k \sum_{m=0}^{k} \left(\frac{1}{2}\right)^m$$

$$= 2^k \left[\frac{1-\left(\frac{1}{2}\right)^{k+1}}{1-1/2}\right]$$

$$= 2^{k+1}\left[1-\left(\frac{1}{2}\right)^{k+1}\right]$$

Example 15 Using convolution theorem, find the inverse Z-transform of

$$E(z) = \frac{z^2}{(z-a)(z-b)}.$$

Solution: Given that

$$E(z) = \frac{z^2}{(z-a)(z-b)}$$

Let us assume $F(z) = \dfrac{z}{z-a}$

$$G(z) = \frac{z}{z-b}$$

Apply the inverse Z-transform,

$$f(k) = a^k$$

$$g(k) = b^k, \quad k \geq 0$$

Convolution of two discrete functions $f(k)$ and $g(k)$ is defined as

$$f(k) * g(k) = \sum_{m=0}^{k} f(m)\, g(k-m)$$

Now, $Z^{-1}[F(z)\, G(z)] = f(k) * g(k)$

$$\Rightarrow \quad Z^{-1}\left[\frac{z^2}{(z-a)(z-b)}\right] = f(k) * g(k)$$

$$= \sum_{m=0}^{k} a^m\, b^{k-m} = b^k \sum_{m=0}^{k} a^m\, b^{-m} = b^k \cdot \sum_{m=0}^{k} \left(\frac{a}{b}\right)^m$$

$$= b^k \frac{\left(\frac{a}{b}\right)^{n+1} - 1}{\frac{a}{b} - 1} = \frac{a^{k+1} - b^{k+1}}{a-b}$$

$$\Rightarrow \quad Z^{-1}\left[\frac{z^2}{(z-a)(z-b)}\right] = \frac{a^{k+1} - b^{k+1}}{a-b}$$

EXERCISE

1. Find Z-transform of sequences given below:

(i) $f(k) = \{1, 2, -3, 4, 6, 0, 5\}$ $\left[\textbf{Ans. } 1 + \frac{2}{z} - \frac{3}{z^2} + \frac{4}{z^3} + \frac{6}{z^4} + \frac{5}{z^6}\right]$

(ii) $f(k) = \{15, 10, 7, \underset{\uparrow}{4}, 1, -1, 0, 3, 6\}$

$\left[\textbf{Ans. } 15z^3 + 10z^2 + 7z + 4 + \frac{1}{z} - \frac{1}{z^2} + \frac{3}{z^4} + \frac{6}{z^5}\right]$

(iii) $f(k) = \frac{1}{5^k}$. $\left[\textbf{Ans. } ... + (5z)^2 + (5z)^1 + 1 + \frac{1}{5z} + \frac{1}{(5z)^2} + ...\right]$

(iv) $f(k) = \{a^k\}$ $k < 0$. $\left[\textbf{Ans. } \frac{-1}{(1 - az^{-1})}\right]$

2. Find the Z-transform of the following (for $k > 0$):

(i) $\sin 5k$. $\left[\textbf{Ans. } \frac{z \sin 5}{z^2 - 2z \cos 5 + 1}\right]$

(ii) $\cos 3k$. $\left[\textbf{Ans. } \frac{z(z - \cos 3)}{z^2 - 2z\cos 3 + 1}\right]$

(iii) $3^k \cos 5k$. $\left[\textbf{Ans. } \frac{z(z - 3\cosh 5)}{z^2 - 2z\cosh 5 + 9}\right]$

(iv) $\sin (5k + 3)$. $\left[\textbf{Ans. } \frac{z^2 \sin 3 + z \sin 2}{z^2 - 2z\cos 5 + 1}\right]$

3. Show that $Z(1/k!) = e^{1/z}$, hence deduce $Z[1/(n + 1)!]$ and $Z[1/(n + 2)!]$.

$[\textbf{Ans. } z(e^{1/z} - 1), z(e^{1/3} - 1 - z^{-1})]$

4. Find inverse Z-transform of

(i) $z/(z + 1)^2$. $[\textbf{Ans. } \{k(-1)^{k-1}\}, k \geq 0]$

(ii) $\frac{z}{3 - z}$, $|z| < 3$. $[\textbf{Ans. } \{3^k\}, k > 0]$

(iii) $\frac{z^3}{z^3 - 27}$, $|z| > 3$. $[\textbf{Ans. } \{3^k\}, k = 0, 3, 6, 9]$

(iv) $\frac{ze^{-\theta}}{(z - e^{-\theta})^2}$, $|z| > e^{-\theta}$ $[\textbf{Ans. } \{ke^{-\theta k}\}, k \geq 0]$

5. Find inverse Z-transform for

(i) $\frac{20z}{(4z - 1)(5z - 1)}$, $|4z| > 1$. $[\textbf{Ans. } \{20(4^{-k} - 5^{-k})U(k)]$

(ii) $\dfrac{2z^2+3z}{(z+2)(z-4)}$, $|z| > 4$. $\left[\textbf{Ans.}\ \left\{\frac{1}{6}(-z)^k+\frac{11}{6}(4^k)\right\}k \ge 0\right]$

(iii) $\dfrac{2z^3-20z}{(z-2)^3(z-4)}$, $|z| > 4$. $\left[\textbf{Ans.}\ \left\{2^{k-1}+k^2z^k-\frac{1}{2}4^k\right\}\ k \ge 0\right]$

(iv) $\dfrac{1}{(z-3)(z-2)}$, $|z| < 2$. $[\textbf{Ans.}\ \{2^{-k}-3^{-k}\}k \ge 1]$

(v) $\dfrac{1}{(z-2)(z-3)}$, $|z| > 3$. $\left[\textbf{Ans.}\ \begin{bmatrix} 3^{k-1}-2^{k-1} & , & k \ge 1 \\ 0 & & k \le 0 \end{bmatrix}\right]$

(vi) $\dfrac{z^2}{(z-1)(z-3)}$ $\left[\textbf{Ans.}\ \left\{\frac{1}{2}(3^{k+1}-1)\right\},\ k \ge 0\right]$

6.9 SOLUTION OF DIFFERENCE EQUATION BY USING Z-TRANSFORM

We can solve a linear difference equation by using Z-transform. This method is quite useful as the Laplace transform method is useful to solve linear differential equation.

Steps to slove a linear difference equation using Z-Transform:

1. Take the Z-transform on both sides of the given linear difference equation by using following:

$$Z[f(k+1)] = z\,F(z) - z\,f(0)$$

$$Z[f(k+2)] = z^2\,F(z) - z^2\,f(0) - z\,f(1) \text{ etc.}$$

2. Apply given condition.
3. Resolve into partial fraction if required to apply inverse Z-transform.
4. Apply inverse Z-transform.

Example 16 Solve the difference equation

$$f(k+2) - 4\,f(k+1) + 3f(k) = 0,\ f(0) = 2a,\ f(1) = 4.$$

Solution: Given that $f(k+2) - 4f(k+1) - 3f(k) = 0$ ...(i)

Applying Z-transform both sides, we get

$$Zf(k+2) - 4\,Z\,f(k+1) - 3\,Z\,f(k) = 0$$

$$\Rightarrow \quad z^2\left[F(z) - f(0) - \frac{f(1)}{z}\right] - 4\,[zF(z) - zf(0)] + 3\,F(z) = 0 \quad \text{...(ii)}$$

Applying given conditions, we get

$$z^2\left[F(z) - 2 - \frac{4}{z}\right] - 4\,[z\,F(z) - z2] + 3\,F(z) = 0$$

$$\Rightarrow \quad F(z)[z^2 - 4z + 3] = 2z^2 + 4z - 8z$$

or $$F(z) = \frac{z(2z-4)}{(z-1)(z-3)} \qquad \text{...(iii)}$$

Resolving into partial fractions:

$$\frac{(2z-4)}{(z-1)(z-3)} = \frac{a}{z-1} + \frac{b}{z-3}$$

We get $\dfrac{2z-4}{(z-1)(z-3)} = \dfrac{1}{z-1} + \dfrac{1}{z-3}$

Therefore, $$F(z) = \frac{(2z-4)}{(z-1)(z-3)} = \frac{z}{z-1} + \frac{z}{z-3} \qquad \text{...(iv)}$$

Applying inverse Z-transform on equation (iv), we get

$$Z^{-1}\{f(z)\} = Z^{-1}\left(\frac{z}{z-1}\right) + Z^{-1}\left(\frac{z}{z-3}\right)$$

$$\Rightarrow \qquad f(k) = 1^k + 3^k \qquad \text{since}\left[Z^{-1}\left(\frac{z}{z-a}\right) = a^k, k \geq 0\right]$$

or $$f(k) = 1 + 3^k$$

Example 17 Solve the difference equation using Z-transform method.

$$f(k+2) - 2f(k+1) + f(k) = 3k + 5$$

Solution: Given that $f(k + 2) - 2f(k + 1) - f(k) = 3k + 5$ ...(i)

Applying Z-transform to the equation (i), we get

$$Z\{f(k+2)\} - 2\,Z\{k+1\} + Z\{(k)\} = 3Z\{(k)\} + 5\,Z\{1\}$$

$$\Rightarrow \qquad z^2\left[F(z) - f(0) - \frac{f(1)}{z}\right] - 2z\,[F(z) - f(0)] + F(z)$$

$$= 3 \cdot \frac{z}{(z-1)^2} + 5 \cdot \frac{z}{z-1}$$

or $$(z-1)^2 F(z) = \frac{5z^2 - 2z}{(z-1)^2} + f(0) + zf(1)$$

$$\Rightarrow \qquad F(z) = \frac{5z^2 - 2z}{(z-1)^4} + \frac{z^2 - 2z}{(z-1)^2} f(0) + \frac{z}{(z-1)^2} f(1)$$

Let $f(0) = c_1$

and $f(1) = c_2$

$$Z\{1\} = \frac{z}{z-1}$$

$$Z\{k\} = \frac{z}{(z-1)^2}$$

$$Z\{k^2\} = \frac{z^2+z}{(z-1)^3}$$

$$Z\{k^3\} = \frac{z^3+4z^2+z}{(z-1)^4}$$

Then, $$F(z) = \frac{5z^2-2z}{(z-1)^4} + \frac{z^2-2z}{(z-1)^2}c_1 + \frac{z}{(z-1)^2}c_2 \qquad ...(ii)$$

Applying inverse Z-transform of the equation (ii), we get

$$Z^{-1}\{F(z)\} = Z^{-1}\left\{\frac{5z^2-2z}{(z-1)^4}\right\} + c_1 Z^{-1}\left\{\frac{z^2-2z}{(z-1)^2}\right\} + x_2 Z^{-1}\left\{\frac{z}{(z-1)^2}\right\} \qquad ...(iii)$$

Then, resolving into partial tractions, we have

$$Z^{-1}\left\{\frac{5z^2-2z}{(z-1)^4}\right\} = Z^{-1}\left\{\frac{z^3+4z^2+z}{2(z-1)^4} + \frac{z^2+z}{(z-1)^3} - \frac{3z}{2(z-1)^2}\right\} \qquad ...(iv)$$

$$Z^{-1}\left\{\frac{z^2-2z}{(z-1)^2}\right\} = Z^{-1}\left\{\frac{z}{(z-1)} - \frac{z}{(z-1)^2}\right\} \qquad ...(v)$$

Using equations (iv) and (v) in equation (iii), we get

$$Z^{-1}(F(z)) = Z^{-1}\left[\frac{z^2+4z^2+z}{2(z-1)^4} + \frac{z^2+z}{(z-1)^3} - \frac{3z}{2(z-1)^2}\right]$$
$$+ c_1 Z^{-1}\left[\frac{z}{(z-1)} - \frac{z}{(z-1)^2}\right] + c_2 Z^{-1}\left[\frac{z}{(z-1)^2}\right]$$

$$\Rightarrow \qquad f(k) = \left(\frac{1}{2}k^3 + 1\cdot k^2 - \frac{3}{2}k\right) + c_1(1-k) + c_2 k.$$

Example 18 Solve the difference equation by using Z-transform,

$$f(k+2) - 4f(k) = 0, f(0) = 0, f(1) = 2$$

Solution: Given that

$$f(k+2) - 4f(k) = 0 \qquad ...(i)$$

Applying Z-transform to equation (i), we get

$$Z\{f(k+2)\} - 4\,Z\{f(k)\} = 0$$

$$\Rightarrow \quad z^2\left[F(z) - f(0) - \frac{f(1)}{z}\right] - 4F(z) = 0 \qquad \text{...(ii)}$$

Applying given conditions, we get

$$z^2\left[F(z) - 0 - \frac{2}{z}\right] - 4F(z) = 0 \quad \Rightarrow \quad (z^2 - 4)\,f(z) = 2z$$

or
$$F(z) = \frac{2z}{z^2 - 4}$$

$$= z\left[\frac{1}{2}\frac{1}{z-2} - \frac{1}{2}\frac{1}{z+2}\right]$$

or
$$F(z) = \frac{1}{2}\cdot\frac{z}{z-2} - \frac{1}{2}\cdot\frac{z}{z+2} \qquad \text{...(iii)}$$

Applying inverse Z-transform on equation (iii), we get

$$Z^{-1}[F(z)] = \frac{1}{2}Z^{-1}\left[\frac{z}{z-2}\right] - \frac{1}{2}Z^{-1}\left[\frac{z}{z+2}\right]$$

or
$$f(k) = \frac{1}{2}2^k - \frac{1}{2}(-2)^k$$

$$\Rightarrow \quad f(k) = \frac{1}{2}\ [2^k - (-\,2)^k$$

Example 19 Solve the difference equation by using Z-transform method $f(k+2) + 6\,f(k+1) + 9f(k) = 2^k$, $f(0) = 0 = f(1)$.

Solution: Given that

$$f(k+2) + 6\,f(k+1) + 9\,f(k) = 2^k \qquad \text{...(i)}$$

Applying Z-transform to the equation (i), we get

$$Z\{f(k+2)\} + 6\,Z\{f(k+1)\} + 9\,Z\{f(k)\} = Z\,\{2^k\}$$

$$z^2\left[F(z) - f(0) + \frac{f(1)}{z}\right] + 6z\left[F(z) - f(0)\right] + 9F(z) = \frac{z}{z-2} \qquad \text{...(ii)}$$

Putting $f(0) = 0 = f(1)$ in equation (ii), we get

$$(z^2 + 6z + 9)\,F(z) = \frac{z}{z-2}$$

or
$$F(z) = \frac{z}{(z-2)(z+3)^2}$$

Resolving into partial fractions, we get

$$F(z) = \frac{1}{25}\left[\frac{z}{z-2} - \frac{z}{z+3} - 5\cdot\frac{z}{(z+3)^2}\right] \qquad \text{...(iii)}$$

Applying inverse Z-transform to equation (iii), we get

$$f(z) = \frac{1}{25}\left[2^k - (-3)^k + \frac{5}{3}k(-3)^k\right]$$

Example 20 Solve the difference equation using Z-transform method.

$f(k+2) - 6f(k+1) + 8f(k) = 2^k + 6k$

Solution: Given that

$$f(k+2) - 6f(k+1) + 8f(k) = 2^k + 6k \qquad \text{...(i)}$$

Applying Z-transform to the equation (i), we get

$$Z\{f(k+2)\} - 6\{f(k+1)\} + 8\{f(k)\} = Z(2^k) + 6Z\{k\}$$

$$\Rightarrow \quad z^2\left[F(z) - f(0) - \frac{f(1)}{z}\right] - 6[zF(z) - zf(0)] + 8F(z) = \frac{z}{z-2} + 6\cdot\frac{z}{(z-1)^2} \qquad \text{...(ii)}$$

Suppose $f(0) = c_1$ ans $f(1) = c_2$.

Then, $(z^2 - 6z + 8)F(z) = (z^2 - 6z)\ c_1 + c_2 z + \dfrac{z}{z-2} + \dfrac{6z}{(z-1)^2}$

or

$$F(z) = \frac{1}{(z-2)(z-4)}\left[\frac{z}{z-2} + \frac{6z}{(z-1)^2} + (z^2 - 6z)c_1 + c_2 z\right]$$

$$= \frac{z}{(z-4)(z-2)^2} + \frac{6z}{(z-2)(z-4)(z-1)^2} + \frac{c_1(z^2+6z)}{(z-2)(z-4)} + \frac{c_2 z}{(z-2)(z-4)}$$

Resolving into partial fraction various terms, we get

$$F(z) = \left[-\frac{1}{4}\cdot\frac{z}{z-2} - \frac{z}{2(z-2)^2} + \frac{z}{4(z-4)}\right]$$

$$+\left[\frac{8}{3}\cdot\frac{z}{z-1} + \frac{1}{3}\frac{z}{z-4} - 3\cdot\frac{z}{z-2} + \frac{2z}{(z-1)^2}\right]$$

$$+ c_1\left[\frac{2z}{z-2} - \frac{z}{z-4}\right] + \frac{c_2}{2}\left[\frac{z}{z-4} - \frac{z}{z-2}\right]$$

$$\Rightarrow \qquad F(z) = a\cdot\frac{z}{z-2}+b\cdot\frac{z}{z-4}-\frac{1}{2}\cdot\frac{z}{(z-2)^2}+\frac{8}{3}\frac{z}{z-1}+\frac{2z}{(z-1)^2}$$

$$a = \frac{24c_1-30c_2-1}{12}$$

$$b = \frac{10c_2-4c_1+27}{2}$$

Applying inverse Z-transform, we get

$$f(k) = a\cdot 2^k+b\,4^k-\frac{1}{2}(k\cdot 2^k)+2k+\frac{8}{3}$$

Example 21 Solve the difference equation using Z-transform

$$f(k + 2) + 2\,f(k + 1) + f(k) = k, f(0) = f(1) = 0.$$

Solution: Given that

$$f(k + 2) + 2\,f(k + 1) + f(k) = k \qquad ...(i)$$

Applying Z-transform to the equation (i), we get

$$Z\{f(k + 2)\} + 2\,Z\{f(k + 1)\} + Z\{f(k)\} = Z\{k\}$$

$$\Rightarrow \qquad z^2\left[F(z)-f(0)-\frac{f(1)}{z}\right]-2\,z\{F(z)-f(0)\}+F(z) = \frac{z}{(z-1)^2}$$

Putting the values $f(0) = 0 = f(1)$, we get

$$z^2\{F(z) - 0 - 0\} + 2z\,\{F(z)\} + F(z) = \frac{z}{(z-1)^2}$$

or

$$F(z) = \frac{z}{(z-1)^2(z+1)^2} \qquad ...(ii)$$

Resolving into partial fraction, we get

$$F(z) = \frac{1}{4}\left[-\frac{z}{z-1}+\frac{z}{(z-1)^2}+\frac{z}{z+1}+\frac{z}{(z+1)^2}\right] \qquad ...(iii)$$

Applying inverse Z-transform to the equation (iii), we get

$$f(k) = \frac{1}{4}\left[-1^k+k+(-1)^k-k(-1)^k\right]$$

$$\Rightarrow \qquad f(k) = \frac{(k+1)}{4}\,\{1-(-1)^k\}$$

EXERCISE

Solve the following difference equations:

1. $f(k + 2) - 4f(k + 1) + 3f(k) = 5^k,\ k \geq 0.$ $\left[\textbf{Ans. } f(k) = c_1 + c_2\, 3^k + \frac{1}{8} 5^k\right]$

2. $f(k + 2) - 4f(k) = 0,\ k \geq 0, f(0) = 0, f(1) = 2.$ $[\textbf{Ans. } f(k) = 2^{k-1} + (-2)^{k-1}]$

3. $6f(k + 2) - 7f(k + 1) + f(k) = 0, f(0) = 1 = f(1).$ $\left[\textbf{Ans. } f(k) = \frac{6}{5}\left[\left(\frac{1}{2}\right)^k - \left(-\frac{1}{3}\right)^k\right]\right]$

4. $f(k + 2) - 2f(k + 1) + f(k) = 3k + 5.$ $\left[\textbf{Ans. } f(k) = \frac{1}{2} k(k - 1)\ (k + 3) + c_0 + c_1 k\right]$

5. $2f(k + 2) - 5f(k + 1) + 3f(k) = 0\ f(0) = 0, f(1) = 1.$

$\left[\textbf{Ans. } f(u) = -\frac{2}{3}\left(\frac{1}{2}\right)^k + \frac{2}{3}2^k\right]$

Chapter 7

Fourier Series

7.1 FOURIER SERIES

One of the most useful methods of mathematical analysis is Fourier Series, named after the French physicist Jean Baptiste Joseph Fourier (1760–1830). A Fourier series is an expansion of a periodic function in the form of an infinite series of sines and cosines. In Applied Sciences and Engineering expansion of function in terms of sine and cosine is useful because it allows us to more easily manipulate functions that are discontinuous or difficult to represent analytically.

Fourier series is helpful for the analysis of periodic and real-valued functions. It can break up a periodic function into an infinite series of sine and cosine. This property makes Fourier series very useful in many applications, for example: study of heat flow in a thin wire fixed at two end points, concentration of chemical, vibration in a string, theory of sound, etc.

7.1.1 Some Useful Results from Definite Integrals

1. $\int_0^{2\pi} \cos nx \, dx = 0, \ (n \neq 0)$

2. $\int_0^{2\pi} \sin nx \, dx = 0, \ (n \neq 0)$

3. $\int_0^{2\pi} \cos^2 nx \, dx = \pi, \ (n \neq 0)$

4. $\int_0^{2\pi} \sin^2 nx \, dx = \pi, \ (n \neq 0)$

5. $\int_0^{2\pi} \cos mx \cos nx \, dx = 0, \ (m \neq n)$

6. $\int_0^{2\pi} \sin mx \cos nx \, dx = 0, (m \neq n)$

7. $\int_0^{2\pi} \sin mx \sin nx \, dx = 0, (m \neq n)$

8. (i) $\sin n\pi = 0$

(ii) $\cos n\pi = (-1)^n$

(iii) $\sin (2n + 1) \pi = (-1)^n$

(iv) $\cos \frac{1}{2} (2n + 1)\pi = 0$

7.1.2 Periodic Functions

A function $f(t)$ is known as periodic function if

$$f(t \pm p) = f(t) \ \forall \ t,$$

for some positive number p, where p is said to be period of $f(t)$.

For example: $\cos t$ is a periodic function with period 2π as $\cos (t + 2\pi) = \cos t$.

Graphically: A function whose graph remains the same if it is shifted on p to the left or right (where p is the period of the given function) is said to be periodic function, or if we place the function over p, it remains the same.

Fourier Series: Fourier series comes into existence when we represent a given periodic function in terms of sine and cosine functions.

Let us assume a periodic function of t, $f(t)$, defined in the interval $[\alpha, \alpha + 2\pi]$. Then $f(t)$ can be represented by trigonometric series

$$f(t) = a_0 + a_1 \cos t + a_2 \cos 2t + a_3 \cos 3t + ... + a_n \cos nt + ...$$
$$b_1 \sin t + b_2 \sin 2t + ... + b_n \sin nt + ...$$

$$\Rightarrow \qquad f(t) = a_0 + \sum_{n=1}^{\infty} (a_n \cos nt + b_n \sin nt) \qquad ...(i)$$

equation (i) is known as Fourier series, where a_0, a_1, ... b_1, b_2 ... are Fourier coefficients.

Formula (i) is known as Fourier-Euler formula.

Determination of the coefficients: In this section we shall calculate the values of a_0, a_n and b_n.

Determination of a_0: Integrate both sides of equation (i) with respect to t, between the limits α and $\alpha + 2\pi$, we get

$$\int_{\alpha}^{\alpha+2\pi} f(t) \, dt = \int_{\alpha}^{\alpha+2\pi} \left[a_0 + \sum_{n=1}^{\infty} (a_n \cos nt + b_n \sin nt) \right] dt$$

$$= \int_{\alpha}^{\alpha+2\pi} a_0\, dt + \sum_{n=1}^{\infty} \left(a_n \int_{\alpha}^{\alpha+2\pi} \cos nt\, dt + b_n \int_{\alpha}^{\alpha+2\pi} \sin nt\, dt \right)$$

$$\Rightarrow \qquad \int_{\alpha}^{\alpha+2\pi} f(t)dt = a_0\, 2\pi + 0 + 0$$

$$\Rightarrow \qquad a_0 = \frac{1}{2\pi} \int_{\alpha}^{\alpha+2\pi} f(t)dt \qquad \text{...(ii)}$$

Determination of a_n: Multiply both sides of equation (i) by cos mt (for positive integer m) and integrate with respect to t between the limits α to $\alpha + 2\pi$, we have

$$\int_{\alpha}^{\alpha+2\pi} f(t)dt \cos mt\, dt = \int_{\alpha}^{\alpha+2\pi} \left[a_0 + \sum_{n=1}^{\infty} (a_n \cos nt + b_n \sin nt) \right] \cos mt\, dt$$

$$= a_0 \int_{\alpha}^{\alpha+2\pi} \cos mt\, dt + \sum_{n=1}^{\infty} \left[a_n \int_{\alpha}^{\alpha+2\pi} \cos nt \cos mt\, dt + b_n \int_{\alpha}^{\alpha+2\pi} \sin nt \cos mt\, dt \right]$$

$$= 0 + a_n\pi + 0 \quad \text{if, } m = n,$$

$$\int_{\alpha}^{\alpha+2\pi} \cos^2 nt\, dt = \pi$$

$$\Rightarrow \qquad a_0 = \frac{1}{\pi} \int_{\alpha}^{\alpha+2\pi} f(t) \cos mt\, dt$$

Determination of b_n: Multiply both sides of equation (i) by sin mt and integrate with respect to t between the limits α to $\alpha + 2\pi$

$$\int_{\alpha}^{\alpha+2\pi} f(t) \sin mt\, dt = \int_{\alpha}^{\alpha+2\pi} \left[a_0 + \sum_{n=1}^{\infty} (a_n \cos nt + b_n \sin nt) \right] \sin mt\, dt$$

$$= a_0 \int_{\alpha}^{\alpha+2\pi} \sin mt\, dt \sum_{n=1}^{\infty} \left[a_n \int_{\alpha}^{\alpha+2\pi} \cos nt \sin mt\, dt + b_n \int_{\alpha}^{\alpha+2\pi} \sin mt \sin nt\, dt \right]$$

$$= a_0 \int_{\alpha}^{\alpha+2\pi} \sin mt\, dt + \sum_{n=1}^{\infty} a_n \int_{\alpha}^{\alpha+2\pi} \cos nt \sin mt\, dt + \sum_{n=1}^{\infty} b_n \int_{\alpha}^{\alpha+2\pi} \sin nt \sin mt\, dt$$

$$= 0 + 0 + b_n\pi$$

$$\text{For } m = n,\ a_0 \int_{\alpha}^{\alpha+2\pi} \sin^2 nt\, dt = \pi$$

Hence, $$b_n = \frac{1}{\pi} \int_{\alpha}^{\alpha+2\pi} f(t)\sin nt\, dt$$

Now, $$f(t) = a_0 + \sum_{n=1}^{\infty} (a_n \cos nt + b_n \sin nt)$$

where $$a_0 = \frac{1}{2\pi} \int_{\alpha}^{\alpha+2\pi} f(t)dt$$

$$a_n = \frac{1}{\pi} \int_{\alpha}^{\alpha+2\pi} f(t) \cos nt\, dt$$

and $$b_n = \frac{1}{\pi} \int_{\alpha}^{\alpha+2\pi} f(t) \sin nt\, dt$$

(i) If the periodic function $f(t)$ is defined in the internal $-\pi$ to π, then

$$f(t) = a_0 + \sum_{n=1}^{\infty} (a_n \cos nt + b_n \sin nt)$$

Here, $$a_0 = \frac{1}{2\pi} \int_{-\pi}^{\pi} f(t)dt$$

$$a_n = \frac{1}{\pi} \int_{-\pi}^{\pi} f(t)\cos nt\, dt$$

and $$= \frac{1}{\pi} \int_{-\pi}^{\pi} f(t)\sin nt\, dt$$

(ii) If the periodic function is defined in the interval 0 to 2π.

Then, $$f(t) = a_0 + \sum_{n=1}^{\infty} (a_n \cos nt + b_n \sin nt)$$

where $$a_0 = \frac{1}{2\pi} \int_{0}^{2\pi} f(t)dt$$

$$a_n = \frac{1}{\pi} \int_{0}^{2\pi} f(t) \cos nt\, dt$$

$$b_n = \frac{1}{\pi} \int_{0}^{2\pi} f(t) \sin nt\, dt$$

(iii) **Dirichlet Conditions:**

A function $f(t)$ can be expressed as a Fourier series if,

(i) $f(t)$ is single valued and finite,

(ii) $f(t)$ is a periodic function,

(iii) $f(t)$ is piecewise continuous,

(iv) $f(t)$ has finite number of discontinuities in the interval, or

(v) $f(t)$ has at the most a finite number of maxima and minima.

(i) At the point of continuity, the Fourier series of the given function $f(t)$ converges to the given function.

(ii) If $f(t)$ is defined in the interval $(\alpha, \alpha + 2\pi)$ such that

$$f(t) = \phi(t), \quad \alpha < t < c$$
$$= \psi(t), \quad c < t < \alpha + 2\pi$$

Then at the point of discontinuity at the point $t = c$, the Fourier series of the given function $f(t)$ converges to the arithmatic mean of the left hand and right hand limits of $f(t)$ at the point $t = c$, or

$$f(t) = \frac{1}{2}[f(c-0) + f(c+0)]$$

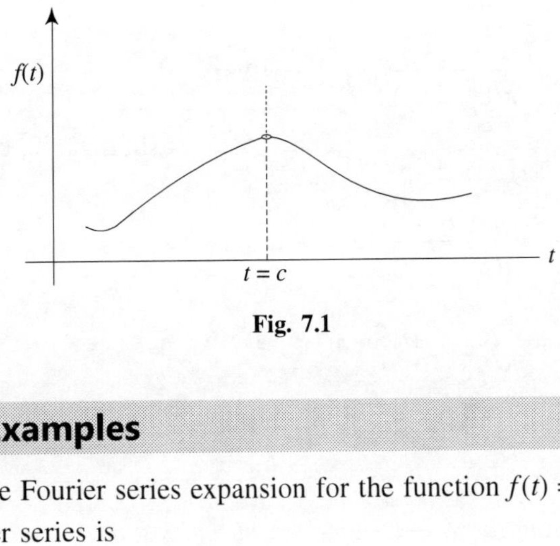

Fig. 7.1

Illustrative Examples

Example 1 Find the Fourier series expansion for the function $f(t) = e^{at}$ in $[0, 2\pi)$:

Solution: The Fourier series is

$$f(t) = a_0 + \sum_{n=1}^{\infty} (a_n \cos nt + b_n \sin nt) \quad ...(i)$$

where
$$a_0 = \frac{1}{2\pi} \int_0^{2\pi} f(t)dt$$

Then,
$$a_0 = \frac{1}{2\pi} \int_0^{2\pi} e^{at}\, dt$$

$$= \frac{1}{2\pi}\frac{1}{a}(e^{at})_0^{2\pi}$$

$$\Rightarrow \qquad a_0 = \frac{1}{2\pi a}(e^{2a\pi} - 1) \qquad \text{...(ii)}$$

and $$a_n = \frac{1}{\pi}\int_0^{2\pi} f(t)\cos nt\, dt$$

$$= \frac{1}{\pi}\int_0^{2\pi} e^{at}\cos nt\, dt$$

But $$\int e^{at}\cos nt\, dt = \frac{e^{at}}{a^2+n^2}(a\cos nt + n\sin nt)$$

Then, $$a_n = \frac{1}{\pi}\left[\frac{e^{at}(a\cos nt + n\sin nt)}{a^2+n^2}\right]_0^{2\pi}$$

$$= \frac{1}{\pi(a^2+n^2)}[e^{2a\pi}(a\cos 2n\pi + n\sin 2n\pi) - e^0(a\cos 0 + n\sin 0)]$$

$$= \frac{1}{\pi(a^2+n^2)}[a e^{2a\pi}\cos 2n\pi - a]$$

(as $\sin 2n\pi = 0$, $\cos 0 = 1 = \cos 2n\pi$)

$$\Rightarrow \qquad a_n = \frac{a}{\pi(a^2+n^2)}[e^{2a\pi} - 1] \qquad \text{...(iii)}$$

Again $$b_n = \frac{1}{\pi}\int_0^{2\pi} f(t)\sin nt\, dt$$

$$= \frac{1}{\pi}\int_0^{2\pi} e^{at}\sin nt\, dt$$

We know that, $$\int e^{at}\sin nt\, dt = \frac{e^{at}}{a^2+n^2}(a\sin nt - n\cos nt)$$

Then, $$b_n = \frac{1}{\pi}\left[\frac{e^{at}(a\sin nt - n\cos nt)}{a^2+n^2}\right]_0^{2\pi}$$

$$= \frac{1}{\pi}\left[\frac{e^{2a\pi}(a\sin 2n\pi - n\cos 2n\pi) - e^0(a\sin 0 - n\cos 0)}{a^2+n^2}\right]$$

$$= \frac{1}{\pi}\left[\frac{e^{2a\pi}(0-n)-e^{0}(a\sin 0 - n\cos 0)}{a^2+n^2}\right]$$

$$= \frac{1}{\pi}\left[\frac{-ne^{2a\pi}+n}{a^2+n^2}\right]$$

$$\Rightarrow \qquad b_n = \frac{n}{\pi(a^2+n^2)}(1-e^{2a\pi})$$

On replacing of the values of a_0, a_n, b_n, we have

$$f(t) = \frac{1}{2\pi}\cdot\frac{e^{2a\pi}-1}{a} + \sum_{n=1}^{\infty}\frac{a(e^{2a\pi}-1)}{\pi(a^2+n^2)}\cos nt + \sum_{n=1}^{\infty}\frac{n(1-e^{2a\pi})}{\pi(a^2+n^2)}\sin nt$$

$$\Rightarrow \qquad f(t) = \frac{1}{\pi}(e^{2a\pi}-1)\left[\frac{1}{2a}+\sum_{n=1}^{\infty}\frac{1}{(a^2+n^2)}(-n\sin nt)\right] + \sum_{n=1}^{\infty}\frac{a(e^{2a\pi}-1)}{\pi}\cdot\frac{\cos nt}{(a^2+n^2)}$$

Ans.

Example 2 Find the Fourier series expansion for the function

$$f(t) = \begin{cases} -\pi, & -\pi < t < 0 \\ t, & 0 < t < \pi \end{cases}$$

hence, find the Fourier expansion for the function

$$\frac{1}{1^2}+\frac{1}{3^2}+\frac{1}{5^2}+\ldots = \frac{\pi^2}{8}$$

Solution: The Fourier series is

$$f(t) = a_0 + \sum_{n=1}^{\infty}(a_n \cos nt + b_n \sin nt)$$

Now,

$$a_0 = \frac{1}{2\pi}\int_{-\pi}^{\pi} f(t)dt$$

$$= \frac{1}{2\pi}\left[\int_{-\pi}^{0} f(t)dt + \int_{0}^{\pi} f(t)dt\right]$$

$$= \frac{1}{2\pi}\int_{-\pi}^{0}(-\pi)dt + \frac{1}{2\pi}\int_{0}^{2\pi} t\,dt$$

$$= \frac{1}{2\pi}\left[-\pi(t)_{-\pi}^{0} + \left(\frac{t^2}{2}\right)_0^{\pi}\right]$$

$$= \frac{1}{2\pi}\left(-\pi^2 + \frac{\pi^2}{2}\right) = -\frac{\pi}{4}$$

$$\Rightarrow \qquad a_0 = -\frac{\pi}{4}$$

and $$a_n = \frac{1}{\pi}\int_{-\pi}^{\pi} f(t)\cos nt\, dt$$

$$= \frac{1}{\pi}\left[\int_{-\pi}^{0} f(t)\cos nt\, dt + \int_{0}^{\pi} f(t)\cos nt\, dt\right]$$

$$= \frac{1}{\pi}\int_{-\pi}^{0} (-\pi)\cos nt\, dt + \frac{1}{\pi}\int_{0}^{\pi} t\cos nt\, dt$$

$$= \frac{1}{\pi}\left[-\pi\left(\frac{\sin nt}{n}\right)_{-\pi}^{0}\right] + \frac{1}{\pi}\left[t\left(\frac{\sin nt}{n}\right)\right]_0^{\pi} - \frac{1}{\pi}\int_0^{\pi} 1\cdot\frac{\sin nt}{n}\, dt\Bigg]$$

$$= -\frac{\pi}{n\pi}(\sin nt)_{-\pi}^{0} + \frac{1}{n\pi}(t\sin t)_0^{\pi} + \frac{1}{\pi n^2}(\sin nt)_0^{\pi}$$

$$= 0 + 0 + \frac{1}{\pi n^2}\cos n\pi - \frac{1}{\pi n^2}$$

$$= 0 + 0 + \frac{1}{\pi n^2}(\cos n\pi - 1)$$

$$\Rightarrow \qquad a_n = \frac{1}{\pi n^2}[(-1)^n - 1]$$

and $$b_n = \frac{1}{\pi}\int_{-\pi}^{\pi} f(t)\sin nt\, dt$$

$$= \frac{1}{\pi}\int_{-\pi}^{0} f(t)\sin nt\, dt + \frac{1}{\pi}\int_{0}^{\pi} f(t)\sin nt\, dt$$

$$= \frac{1}{\pi}\int_{-\pi}^{0} (-\pi)\sin nt\, dt + \frac{1}{\pi}\int_{0}^{\pi} t\sin nt\, dt$$

$$= \left[-\pi\frac{-\cos nt}{n}\right]_{-\pi}^{0} + \frac{1}{\pi}\left[t.\frac{-\cos nt}{n}\right]_0^{\pi} - \frac{1}{\pi}\int_0^{\pi} 1\cdot\frac{-\cos nt}{n}\, dt$$

$$= -\left[\frac{\cos nt}{n}\right]_{-\pi}^{0} - \frac{1}{\pi}\left[\frac{t\cos nt}{n}\right]_{0}^{\pi} + \left(\frac{1}{\pi}\frac{\sin t}{n^2}\right)_{0}^{\pi}$$

$$= \frac{1}{\pi}\left[\frac{\pi}{n} - \frac{\pi}{n}\cos n\pi - \frac{\pi}{n}\cos n\pi\right]$$

$$= \frac{1}{n}(1 - 2\cos n\pi)$$

$$\Rightarrow \qquad b_n = \frac{1}{n}[1 - 2(-1)^n]$$

On replacing the values of a_0, a_n, b_n in the Fourier series, we get

$$f(t) = a_0 + \sum_{n=1}^{\infty} \frac{1}{\pi n^2}[(-1)^n - 1]\cos nt + \frac{1}{n}[1 - 2(-1)^n]\sin nt$$

$$= \frac{-\pi}{4} + \sum_{n=1}^{\infty} \frac{1}{\pi n^2}[(-1)^n - 1]\cos nt + \sum_{n=1}^{\infty} \frac{1}{n}[1 - 2(-1)^n]\sin nt$$

$$= -\frac{\pi}{4} + \left(-\frac{2}{\pi}\cos t - \frac{2}{\pi}\frac{\cos 3t}{3^2} - \frac{2}{\pi}\frac{\cos 5t}{5^2} + \ldots\right) + \left(3\cdot\frac{\sin t}{1} - \frac{\sin 2t}{2} + 3\frac{\sin 3t}{3} - \frac{\sin 4t}{5^2} + \ldots\right)$$

$$= -\frac{\pi}{4} - \frac{2}{\pi}\left(\cos t + \frac{\cos 3t}{3^2} - \frac{\cos 5t}{5^2} + \ldots\right) + \left(3\sin t - \frac{\sin 2t}{2} + \frac{3\sin 3t}{3} - \ldots\right)$$

The above expression on the right-hand side represents $f(t)$ for all the values of t in the interval $-\pi < t < \pi$ except at $t = 0$.

Now, for $t = 0$, sum of the series

$$= \frac{1}{2}[f(0 + 0) + f(0 - 0)]$$

$$= \frac{1}{2}(0 - \pi) = -\pi/2$$

On replacing $t = 0$, we get

$$-\frac{\pi}{2} = -\frac{\pi}{4} - \frac{2}{\pi}\left[1 + \frac{1}{3^2} + \frac{1}{5^2} + \ldots\right] + 0$$

$$-\frac{\pi}{2} + \frac{\pi}{4} = -\frac{2}{\pi}\left[1 + \frac{1}{3^2} + \frac{1}{5^2} + \ldots\right]$$

$$\Rightarrow \qquad \frac{\pi^2}{8} = 1 + \frac{1}{3^2} + \frac{1}{5^2} + \ldots$$

Example 3 Find the Fourier series for $F(t) = t + t^2, -\pi < t < \pi$,
Hence, show that

$$\frac{\pi^2}{6} = 1 + \frac{1}{2^2} + \frac{1}{3^2} + \ldots$$

Solution: We know that,

$$f(t) = a_0 + \sum_{n=1}^{\infty} (a_n \cos nt + b_n \sin nt)$$

Then,
$$a_0 = \frac{1}{2\pi} \int_{-\pi}^{\pi} f(t)dt$$

$$= \frac{1}{2\pi} \int_{-\pi}^{\pi} (t + t^2)dt$$

$$= \frac{1}{2\pi} \left[\int_{-\pi}^{\pi} tdt + \int_{-\pi}^{\pi} t^2 dt \right]$$

$$= \frac{1}{2\pi} \left[0 + 2\int_{0}^{\pi} t^2 dt \right]$$

$$= \frac{1}{\pi} \int_{0}^{\pi} t^2 dt = \frac{1}{\pi} \left(\frac{t^3}{3} \right)_0^{\pi}$$

$$= \frac{\pi^3}{3}$$

and
$$a_n = \frac{1}{\pi} \int_{-\pi}^{\pi} (t + t^2) \cos nt \, dt$$

$$= \frac{1}{\pi} \left[\int_{-\pi}^{\pi} t \cos nt \, dt + \int_{-\pi}^{\pi} t^2 \cos nt \, dt \right]$$

$$= \frac{1}{\pi} \left[0 + \int_{-\pi}^{\pi} t^2 \cos nt \, dt \right]$$

$$= \frac{1}{\pi} \int_{-\pi}^{\pi} t^2 \cos nt \, dt$$

$$= \frac{2}{\pi}\int_0^\pi t^2 \cos nt\, dt$$

$$= \frac{2}{\pi}\left[\left(\frac{t^2 \sin nt}{n}\right)_0^\pi - \int_0^\pi 2t\,\frac{\sin nt}{n}\,dt\right]$$

$$= \frac{2}{\pi}\left[-\frac{2}{n}\int_0^\pi t \sin nt\, dt\right]$$

$$= \frac{2}{\pi}\left[-\frac{2}{n}\left(\frac{t\cos nt}{-n}\right)_0^\pi + \frac{2}{n}\int_0^\pi \frac{\cos nt}{-n}\,dt\right]$$

$$= \frac{2}{\pi}\left[-\frac{2}{n^2}(\pi \cos n\pi) - \frac{2}{n^2}\left(\frac{\sin nt}{n}\right)_0^\pi\right]$$

$$= \frac{4}{n^2}\cos n\pi \qquad [\cos n\pi = (-1)^n \quad \sin n\pi = 0]$$

$$\Rightarrow \qquad a_n = \frac{4}{n^2}(-1)^n$$

Now,

$$b_n = \frac{1}{\pi}\int_{-\pi}^{\pi}(t + t^2)\sin nt\, dt$$

$$= \frac{2}{\pi}\int_{-\pi}^{\pi} t \sin nt\, dt \quad \text{as} \quad \int_{-\pi}^{\pi} t^2 \sin t\, dt = 0$$

$$= \frac{2}{\pi}\left[\left(\frac{t\cos nt}{-n}\right)_0^\pi - \int_0^\pi \frac{\cos nt}{-n}\,dt\right]$$

$$= \frac{2}{\pi}\left[\frac{\pi \cos n\pi}{-n} + \frac{1}{n}\left(\frac{\sin nt}{n}\right)_0^\pi\right]$$

$$= \frac{-2}{n}(-1)^n, \quad \text{as} \quad \sin n\pi = 0$$

On replacing values of a_0, a_n and b_n in the Fourier series,

$$f(t) = a_0 + \sum_{n=1}^{\infty}(a_n \cos nt + b_n \sin nt)$$

$$\Rightarrow \qquad t + t^2 = \frac{\pi^2}{3} - 4\left(\cos t - \frac{\cos 2t}{2^2} + \frac{\cos 3t}{3^2} \ldots\right) + 2\left(\sin t - \frac{1}{2}\sin 2t + \frac{1}{3}\sin 3 + \ldots\right)$$

Function $f(t) = t + t^2$ represents a series on RHS by the above expression in the interval of $-\pi < t < \pi$ except at end point π and $-\pi$.

$\Rightarrow$ Sum of the series on right-hand side of the above expression

$$= \frac{1}{2}\,[f(-\pi + 0) + f(-\pi - 0)]$$

$$= \frac{1}{2}\,[f(-\pi + 0) + f(\pi - 0)] = \frac{1}{2}\,[-\pi + \pi^2 + \pi + \pi^2]$$

$$= \pi^2$$

Putting $x = \pi$ in the equation,

$$\pi^2 = \frac{\pi^2}{3} - 4\left(\cos\pi - \frac{\cos 2\pi}{2^2} + \frac{\cos 3\pi}{3^2} - \ldots\right) + 0$$

$$\Rightarrow \qquad \pi^2 - \frac{\pi^2}{3} = 4\left(1 + \frac{1}{2^2} + \frac{1}{3^2} + \frac{1}{4^2} + \ldots\right)$$

$$\Rightarrow \qquad \frac{2\pi^2}{3} = 4\left(1 + \frac{1}{2^2} + \frac{1}{3^2} + \frac{1}{4^2} + \ldots\right)$$

$$\Rightarrow \qquad \frac{\pi^2}{6} = 1 + \frac{1}{2^2} + \frac{1}{3^2} + \frac{1}{4^2} + \ldots$$

Example 4 Find the Fourier series expansion for the function

$$f(t) = 0 \quad -\pi \le t < 0$$
$$= \sin t \quad 0 < t \le \pi$$

Hence, show that

$$\frac{1}{2} = \frac{1}{1.3} + \frac{1}{3.5} + \frac{1}{5.7} + \ldots$$

and
$$\frac{\pi - 2}{4} = \frac{1}{1.3} + \frac{1}{3.5} + \frac{1}{5.7} + \ldots$$

Solution: Given that

$$f(t) = \begin{matrix} 0, & -\pi \le t < 0 \\ \sin t, & 0 < t \le \pi \end{matrix}$$

and, Fourier series is

$$f(t) = a_0 + \sum_{n=1}^{\infty} (a_n \cos nt + b_n \sin nt) \qquad \text{...(i)}$$

Now,
$$a_0 = \frac{1}{2\pi} \int_{-\pi}^{\pi} f(t)\, dt$$

$$= \frac{1}{2\pi} \left[\int_{-\pi}^{0} f(t)dt + \int_{0}^{\pi} f(t)dt \right]$$

$$= \frac{1}{2\pi} \int_{0}^{\pi} \sin t\, dt$$

$$= \frac{1}{2\pi} (-\cos t)_0^{\pi}$$

$$= \frac{1}{2\pi} (-\cos t) = \frac{2}{2\pi}$$

$\Rightarrow$
$$a_0 = \frac{1}{\pi} \qquad \text{...(ii)}$$

$$a_n = \frac{1}{\pi} \int_{-\pi}^{\pi} f(t) \cos nt\, dt$$

$$= \frac{1}{\pi} \left[\int_{-\pi}^{0} f(t) \cos nt\, dt + \int_{0}^{\pi} f(t) \cos nt\, dt \right]$$

$$= \frac{1}{\pi} \int_{0}^{\pi} \sin t \cos nt\, dt$$

$$= \frac{1}{\pi} \int_{0}^{\pi} \frac{1}{2} (2 \sin t \cos nt\, dt)$$

$$= \frac{1}{\pi} \int_{0}^{\pi} \frac{1}{2} [\sin(n + 1)t - \sin(n - 1)]dt$$

$$= \frac{1}{2\pi} \left[\int_{0}^{\pi} \sin(n+1)t\, dt - \int_{0}^{\pi} \sin(n-1)t\, dt \right]$$

$$= \frac{1}{2\pi} \left[-\frac{\cos (n+1)t}{(n+1)} + \frac{\cos (n-1)t}{(n-1)} \right]_0^{\pi}$$

$$= \frac{1}{2\pi}\left[\frac{\cos(n+1)\pi}{(n+1)} + \frac{\cos(n-1)}{(n-1)} + \frac{1}{(n+1)} - \frac{1}{(n-1)}\right]$$

$$= \frac{1}{2\pi}\left[\frac{\cos n\pi \cos\pi - \sin n\pi \sin \pi}{(n+1)} + \frac{\cos n\pi \cos\pi + \sin n\pi \sin\pi}{(n-1)} + \frac{1}{n+1} - \frac{1}{n-1}\right]$$

$$= \frac{1}{2\pi}\left[\frac{\cos n\pi}{(n+1)} - \frac{\cos n\pi}{(n-1)} - \frac{2}{n^2-1}\right] \qquad \sin n\pi = 0$$

$$= \frac{1}{2\pi}\left[\frac{(n-1)\cos n\pi - (n+1)\cos\pi - 2}{n^2-1}\right]$$

$$= \frac{1}{2\pi}\left[\frac{-2-2\cos n\pi}{n^2-1}\right]$$

$$\Rightarrow \qquad a_n = \frac{(1+\cos n\pi)}{\pi(1-n^2)}, \quad n \neq 1 \qquad \text{...(iii)}$$

For $n = 1$,

$$a_1 = \frac{1}{\pi}\int_{-\pi}^{\pi} f(t)\cos t\ at,\ n = 1$$

$$= \frac{1}{\pi}\int_{0}^{\pi} \sin t \cos t\ dt$$

$$= \frac{1}{2\pi}\int_{0}^{\pi} \sin 2t\ dt$$

$$= \frac{1}{2\pi}\left[-\frac{\cos 2t}{2}\right]_0^{\pi} = \frac{-1}{4\pi}(\cos 2t)_0^{\pi}$$

$$= 0 \qquad \text{...(iv)}$$

$$b_n = \frac{1}{\pi}\int_{-\pi}^{\pi} f(t)\sin nt\ dt$$

$$= \frac{1}{\pi}\int_{0}^{\pi} \sin t \sin nt\ dt$$

$$= \frac{1}{2\pi}\int_{0}^{\pi} 2\sin t \sin nt\ dt \qquad [\cos(x-y) - \cos(x+y) = 2\sin x \sin y]$$

$$= \frac{1}{2\pi}\int_0^{\pi} [\cos(n-1)t - \cos(n+1)t]\, dt$$

$$= \frac{1}{2\pi}\left[\frac{\sin(n-1)t}{n-1} - \frac{\sin(n+1)t}{t}\right]_0^{\pi}$$

$$= \frac{1}{2\pi}\left[\frac{\sin nt\cos t - \cos nt\sin t}{(n-1)} - \frac{\sin nt\cos t + \cos nt\sin t}{(n+1)}\right]_0^{\pi} \quad ...(v)$$

$\Rightarrow$ $\qquad b_n = 0,\ n \neq 1.$

Then, $\qquad b_1 = \frac{1}{\pi}\int_{-\pi}^{\pi} f(t)\sin nt\, dt$

$$= \frac{1}{\pi}\int_0^{\pi} \sin^2 t\, dt$$

$$= \frac{1}{2\pi}\int_0^{\pi} 2\sin^2 t\, dt$$

$$= \frac{1}{2\pi}\int_0^{\pi} (1-\cos 2t)dt$$

$$= \frac{1}{2\pi}\left[t - \frac{\sin 2t}{2}\right]_0^{\pi}$$

$$= \frac{1}{2\pi}[(\pi - 0) - 0] = \frac{1}{2} \quad ...(vi)$$

Now, replacing the values of a_0, a_n, b_n, a_1, b_1 in equation (i)

$$f(t) = \frac{1}{\pi} + 0 + \sum_{n=2}^{\infty} \frac{(1+\cos n\pi)}{\pi(1-n^2)}\cos nt + \frac{1}{2}\sin t + 0$$

$$= \frac{1}{\pi} + \left[\frac{2}{\pi(1-2^2)}\cos 2t + \frac{2}{\pi(1-4^2)}\cos 4t + \frac{2}{\pi(1-6^2)}\cos 6t + ... + \frac{1}{2}\sin t\right] \quad ...(vii)$$

Equation (vii) represents a series for $f(t)$ in the interval $-\pi \le t \le \pi$ except at $t = 0$, since $f(t)$ is discontinuous at $t = 0$.

$\therefore$ Sum of the series represented by (vii) on R.H.S., for $t = 0$, we get

$$\frac{1}{2}[f(0+0) + f(0-0)]$$

$$= \frac{1}{2}[\sin 0 + 0] = 0$$

On replacing $t = 0$ in equation (vii), we have

$$0 = \frac{1}{\pi} + \frac{2}{\pi}\left[\frac{1}{(1+2)(1-2)} + \frac{1}{(1-4)(1+4)} + \frac{1}{(1-6)(1+6)} + \ldots\right]$$

$$\Rightarrow \quad \frac{1}{\pi} = \frac{2}{\pi}\left[\frac{1}{1.3} + \frac{1}{3.5} + \frac{1}{5.7} + \ldots\right]$$

$$\Rightarrow \quad \frac{1}{2} = \frac{1}{1.3} + \frac{1}{3.5} + \frac{1}{5.7} + \ldots$$

and putting $t = \pi/2$ in the equation (vii), we have

$$\sin \pi/2 = \frac{1}{\pi} + \frac{2}{\pi}\left[\frac{1}{(1+2)(1-2)}(-1) + \frac{1}{(1-4)(1+4)}(1) + \frac{1}{(1-6)(1+6)}(-1)\right] + \frac{1}{2}\sin \pi/2$$

$$\Rightarrow \quad 1 - \frac{1}{2} = \frac{1}{\pi} + \frac{2}{\pi}\left[\frac{1}{1.3} - \frac{1}{3.5} + \frac{1}{5.7} - \ldots\right]$$

$$\Rightarrow \quad \frac{1}{2} - \frac{1}{\pi} = \frac{2}{\pi}\left[\frac{1}{1.3} - \frac{1}{3.5} + \frac{1}{5.7} - \ldots\right]$$

$$\Rightarrow \quad \frac{\pi - 2}{4} = \frac{2}{\pi}\left[\frac{1}{1.3} - \frac{1}{3.5} + \frac{1}{5.7} - \ldots\right]$$

Example 5 Find the Fourier series expansion for the function

$$f(t) = \begin{cases} -1, & -\pi \le t < 0 \\ 0, & t = 0 \\ 1, & 0 < t \le \pi \end{cases}$$

Hence, prove that

$$\frac{\pi}{4} = 1 - \frac{1}{3} + \frac{1}{5} - \frac{1}{7} + \ldots$$

Solution: The given function is

$$f(t) = \begin{cases} -1, & -\pi \le t < 0 \\ 0, & t = 0 \\ 1, & 0 < t \le \pi \end{cases}$$

and, Fourier series expansion for the function $f(t)$ can be defined as

$$f(t) = a_0 + \sum_{n=1}^{\infty} (a_n \cos nt + b_n \sin nt) \quad \text{...(i)}$$

Now,

$$a_0 = \frac{1}{2\pi} \int_{-\pi}^{\pi} f(t)\, dt$$

$$= \frac{1}{2\pi} \left[\int_{-\pi}^{0} f(t)dt + \int_{0}^{\pi} f(t)dt \right]$$

$$= \frac{1}{2\pi} \left[\int_{-\pi}^{0} -1\, dt + \int_{0}^{\pi} 1 \cdot dt \right]$$

$$= \frac{1}{2\pi} \left[-\int_{-\pi}^{0} dt + \int_{0}^{\pi} dt \right]$$

$$\Rightarrow \quad a_0 = \frac{1}{2\pi} [-\pi + \pi] = 0 \quad \text{...(ii)}$$

$$a_n = \frac{1}{\pi} \left[\int_{-\pi}^{0} f(t) \cos nt\, dt + \int_{0}^{\pi} f(t) \cos nt\, dt \right]$$

$$= \frac{1}{\pi} \left[\int_{-\pi}^{0} (-1) \cos nt\, dt + \int_{0}^{\pi} \cos nt\, dt \right]$$

$$= \frac{1}{\pi} \left[\left(\frac{\sin nt}{n} \right)_{-\pi}^{0} + \left(\frac{\sin nt}{n} \right)_{0}^{\pi} \right]$$

$$\Rightarrow \quad a_n = 0, \quad \text{as} \quad \sin n\pi = 0 = \sin 0 \quad \text{...(iii)}$$

and

$$b_n = \frac{1}{\pi} \int_{-\pi}^{\pi} f(t) \sin nt\, dt$$

$$= \frac{1}{\pi} \left[-\int_{-\pi}^{0} \sin nt\, dt + \int_{0}^{\pi} \sin nt\, dt \right]$$

$$= \frac{1}{\pi} \left[\left(\frac{\cos nt}{n} \right)_{-\pi}^{0} + \left(\frac{\cos nt}{n} \right)_{0}^{\pi} \right]$$

$$= \frac{1}{\pi} \left[\frac{1}{n} [\cos 0 - (\cos - n\pi)] + \frac{1}{n} [1 - \cos n\pi] \right]$$

$$= \frac{2}{n\pi}(1 - \cos n\pi)$$

$$= \frac{2}{n\pi}[(1 - (-1)^n] \qquad \text{...(iv)}$$

On replacing the values of a_0, a_n, b_n in equation (i), we get

$$f(t) = 0 + \sum_{n=1}^{\infty}\left[0 + \frac{2}{n\pi}[1 \quad (-1)^n]\right]\sin nt$$

$$= \frac{2}{n\pi}\left[2\sin t + \frac{2\sin 3t}{3} + \frac{2\sin 5t}{5} + ...\right]$$

$$= \frac{4}{\pi}\left[\sin t + \frac{\sin 3t}{3} + \frac{\sin 5t}{5} + ...\right]$$

On putting $x = \pi/2$

$$1 = \frac{4}{\pi}\left[1 - \frac{1}{3} + \frac{1}{5} - \frac{1}{7} + ...\right]$$

$$\frac{\pi}{4} = 1 - \frac{1}{3} + \frac{1}{5} - \frac{1}{7} + ...$$

7.1.3 Even and Odd Functions

Even Function: A function $f(t)$ is said to be even function if $f(-t) = f(t)$ for all values of t.

Important Properties of Even Function:

1. The product of two even functions is an even function.
   If $f(x)$ and $\phi(x)$ are two even functions then $F(x) = f(x).\ \phi(x)$ is also an even function.
2. The addition of two even functions is an even function.
   If $f(x)$ and $\phi(x)$ are two even functions, then $F(x) = \phi(x) + f(x)$ is also an even function.
3. The graph of an even function is always symmetrical about y-axis.
4. An algebraic expression in an independent variable will always be in even powers.
5. For an even function, $f(t)$:

$$\int_{-a}^{a} f(t)dt = 2\int_{0}^{a} f(t)dt$$

Odd Function: A function $f(t)$ is said to be an odd function if $f(-t) = -f(t)$.

(i) The product of two odd functions is an even function.

(ii) The product of an even function and an odd function is always an odd function.

(iii) An algebraic odd function always has odd powers.

(iv) The sum of two odd functions is always an odd function.

(v) For an odd function, $\int_{-a}^{a} f(t)dt = 0$.

Fourier Series for Even and Odd Functions

For a periodic function, $f(t)$, having period 2π between interval $-\pi$ and π, we can define Fourier series as below.

$$f(t) = a_0 + \sum_{n=1}^{\infty} (a_n \cos nt + b_n \sin nt)$$

where,

$$a_0 = \frac{1}{2\pi} \int_{-\pi}^{\pi} f(t)\, dt$$

$$a_n = \frac{1}{\pi} \int_{-\pi}^{\pi} f(t) \cos nt\, dt$$

$$b_n = \frac{1}{\pi} \int_{-\pi}^{\pi} f(t) \sin nt\, dt$$

Case 1: *For an even function, $f(t)$*: In between the interval $-\pi$ and π, then

$$a_0 = \frac{1}{2\pi} \int_{-\pi}^{\pi} f(t)\, dt$$

$$= \frac{1}{2\pi} \cdot 2\int_{0}^{\pi} f(t)\, dt = \frac{1}{\pi} \int_{0}^{\pi} f(t)\, dt$$

$$\Rightarrow \quad a_0 = \frac{1}{\pi} \int_{0}^{\pi} f(t)\, dt$$

and

$$a_n = \frac{1}{\pi} \int_{-\pi}^{\pi} f(t) \cos nt\, dt$$

Product of two even functions, $f(t)$ and $\cos nt$, is even. Then

$$a_n = \frac{2}{\pi} \int_{0}^{\pi} f(t) \cos nt\, dt$$

and

$$b_n = \frac{1}{\pi} \int_{-\pi}^{\pi} f(t) \sin nt\, dt$$

$$= 0 \quad \text{(since product of an even function and odd function is an odd function)}$$

Therefore, for an even function, Fourier series

$$f(t) = a_0 + \sum_{n=1}^{\infty} a_n \cos nt$$

Thus, the series is called Fourier cosine series.

Case 2: For an odd function, *f*(*t*):

$$a_0 = \frac{1}{2\pi} \int_{-\pi}^{\pi} f(t)\, dt$$

$$\Rightarrow \qquad a_0 = 0, \quad \text{as} \quad \int_{-\pi}^{\pi} f(t)\, dt = 0$$

$$a_n = \frac{1}{\pi} \int_{-\pi}^{\pi} f(t) \cos nt\, dt$$

$\Rightarrow \qquad a_n = 0 \qquad$ ($\because$ $f(t) \cos nt$ product of an even and an odd function is an odd function).

$$b_n = \frac{1}{\pi} \int_{-\pi}^{\pi} f(t) \sin nt\, dt$$

$$= \frac{2}{\pi} \int_{0}^{\pi} f(t) \sin nt\, dt$$

(Product of two odd functions is an even function.)

Then, Fourier series can be written as

$$f(t) = \sum_{n=1}^{\infty} b_n \cos nt$$

Hence, the series is known as sine Fourier series.

Example 6 Find the Fourier series for the function

$$f(t) = t \cos t, -\pi < t < \pi$$

Solution: Given function is

$$f(t) = t \cos t \qquad \text{...(i)}$$

Since $f(t)$ is a product of an even and an odd functions, therefore, $f(t)$ is an odd function. Therefore,

$$f(t) = \sum_{n=1}^{\infty} b_n \sin nt$$

$$b_n = \frac{1}{\pi} \int_{-\pi}^{\pi} f(t) \sin nt\, dt$$

$$= \frac{2}{\pi}\int_0^{\pi} t \cos t \sin nt \, dt$$

$$= \frac{1}{\pi}\int_0^{\pi} t(2 \cos t \sin nt) \, dt$$

$$= \frac{1}{\pi}\int_0^{\pi} t[\sin(n + 1)t + \sin(n - 1)t] \, dt$$

$$= \frac{1}{\pi}\left[-\frac{t\cos(n+1)t}{n+1} + \frac{\sin(n+1)t}{(n+1)^2} - \frac{t\cos(n-1)t}{(n-1)} + \frac{\sin(n-1)t}{(n-1)^2}\right]_0^{\pi}$$

$$= \frac{1}{\pi}\left[\frac{-t(\cos nt\cos t - \sin nt\sin t)}{(n+1)} + \frac{(\sin nt\cos t + \cos nt\sin t)}{(n+1)^2}\right.$$

$$\left.\frac{-t(\cos nt\cos t + \sin nt\sin t)}{(n+1)} + \frac{\sin nt\cos t - \cos nt\sin t)}{(n-1)^2}\right]_0^{\pi} \quad n \neq 1$$

$$= \frac{1}{\pi}\left[-\frac{\pi\cos\pi\cos n\pi}{(n+1)} + 0 - \frac{\pi\cos n\pi\cos\pi}{(n-1)} + 0\right]$$

$$= \frac{1}{\pi}\left[\frac{\pi(-1)^n}{(n+1)} + \frac{\pi(-1)^n}{(n-1)}\right], \; n \neq 1$$

$$= \frac{(-1)^n 2n}{n^2 - 1}$$

$$\Rightarrow \qquad b_n = \frac{(-1)^n 2n}{n^2 - 1}, \; n \neq 1$$

For $n = 1$

$$b_1 = \frac{2}{\pi}\int_0^{\pi} \cdot \cos t \sin t \, dt$$

$$= \frac{1}{\pi}\int_0^{\pi} t \sin 2t \, dt$$

$$= \frac{1}{\pi}\left[-\frac{t\cos 2t}{2} + \int \frac{\cos 2t}{2} dt\right]_0^{\pi}$$

$$= \frac{1}{\pi}\left[-\frac{t\cos 2t}{2} + \frac{\sin 2t}{2^2}\right]_0^{\pi}$$

$$= \frac{1}{\pi}\left[\left(-\frac{\pi}{2}\right)\right] = -\frac{1}{2}$$

Replacing the values of b_1 and b_n in the Fourier series,

$$f(t) = t \cos t = -\frac{1}{2}\sin t + \sum_{n=2}^{\infty} \frac{2(-1)^n n}{n^2 - 1} \sin nt$$

or $$t \cos t = -\frac{1}{2}\sin t + 2\left[\frac{2\sin 2t}{2^2-1} - \frac{3\sin 3t}{3^2-1} + \frac{4\sin 4t}{4^2-1} - \ldots\right]$$ ***Ans.***

Example 7 Find the Fourier series for the function

$$f(t) = |\cos t|, -\pi < t < \pi$$

Solution: Given function

$$f(t) = |\cos t| \quad \ldots\text{(i)}$$

Since $$f(t) = f(-t) = |\cos t|$$

$\Rightarrow$ $f(t)$ is an even function

Therefore, $b = 0$

Hence, Fourier series for the given function (i)

$$f(t) = a_0 + \sum_{n=1}^{\infty} b_n \cos nt$$

Now, $$a_0 = \frac{1}{2\pi}\int_{-\pi}^{\pi} f(t)\, dt$$

$$= \frac{1}{2\pi} \cdot 2 \cdot \int_0^{\pi} f(t)\, dt$$

$$= \frac{1}{\pi}\int_0^{\pi} f(t)\, dt$$

$$= \frac{1}{\pi}\left[\int_0^{\pi/2} \cos t\, dt + \int_{\pi/2}^{\pi} (-\cos t)\, dt\right]$$

$$= \frac{1}{\pi}\left[(\sin t)_0^{\pi/2} - (\sin t)_{\pi/2}^{\pi}\right]$$

$\Rightarrow$ $$a_0 = \frac{1}{\pi}[(1-0) - (1-1)]$$

$$= \frac{1}{\pi}[1 + 1]$$

$$\Rightarrow \qquad a_0 = \frac{2}{\pi}$$

and, $$a_n = \frac{1}{\pi}\int_{-\pi}^{\pi} f(t)\cos nt\, dt$$

$$= \frac{2}{\pi}\int_0^{\pi} f(t)\cos nt\, dt$$

$$= \frac{2}{\pi}\left[\int_0^{\pi/2} 2\cos t \cos nt\, dt + \int_{\pi/2}^{\pi} (-\cos t)\cos nt\, dt\right]$$

$$= \frac{1}{\pi}\left[\int_0^{\pi/2} 2\cos t \cos nt\, dt - \int_{\pi/2}^{\pi} 2\cos t \cos nt\, dt\right]$$

$$= \frac{1}{\pi}\left[\int_0^{\pi/2} [\cos(n+1)t + \cos(n-1)t]dt - \int_{\pi/2}^{\pi} [\cos(n+1)t + \cos(n-1)t]dt\right]$$

$$= \frac{1}{\pi}\left[\left(\frac{\sin(n+1)t}{(n+1)} + \frac{\sin(n-1)}{(n-1)}\right)_0^{\pi/2} - \left(\frac{\sin(n+1)t}{(n+1)} + \frac{\sin(n-1)t}{(n-1)}\right)_{\pi/2}^{\pi}\right]$$

$$= \frac{1}{\pi}\left[\left(\frac{\sin(n+1)\pi/2}{(n+1)} + \frac{\sin(n-1)\pi/2}{(n-1)}\right) + \left(\frac{\sin(n+1)\pi/2}{n+1} + \frac{\sin(n-1)\pi/2}{(n-1)}\right)\right]$$

$$= \frac{1}{\pi}\left[\frac{\cos n\pi/2}{(n+1)} - \frac{\cos(n\pi)/2}{(n-1)} + \frac{\cos n\pi/2}{(n+1)} - \frac{\cos n\pi/2}{(n-1)}\right]$$

$$= \frac{2}{\pi}\left[\frac{\cos\frac{n\pi}{2}}{(n+1)} - \frac{\cos\frac{n\pi}{2}}{(n-1)}\right]$$

$$= -\frac{4\cos n\pi/2}{\pi(n^2-1)}$$

$$= \frac{-4}{\pi(n^2-1)}\cos n\pi/2, \quad n \neq 1$$

Therefore, $$a_1 = \frac{2}{\pi}\int_0^{\pi} f(t)\cos t\, dt$$

$$= \frac{2}{\pi}\int_0^{\pi} |\cos t| \cos t\, dt$$

$$= \frac{2}{\pi}\left[\int_0^{\pi/2} \cos t + \cos t\, dt + \int_{\pi/2}^{\pi} -\cos t \cos t\, dt\right]$$

$$= \frac{2}{\pi}\left[\frac{1}{2}\int_0^{\pi/2} (1+\cos 2t)dt + \frac{1}{2}\int_{\pi/2}^{\pi} (1+\cos 2t)dt\right]$$

$$= \frac{1}{\pi}\left[\left(t+\frac{\sin 2t}{2}\right)_0^{\pi/2} - \left(t+\frac{\sin 2t}{2}\right)_{\pi/2}^{\pi}\right]$$

$$= \frac{1}{\pi}\left[\frac{\pi}{2}+\frac{\sin\pi}{2}\right] - 0 - \left(\pi+\frac{\sin 2\pi}{2}\right)+\left(\frac{\pi}{2}+\frac{\sin\pi}{2}\right]$$

$$= \frac{1}{\pi}\left[\frac{\pi}{2}-\pi+\frac{\pi}{2}+0+0\right]$$

$\Rightarrow \qquad a_1 = 0$

Substituting the values of a_0, a_1 and a_n in the given Fourier series, we get

$$f(t) = |\cos t| = \frac{2}{\pi} + 0 + \sum_{n=2}^{\infty} - \frac{-4}{\pi(n^2-1)} \cos\frac{n\pi}{2} \cos nt$$

$$\Rightarrow \qquad |\cos t| = \frac{2}{\pi} + \frac{4}{\pi}\left[\frac{\cos 2t}{3} - \frac{\cos 4t}{15} + \frac{\cos 6t}{35} - \ldots\right]$$

Example 8 Find the Fourier series to represent

$$f(t) = |t|,\ -\pi \le t \le \pi.$$

Solution: The given function is

$$f(t) = |t|$$

Since, $\qquad f(t) = |-t| = |t| = f(t)$

$\Rightarrow \qquad f(t) = |t|$ is an even function

Hence, $\qquad b_n = 0$

Therefore, Fourier series for the given function can be written as

$$f(t) = a_0 + \sum_{n=1}^{\infty} a_n \cos nt$$

Then, $\qquad a_0 = \dfrac{1}{2\pi}\displaystyle\int_{-\pi}^{\pi} f(t)\, dt$

$$= \frac{1}{2\pi}\cdot 2\cdot \int_0^{\pi} f(t)\, dt, \text{ as } f(t) \text{ is an even function}$$

$$= \frac{1}{\pi} \int_0^{\pi} f(t)\, dt$$

$$= \frac{1}{\pi} \int_0^{\pi} |x|\, dx = \frac{1}{\pi} \int_0^{\pi} x\, dx = \frac{\pi}{2}$$

$$\Rightarrow \qquad a_0 = \frac{\pi}{2}$$

and

$$a_n = \frac{1}{\pi} \int_{-\pi}^{\pi} f(t) \cos nt\, dt$$

$$= \frac{1}{\pi} \cdot 2 \cdot \int_0^{\pi} |t| \cos nt\, dt$$

$$= \frac{2}{\pi} \int_0^{\pi} t \cos nt\, dt$$

$$\Rightarrow \qquad a_n = \frac{2}{\pi} \left[\frac{t \sin nt}{n} + \frac{\cos nt}{n^2} \right]_0^{\pi}$$

$$= \frac{2}{\pi} \left[0 + \frac{\cos n\pi - \cos 0}{n^2} \right]$$

$$= \frac{2}{\pi n^2} [(-1)^n - 1]$$

On replacing the values of a_0 and a_n, we get

$$f(t) = |t| = \frac{\pi}{2} \sum_{n=1}^{\infty} \frac{2}{\pi n^2} [(-1)^n - 1] \cos nt$$

$$\Rightarrow \qquad |t| = \frac{\pi}{2} + \frac{2}{\pi} \left[\frac{2}{1^2} \cos t - \frac{2}{3^2} \cos 3t - \frac{2}{5^2} \cos 5t - \ldots \right]$$

or

$$|t| = \frac{\pi}{2} - \frac{4}{\pi} \left[\frac{\cos t}{1^2} + \frac{\cos 3t}{3^2} + \frac{\cos 5t}{5^2} + \ldots \right]$$

Example 9 Find the Fourier series for the function

$$f(x) = t^2, \; -\pi \le t \le \pi$$

hence, show that

$$\frac{\pi^2}{6} = 1 + \frac{1}{2^2} + \frac{1}{3^2} + \frac{1}{4^2} + \ldots$$

$$\frac{\pi^2}{12} = 1 - \frac{1}{2^2} + \frac{1}{3^2} - \frac{1}{4^2} + \ldots$$

and
$$\frac{\pi^2}{8} = 1 + \frac{1}{3^2} + \frac{1}{5^2} + \frac{1}{7^2} + \ldots$$

Solution: Given function

$$f(t) = t^2$$

which is an even function, therefore,

$$b_n = 0$$

Thus,
$$f(t) = a_0 + \sum_{n=1}^{\infty} a_n \cos nt$$

Now,
$$a_0 = \frac{1}{2\pi}\int_{-\pi}^{\pi} f(t)\, dt$$

$$= \frac{1}{\pi}\int_0^{\pi} f(t)\, dt$$

$$= \frac{1}{\pi}\int_0^{\pi} t^2\, dt$$

$\Rightarrow$
$$a_0 = \frac{1}{3}\pi^2$$

and
$$a_n = \frac{2}{\pi}\int_0^{\pi} f(t) \cos nt\, dt$$

$$= \frac{2}{\pi}\int_0^{\pi} t^2 \cos nt\, dt$$

$$= \frac{2}{\pi}\left[t^2 \frac{\sin nt}{n} - 2\int t \frac{\sin nt}{n} dt\right]_0^{\pi}$$

$$= \frac{2}{\pi}\left[t^2 \frac{\sin nt}{n} + 2t\frac{\cos nt}{n^2} - 2\frac{\sin nt}{n^3}\right]_0^{\pi}$$

$$= \frac{2}{\pi}\left[0 + \frac{2\pi \cos n\pi}{n^2} + 0\right]$$

$$= \frac{4}{n^2}\cos n\pi$$

$$= \frac{4}{n^2}(-1)^n$$

On replacing the values of a_0 and a_n, we get

$$f(t) = \frac{1}{3}\pi^2 + \sum_{n=1}^{\infty}\frac{4}{n^2}(-1)^n \cos nt$$

$$\Rightarrow \qquad t^2 = \frac{1}{3}\pi^2 + 4\left[-\frac{\cos t}{1^2} + \frac{\cos 2t}{2^2} - \frac{\cos 3t}{3^2} + \frac{\cos 4t}{4^2} \cdots\right]$$

or
$$t^2 = \frac{\pi^2}{3} - 4\left[\frac{\cos t}{1^2} - \frac{\cos 2t}{2^2} + \frac{\cos 3t}{3^2} - \frac{\cos 4t}{4^2} + \cdots\right]$$

Replacing $t = \pi$, we get

$$\pi^2 = \frac{\pi^2}{3} - 4\left[-\frac{1}{1^2} - \frac{1}{2^2} - \frac{1}{3^2} - \frac{1}{4^2} \cdots\right]$$

$$\frac{\pi^2}{6} = 1 + \frac{1}{2^2} - \frac{1}{3^2} - \frac{1}{4^2} \cdots \qquad \ldots(1)$$

and, on replacing $t = 0$, we get

$$0 = \frac{\pi^2}{3} - 4\left[-\frac{1}{1^2} - \frac{1}{2^2} + \frac{1}{3^2} - \frac{1}{4^2} + \cdots\right]$$

$$\Rightarrow \qquad \frac{\pi^2}{12} = 1 - \frac{1}{2^2} + \frac{1}{3^2} - \frac{1}{4^2} + \cdots \qquad \ldots(2)$$

On adding equations (1) and (2), we get

$$\frac{\pi^2}{6} + \frac{\pi^2}{12} = \left(1 + \frac{1}{2^2} + \frac{1}{3^2} + \frac{1}{4^2} + \cdots\right) + \left(1 - \frac{1}{2^2} + \frac{1}{3^2} - \frac{1}{4^2} + \cdots\right)$$

$$\Rightarrow \qquad \frac{\pi^2}{8} = 1 + \frac{1}{3^2} + \frac{1}{5^2} + \frac{1}{7^2} + \cdots$$

Example 10 Find the Fourier series corresponding to the function $f(t) = |\sin t|$ in the interval $-\pi < t < \pi$.

Solution: Given function $f(t) = |\sin t|$
is an even function, thus, $b_n = 0$

Then, $$f(t) = a_0 + \sum_{n=1}^{\infty} a_n \cos nt$$

Now, $$a_0 = \frac{1}{2\pi}\int_{-\pi}^{\pi} f(t)dt$$

$$= \frac{1}{\pi}\int_{0}^{\pi} f(t)dt$$

$$= \frac{1}{\pi}\int_{0}^{\pi} |\sin t|\, dt$$

$$= \frac{1}{\pi}\int_0^{\pi} \sin t\, dt$$

$$= \frac{1}{\pi}[-\cos t]_0^{\pi}$$

$$= \frac{2}{\pi}$$

and $$a_n = \frac{1}{\pi}\int_{-\pi}^{\pi} f(t)\cos nt\, dx$$

$$= \frac{2}{\pi}\int_0^{\pi} f(t)\cos nt\, dx$$

$$= \frac{2}{\pi}\int_0^{\pi} \sin t\cos nt\, dx$$

$$= \frac{1}{\pi}\int_0^{\pi} 2\sin t\cos nt\, dx$$

$$= \frac{1}{\pi}\int_0^{\pi} [\sin (n+1)t - \sin(n-1)]\, dt$$

$$= \frac{1}{\pi}\left[-\frac{\cos(n+1)t}{(n+1)} + \frac{\cos(n-1)t}{(n-1)}\right]_0^{\pi},$$

$$= \frac{1}{\pi}\left[-\frac{(\cos nt + \cos t - \sin nt \sin t)}{(n+1)} + \frac{(\cos nt + \cos t + \sin nt \sin t)}{(n-1)}\right]_0^{\pi}$$

$$= \frac{1}{\pi}\left[(1+\cos n\pi)\left(\frac{1}{n+1} - \frac{1}{n-1}\right)\right]$$

$$= \frac{1}{\pi}(1+\cos n\pi)\frac{(-2)}{n^2-1}$$

$\Rightarrow$ $$a_n = \frac{2}{\pi(1-n^2)}\,[1 + (-1)^n],\ n \neq 1$$

Thus, $$a_n = \begin{cases} 0 & \text{For } n \text{ is an odd number} \\ \dfrac{4}{\pi\,(1-n)^2} & \text{For } n \text{ is an even number} \end{cases}$$

and $$a_1 = \frac{2}{\pi}\int_0^{\pi} f(x) \cos t \, dx$$

$$= \frac{2}{\pi}\int_0^{\pi} \sin t \cos t \, dx$$

$$= \frac{1}{\pi}\int_0^{\pi} \sin 2t \, dt$$

$$= \frac{1}{\pi}\left[\frac{-\cos 2t}{2}\right]_0^{\pi}$$

$\Rightarrow$ $$a_1 = \frac{1}{\pi}\left(\frac{-1+1}{2}\right) = 0$$

$\Rightarrow$ $$a_1 = 0$$

Thus, $$f(x) = |\sin t|$$

$$= \frac{2}{\pi} + 0 + \sum_{n=2}^{\infty} \frac{4}{\pi(1-n^2)} \cos nt$$

$\Rightarrow$ $$|\sin t| = \frac{2}{\pi} - \frac{4}{\pi}\left[\frac{\cos 2t}{3} + \frac{\cos 4\pi}{15} + \frac{\cos 6\pi}{35} + \ldots\right]$$

Example 11 Find Fourier series for the function

$$f(t) = t \sin t, \; 0 < t < 2\pi$$

Solution: Given that $f(t) = t \sin t$

is an even function, thus $b_n = 0, \quad \text{for } n \neq 1$

and $$f(t) = a_0 + \sum_{n=1}^{\infty} a_n \cos nt + b_1$$

Thus, $$a_0 = \frac{1}{2\pi}\int_0^{2\pi} f(t) \, dt$$

$$= \frac{1}{2\pi}\int_0^{2\pi} t \sin t dt$$

$$= \frac{1}{2\pi}[-t \cos t + \sin t]_0^{2\pi}$$

$$= \frac{1}{2\pi}[-2\pi \cos 2\pi + \sin 2\pi - 0]$$

$$= -1$$

and
$$a_n = \frac{1}{\pi}\int_0^{2\pi} t \sin t + \cos nt \, dt$$

$$= \frac{1}{2\pi}\int_0^{2\pi} t \cdot 2 \sin t + \cos nt \, dt$$

$$= \frac{1}{2\pi}\int_0^{2\pi} t[\sin (n+1)\, t + \sin (n-1)\, t]dt$$

$$= \frac{1}{2\pi}\left[-\frac{t \cos (n+1)t}{n+1} + \frac{\sin (n+1)t}{(n+1)^2}\right.$$

$$\left. - \left(\frac{t \cos (n-1)t}{n-1} + \frac{\sin (n-1)t}{(n-1)^2}\right)\right]_0^{2\pi}$$

$$= \frac{1}{2\pi}\left[-\frac{2\pi}{n+1} + \frac{2\pi}{n-1}\right]$$

$$= \frac{2}{n^2 - 1}, \; n \neq 1$$

For $n = 1$, we have

$$a_1 = \frac{1}{\pi}\int_0^{2\pi} t \sin t + \cos t \, dt$$

$$= \frac{1}{2\pi}\int_0^{2\pi} t \sin 2t \, dt$$

$$= \frac{1}{2\pi}\left[-\frac{t \cos 2t}{2} - \left(-\frac{\sin 2t}{4}\right)\right]_0^{2\pi}$$

$$= \frac{1}{2\pi}\left[-\frac{2\pi}{2}\right]$$

$$= -\frac{1}{2}$$

$$b_n = \frac{1}{\pi}\int_0^{2\pi} t \sin t + \sin nt \, dt$$

$$= \frac{1}{2\pi}\int_0^{2\pi} t \cdot 2 \sin t \sin nt \, dt$$

$$= \frac{1}{2\pi}\int_0^{2\pi} t[\cos (n-1)\, t - \cos (n+1)\, t]dt$$

$$= \frac{1}{2\pi}\int_0^{2\pi} t\cos(n-1)t\,dt - \frac{1}{2\pi}\int_0^{2\pi} t\cos(n+1)t\,dt$$

$$= \frac{1}{2\pi}\left[\left(\frac{t\sin(n-1)t}{n-1} + \frac{\cos(n-1)t}{(n-1)^2}\right)\right.$$

$$\left. - \left(\frac{t\sin(n+1)t}{n+1} + \frac{\cos(n+1)t}{(n+1)^2}\right)\right]_0^{2\pi}$$

$$= \frac{1}{2\pi}\left[\frac{1}{(n-1)^2}[\cos(n-1)2\pi - 1]\right.$$

$$\left. - \frac{1}{(n+1)^2}[\cos(n+1)2\pi - 1]\right]$$

$$= 0,\ n \neq 1$$

For $n = 1$, we have

$$b_1 = \frac{1}{\pi}\int_0^{2\pi} t\sin t\sin t\,dt$$

$$= \frac{1}{2\pi}\int_0^{2\pi} t(1 - \cos 2t)\,dt$$

$$= \frac{1}{2\pi}\left[\frac{t^2}{2} - \left(\frac{t\sin 2t}{2} + \frac{\cos 2t}{2^2}\right)\right]_0^{2\pi}$$

$$= \frac{1}{2\pi}\left[\frac{1}{2}(2\pi)^2\right]$$

$$= \pi$$

Thus,

$$f(t) = t\sin t$$

$$= a_0 + \sum_{n=1}^{\infty} a_n\cos nt + b_n\sin nt$$

$$= -1 + \left(-\frac{1}{2}\right)\cos t + 2\left(\frac{\cos 2t}{2^2-1} + \frac{\cos 3t}{3^2-1} + \ldots\right) + \pi\sin t$$

$$= -1 + \pi\sin t - \frac{1}{2}\cos t + 2\left(\frac{\cos 2t}{2^2-1} + \frac{\cos 3t}{3^2-1} + \ldots\right).$$

7.2 FOURIER SERIES IN ARBITRARY INTERVAL (0, 2*c*) OR (–*c*, *c*)

We have studied the problems having interval 2π. In this section, we shall study the problems having interval other than 2π. In this case we shall make the suitable transformation to solve the problem. Let the given interval be $2c$. If we apply the Fourier series to the function of period $2c$. This interval must be converted by a suitable change of variable t into u.

Suppose $f(t)$ be a periodic function defined between the interval α to $\alpha + 2c$, we shall change the problem to the period 2π.

Let $f(t)$ be defined in the interval $(-l, l)$ and suppose that

$$x = \frac{\pi t}{l}$$

$$\Rightarrow \qquad f(t) = \int\left(\frac{lx}{\pi}\right)$$

Thus, the function $F(x) = a_0 + \sum_{n=1}^{\infty} a_n \cos nx + b_n \sin nx$

where,
$$a_0 = \frac{1}{2\pi}\int_{-\pi}^{\pi} F(x)\, dx$$

$$a_n = \frac{1}{\pi}\int_{-\pi}^{\pi} F(x) \cos nx\, dx$$

$$b_n = \frac{1}{\pi}\int_{-\pi}^{\pi} F(x) \sin nx\, dx$$

In this case $F(x)$ is a periodic function with period 2π.

Now, to find Fourier series corresponding to $f(t)$ on the interval $(-c, c)$, we replace the variable with $\pi t/c$, we have

$$f(t) = a_0 + \sum_{n=1}^{\infty} a_n \cos \frac{n\pi t}{c} + b_n \sin \frac{n\pi t}{c}$$

where,

$$a_0 = \frac{1}{2c}\int_{-c}^{c} f(t)\, dt$$

$$a_n = \frac{1}{c}\int_{-c}^{c} f(t) \cos \frac{n\pi}{c} t\, dt$$

$$b_n = \frac{1}{c}\int_{-c}^{c} f(t) \sin \frac{n\pi}{c} t\, dt$$

For a function defined in the interval $(0, 2c)$, we have

$$a_0 = \frac{1}{2c}\int_{0}^{2c} f(t)\, dt$$

$$a_n = \frac{1}{c}\int_{0}^{2c} f(t) \cos \frac{n\pi}{c} t\, dt$$

$$b_n = \frac{1}{c}\int_{0}^{2c} f(t) \sin \frac{n\pi t}{c}\, dt$$

Example 12 Find the Fourier series expansion to the function

$$f(t) = t - t^2 \text{ in the interval } (-1, 1).$$

Solution: The function

$$f(t) = t - t^2$$

defined in the interval (–1, 1), therefore $c = 1$. Therefore, Fourier series expansion can be written as

$$t - t^2 = a_0 + \left[a_n \cos\frac{n\pi t}{c} + b_n \frac{n\pi t}{c}\right] \quad ...(i)$$

Then,
$$a_0 = \frac{1}{2.1}\int_{-1}^{1} (t - t)^2 \, dt$$

$$= \frac{1}{2}\int_{-1}^{1} tdt - \frac{1}{2}\int_{-1}^{1} t^2 dt$$

$\Rightarrow$
$$a_0 = 0 - \frac{1}{3} = \frac{-1}{3} \quad ...(ii)$$

$$= \frac{1}{1}\int_{-1}^{1} (t - t^2) \cos \frac{n\pi t}{1} \, dt$$

$$= \int_{-1}^{1} (t - t^2) \cos n\pi t \, dt$$

$$= \int_{-1}^{1} t \cos n\pi t \, dt - \int_{-1}^{1} t^2 \cos n\pi t \, dt$$

$$= 2\int_{-1}^{1} t^2 \cos n\pi t \, dt$$

$$= -2\left[\frac{t^2}{n\pi}\sin n\pi t + \frac{2t}{n^2\pi^2}\cos n\pi t - \frac{2}{n^3\pi^3}\sin n\pi t\right]_0^1$$

$$= -2\left[\frac{2}{n^2\pi^2}\cos nx\right]$$

$$= \frac{-4}{n^2\pi^2}(-1)^n$$

$$= \frac{4}{n^2\pi^2}(-1)^{n+1}$$

$$\Rightarrow \qquad Q_1 = \frac{4}{\pi^2},\ a_2 = \frac{-4}{2^2\pi^2},\ a_3 = \frac{4}{3^2\pi^2},\ldots$$

$$b_n = \frac{1}{1}\int_{-1}^{1} (t - t^2) \sin n\pi t\, dt$$

$$= 2\int_0^1 t \sin n\pi t\, dt$$

$$= 2\left[-\frac{t}{n\pi}\cos n\pi t + \frac{\sin n\pi t}{n^2\pi^2}\right]_0^1$$

$$= -2\cdot\frac{\cos n\pi}{n\pi}$$

$$= \frac{-2}{n\pi}(-1)^n$$

$$\Rightarrow \qquad b_n = \frac{2}{n\pi}(-1)^{n+1}$$

$$b_1 = \frac{2}{\pi},\ b_2 = \frac{-2}{2\pi},\ b_3 = \frac{2}{3\pi},\ldots$$

Therefore, Fourier series is

$$t - t^2 = -\frac{1}{3} + \frac{4}{\pi^2}\left[\frac{\cos \pi t}{1^2} - \frac{\cos 2\pi t}{2^2} + \frac{\cos 3\pi t}{3^2} - \ldots\right]$$

Example 13 Find the Fourier series to the given below

$$f(t) = e^{-t},\ -a < t < a.$$

Solution: The given function

$$f(t) = e^{-t},\ -a < t < a$$

Here, $c = a$

Fourier series corresponding to the function $f(t) = e^{-t}$

$$e^{-t} = a_0 + \sum_{n=1}^{\infty}\left(a_n \cos\frac{n\pi t}{a} + b_n \sin\frac{n\pi t}{a}\right)$$

Now, $$a_0 = \frac{1}{2a}\int_{-a}^{a} e^{-t}\, dt$$

$$= \frac{-1}{2a}[e^{-t}]_{-a}^{a} = -\frac{1}{2a}[e^{-a} - e^{a}] = \frac{1}{a}\sinh a$$

$$b_n = \frac{1}{a} \int_{-a}^{a} e^{-t} \cos \frac{n\pi t}{a}\, dt$$

$$= \frac{1}{a} \left[\frac{1}{1+\frac{n^2\pi^2}{a^2}} . e^{-t} \left(-\cos\frac{n\pi t}{a} + \frac{n\pi}{a} \sin\frac{n\pi t}{a} \right) \right]_{-a}^{a}$$

$$= \frac{a}{a^2 + n^2\pi^2} \left[e^{-a} \left(-\cos n\pi + \frac{n\pi}{a} \sin n\pi \right) - e^{a} \left(-\cos np - \frac{n\pi}{a} \sin n\pi \right) \right]$$

$$= \frac{a}{a^2 + n^2\pi^2} [(e^a - e^{-a}) \cos n\pi]$$

$$= \frac{2a}{a^2 + n^2\pi^2} \cos n\pi \sinh a$$

Further $$b_n = \frac{1}{a} \int_{-a}^{a} e^{-t} \sin\frac{n\pi t}{a} dt$$

$$= \frac{1}{a} \left[\frac{1}{1+\frac{n^2\pi^2}{a^2}} . e^{-t} \left(-\sin\frac{n\pi t}{a} - \frac{n\pi}{a} \cos\frac{n\pi t}{a} \right) \right]_{-a}^{a}$$

$$= \frac{1}{a^2 + n^2\pi^2} \left[\frac{n\pi}{a} \cos n\pi (e^a - e^{-a}) \right]$$

$$= \frac{2n\pi}{a^2 + n^2\pi^2} \cos n\pi \sinh a$$

On putting the value of a_0, a_n and b_n, we get

$$e^{-t} = \frac{1}{a} \sinh a + \sum_{n=1}^{\infty} \left[\frac{1}{a^2 + \pi^2} \cos\frac{\pi t}{a} - \frac{1}{a^2 + 4\pi^2} \cos\frac{2\pi t}{a} + \ldots \right]$$

$$- 2\pi \sinh a \left[\frac{1}{a^2 + \pi^2} \sin\frac{\pi t}{a} - \frac{2}{a^2 + 4\pi^2} \sin\frac{2\pi t}{a} + \ldots \right]$$

Fourier series for even and odd functions in the interval $(-c, c)$ of length $2c$.

Case 1: Let us consider a function $f(t)$ which is defined in the interval $(-c, c)$ and is even. The Fourier series can be given as below to the function:

$$f(t) = a_0 + \sum_{n=1}^{\infty} a_n \cos \frac{n\pi x}{c}$$

where,
$$a_0 = \frac{1}{c} \int_0^c f(t)\, dt$$

$$a_n = \frac{2}{c} \int_0^c f(t) \cos\left(\frac{n\pi t}{c}\right) dt$$

Case 2: Let $f(t)$ is an odd function defined in the interval $(-c, c)$, then the Fourier series corresponding to the function can be given as

$$f(t) = \sum_{n=1}^{\infty} b_n \sin \frac{n\pi x}{c}$$

where,
$$b_n = \frac{2}{c} \int_0^c f(t) \sin\left(\frac{n\pi t}{c}\right) dt$$

7.3 HALF RANGE SERIES

Many times it is required to obtain the Fourier expansion of a non-periodic function $f(x)$ defined in the interval $(0, c)$ of length c, which is half of the period of Fourier series as given in Fig. 7.2. This function, $f(x)$, can be considered the displacement of a violin string of length c or temperature of a metal bar of given length, c. The Fourier expansion of these type of functions are known as **half range expansion** or **half range Fourier series**.

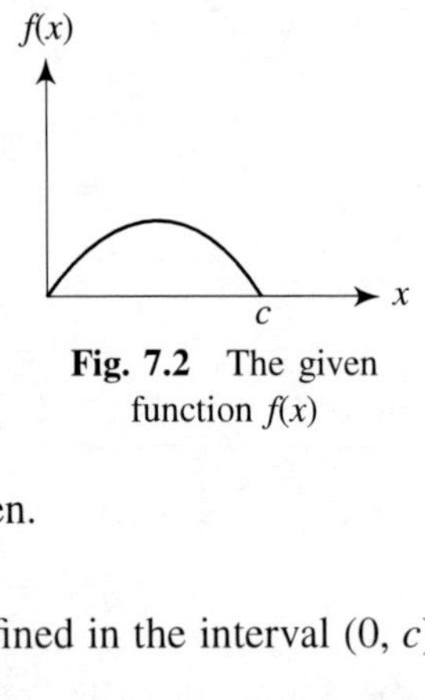

Fig. 7.2 The given function $f(x)$

It is immaterial whatever the function will outside the given interval $(0, c)$. We extend the given function to cover the interval $(-c, c)$ of length $2c$, such that the new function may be odd or even.

Case I: Fourier cosine series

Let the given $f(x)$ is neither periodic function nor even nor odd, defined in the interval $(0, c)$ of length, c. Since $f(x)$ is not defined in the interval $(-c, 0)$.

We define a new function $g(x)$ such that

(i) $g(x) = f(x)$ in the interval $(0, c)$.

(ii) We extend the function $f(x)$ by reflecting it about the y-axis so that $f(-x) = f(x)$ in the other half range $(-c, 0)$.

Then $g(x)$ is an even function in the interval $(-c, c)$ with period $2c$ as in Fig. 7.3 and gives the required Fourier cosine series in the interval $(-c, c)$.

$$g(x) = a_0 + \sum_{n=1}^{\infty} a_n \cos \frac{n\pi x}{c} \qquad ...(1)$$

where, $$a_0 = \frac{1}{c}\int_0^c f(x)\,dx \qquad ...(2)$$

$$a_n = \frac{2}{c}\int_0^c f(x)\cos\frac{n\pi x}{c}\,dx \qquad ...(3)$$

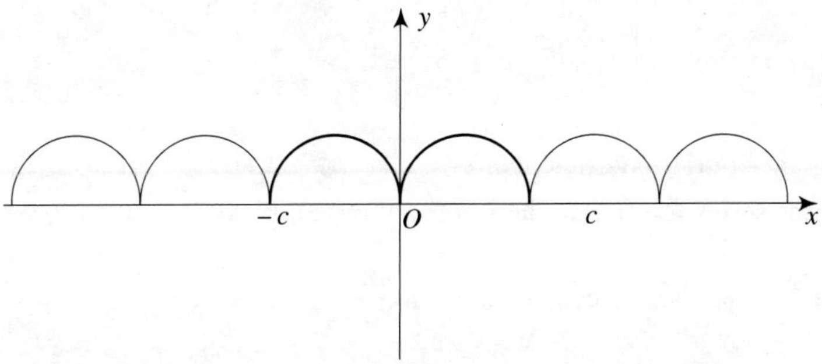

Fig. 7.3 $f(x)$ extended as an even periodic function of period $2c$.

Since $f(x)$ and $g(x)$ are equal in $(0, c)$, then the required half range Fourier sine series of $f(x)$ in the interval $(0, c)$ is given by equation (1) with a_0 and a_n given by equation (2) and equation (3).

Case II: Fourier sine series

Let the given function $f(x)$ is neither periodic function nor even nor odd, defined in the interval $(0, c)$ of length c. Since $f(x)$ is not defined in the interval $(-c, 0)$.

We define a new function $g(x)$ such that:

(i) $g(x) = f(x)$ in the interval $(0, c)$, and

(ii) we extend the given function $f(x)$ by reflecting it about the y-axis so that $f(-x) = -f(x)$ in the other half range $(-c, 0)$.

Then $g(x)$ is an odd function in the interval $(-c, c)$ with period $2c$ as in Fig. 7.4 and given the required **Fourier sine series** in the interval $(-c, c)$.

$$g(x) = \sum_{n=1}^{\infty} b_n \sin\frac{n\pi x}{c} \qquad ...(4)$$

where, $$b_0 = \frac{2}{c}\int_0^c f(x)\sin\frac{n\pi x}{c}\,dx \qquad ...(5)$$

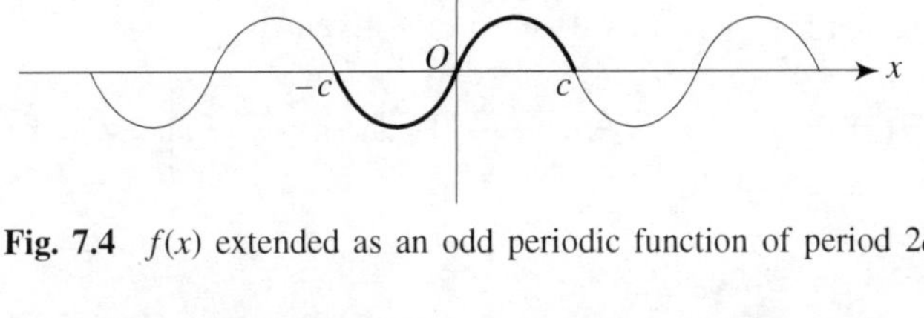

Fig. 7.4 $f(x)$ extended as an odd periodic function of period $2c$

Since $f(x)$ and $g(x)$ are equal in $(0, c)$, then the required half range Fourier sine series of $f(x)$ in the interval $(0, c)$ is given by equation (4) with b_n given by equation (5).

Hence, a given non-periodic function $f(x)$ can be represented in **cosine series** or **sine series**.

7.4 PARSEVAL'S THEOREM

Let $f(t)$ be a periodic function with period $2c$, defined in the interval $(-c, c)$. Then

$$\int_{-c}^{c} |f(t)|^2\, dx - c\left[a_0^2 + \sum_{n=1}^{\infty}(a_n^2 + b_n^2)\right]$$

where, a_0, a_n, b_n are Fourier coefficients.

Proof: The Fourier series for $f(t)$ defined in the interval, $(-c, c)$ is given by

$$f(t) = a_0 + \sum_{n=1}^{\infty}\left(a_n \cos\frac{n\pi t}{c} + b_n \sin\frac{n\pi t}{c}\right)$$

where, a_0, a_n and b_n are Fourier constants. On multiplying both sides by $f(t)$ and then integrating term by term from $-c$ to c, we have

$$\int_{-c}^{c} |f(t)|^2\, dt = a_0 \int_{-c}^{c} f(t)dt + \sum_{n=1}^{\infty}\left[a_n \int_{-c}^{c} f(t)\cos\frac{n\pi t}{c}dt\right] + \sum_{n=1}^{\infty}\left[b_n \int_{-c}^{c} f(t)\sin\frac{n\pi t}{c}dt\right]$$

$$\int_{-c}^{c} |f(t)|\, dt = a_0.c.a_0 + \sum_{n=1}^{\infty} a_n(ca_n) + \sum_{n=1}^{\infty} b_n(cb_n)$$

$$\int_{-c}^{c} |f(t)|^2\, dt = c\left[a_0^2 \sum_{n=1}^{\infty}(a_n^2 + b_n^2)\right]$$

Note: Assume $f(t)$ be a function, which is defined in the interval (a, b), then

$$\left[\left(\frac{1}{b-a}\right)\int_{0}^{b} |f(t)|^2\, dt\right]^{1/2}$$

is called the root mean square or effective value of $f(t)$ and denoted by the symbol, $|f(t)|_{\text{rms}}$

$$|f(t)|_{\text{rms}}^2 = \frac{1}{b-a}\int_{a}^{b} [f(t)]^2\, dt$$

Example 14 Find the Fourier series representation of the function $f(x) = e^{-x}$, $-a < x < a$.

Solution: Fourier series for the interval $-c$ to c is given by

$$f(x) = a_0 + \sum_{n=1}^{\infty}\left[a_n \cos\left(\frac{n\pi x}{c}\right) + b_n \sin\left(\frac{n\pi x}{c}\right)\right], \ -c < x < c$$

since $c = a$

$$\therefore \qquad f(x) = a_0 + \sum_{n=1}^{\infty}\left[a_n \cos\left(\frac{n\pi x}{a}\right) + b_n \sin\left(\frac{n\pi x}{a}\right)\right] \qquad \text{...(i)}$$

where, $a_0 = \dfrac{1}{2a}\int_{-a}^{a} f(x)dx = \dfrac{1}{2a}\int_{-a}^{a} e^{-x}dx$

$$= \frac{1}{2a}(-e^{-x})_{-a}^{a} = -\frac{1}{2a}(e^{-a} - e^{a})$$

$$\Rightarrow \qquad a_0 = \frac{1}{a}\sinh a \qquad \text{...(ii)}$$

Also $a_n = \dfrac{1}{a}\int_{-a}^{a} f(x)\cos\left(\dfrac{n\pi x}{a}\right)dx$

$$= \frac{1}{a}\int_{-a}^{a} e^{-x}\cos\left(\frac{n\pi x}{a}\right)dx$$

$$= \frac{1}{a}\left[\frac{e^{-x}}{1^2 + \frac{n^2\pi^2}{a^2}}\left[-1.\cos\left(\frac{n\pi x}{a}\right) + \left(\frac{n\pi}{a}\right)\sin\left(\frac{n\pi x}{a}\right)\right]\right]_{-a}^{a}$$

$$\left[\because \int e^{\alpha x}\cos bx\, dx = \frac{e^{ex}}{a^2+b^2}(a\cos bx + b\sin bx)\right]$$

$$\therefore \qquad a_n = \frac{1}{a}\left[\frac{a^2}{a^2+n^2\pi^2}\left\{e^{-a}\left(-\cos n\pi + \frac{n\pi}{a}\sin n\pi\right) - e^{a}\left(-\cos n\pi - \frac{n\pi}{a}\sin n\pi\right)\right\}\right]$$

$$= \frac{a}{a^2+n^2\pi^2}\left[\cos n\pi(e^{a} - e^{-a})\right] \qquad [\because \sin n\pi = 0 \text{ for } n = 1, 2, 3...]$$

$$= \frac{2a}{a^2+n^2\pi^2}\cos n\pi \sinh a$$

$$\Rightarrow \qquad a_n = \frac{2a}{a^2+n^2\pi^2}(-1)^n \sinh a$$

Again $b_n = \dfrac{1}{a}\int_{-a}^{a} f(x)\sin\left(\dfrac{n\pi x}{a}\right)dx$

$$= \frac{1}{a}\int_{-a}^{a} e^{-x}\sin\left(\frac{n\pi x}{a}\right)dx$$

$$= \frac{1}{a}\left[\frac{e^{-x}}{(-1)^2 + \frac{n^2\pi^2}{a^2}}\left[-\sin\left(\frac{n\pi x}{a}\right) - \frac{n\pi}{a}\cos\left(\frac{n\pi x}{a}\right)\right]\right]_{-a}^{a}$$

$$\left[\because \int e^{ax}\sin bx\,dx = -\frac{e^{ax}}{a^2+b^2}(a\sin bx - b\cos bx)\right]$$

$$\therefore \qquad b_n = \frac{a}{a^2+n^2\pi^2}\left[e^{-a}\left(-\sin n\pi - \frac{n\pi}{a}\cos n\pi\right) - e^{a}\left(\sin n\pi - \frac{n\pi}{a}\cos n\pi\right)\right]$$

$$= \frac{a}{a^2+n^2\pi^2}\left[\frac{n\pi}{a}\cos n\pi(e^a - e^{-a})\right]$$

$$= \frac{2an\pi}{(a^2+n^2\pi^2)a}\cos n\pi \sinh a$$

$$\Rightarrow \qquad b_n = \frac{2n\pi(-1)^n}{a^2+n^2\pi^2}\sinh a$$

Substituting the values of a_0, a_n and b_n in equation (i), we have

$$f(x) = e^{-x} = \frac{\sinh a}{a} + \sum_{n=1}^{\infty}\frac{2a(-1)^n}{a^2+n^2\pi^2}\sinh a\cos\left(\frac{n\pi x}{a}\right) + \sum_{n=1}^{\infty}\frac{2n\pi(-1)^n}{a^2+n^2\pi^2}\sinh a\sin\left(\frac{n\pi x}{a}\right)$$

$$e^{-x} = \frac{\sinh a}{a} - 2a\sinh a\left[\frac{1}{a^2+\pi^2}\cos\frac{\pi x}{a} - \frac{1}{a^2+4\pi^2}\cos\frac{2\pi x}{a} + \ldots\right]$$

$$- 2\pi\sinh a\left[\frac{1}{a^2+\pi^2}\sin\frac{\pi x}{a} - \frac{1}{a^2+4\pi^2}\sin\frac{2\pi x}{a} + \ldots\right] \qquad \textit{Ans.}$$

Example 15 Find the Fourier series representation for the function

$$f(x) = \begin{cases} \frac{\lambda}{4} & -\lambda < x < -\frac{\lambda}{2} \\ \frac{x^2}{\lambda} & -\frac{\lambda}{2} < x < \frac{\lambda}{2} \\ \frac{\lambda}{4} & \frac{\lambda}{2} < x < \lambda \end{cases}$$

Solution: Here the length of interval $2c = 2\lambda$

$$\therefore \qquad c = \lambda$$

Hence, the Fourier series is

$$f(x) = a_0 + \sum_{n=1}^{\infty}\left[a_n\cos\left(\frac{n\pi x}{\lambda}\right) + b_n\sin\left(\frac{n\pi x}{\lambda}\right)\right], \; -\lambda < x < \lambda \qquad \ldots(i)$$

where, $$a_0 = \frac{1}{2\lambda}\int_{-\lambda}^{\lambda} f(x)dx = \frac{1}{2\lambda}\left[\int_{-\lambda}^{-\lambda/2} f(x)dx + \int_{-\lambda/2}^{\lambda/2} f(x)dx + \int_{\lambda/2}^{\lambda} f(x)dx\right]$$

$$= \frac{1}{2\lambda}\left[\int_{-\lambda}^{\lambda} \frac{\lambda}{4}dx + \int_{-\lambda/2}^{\lambda/2} \frac{x^2}{4}dx + \int_{\lambda/2}^{\lambda} \frac{\lambda}{4}dx\right]$$

$$= \frac{1}{2\lambda}\left[\left(\frac{\lambda x}{4}\right)_{-\lambda}^{-\lambda/2} + \left(\frac{x^3}{3\lambda}\right)_{-\lambda/2}^{\lambda/2} + \left(\frac{\lambda x}{4}\right)_{\lambda/2}^{\lambda}\right]$$

$$= \frac{1}{2\lambda}\left[-\frac{\lambda^2}{8} + \frac{\lambda^2}{4} + \frac{\lambda^2}{24} + \frac{\lambda^2}{24} + \frac{\lambda^2}{4} - \frac{\lambda^2}{8}\right]$$

$$= \frac{1}{2\lambda}\left[\frac{\lambda^2}{3}\right]$$

$$\Rightarrow \quad a_0 = \frac{\lambda}{6} \qquad \text{...(ii)}$$

Also $$a_0 = \frac{1}{\lambda}\int_{-\lambda}^{\lambda} f(x)\cos\left(\frac{n\pi x}{\lambda}\right)dx$$

$$= \frac{1}{\lambda}\left[\int_{-\lambda}^{-\lambda/2} f(x)\cos\left(\frac{n\pi x}{\lambda}\right)dx + \int_{-\lambda/2}^{\lambda/2} f(x)\cos\left(\frac{n\pi x}{\lambda}\right)dx + \int_{\lambda/2}^{\lambda} f(x)\cos\left(\frac{n\pi x}{\lambda}\right)dx\right]$$

$$= \frac{1}{\lambda}\left[\int_{-\lambda}^{-\lambda/2} \frac{\lambda}{4}\cos\left(\frac{n\pi x}{\lambda}\right)dx + \int_{-\lambda/2}^{\lambda/2} \frac{x^2}{\lambda}\cos\left(\frac{n\pi x}{\lambda}\right)dx + \int_{\lambda/2}^{\lambda} \frac{\lambda}{4}\cos\left(\frac{n\pi x}{\lambda}\right)dx\right]$$

$$= \frac{1}{\lambda}\left[\left(\frac{\lambda}{4}\cdot\frac{\sin\frac{n\pi x}{\lambda}}{\frac{n\pi}{\lambda}}\right)_{-\lambda}^{-\lambda/2} + \frac{1}{\lambda}\left\{\left(x^2\frac{\sin\frac{n\pi x}{\lambda}}{\frac{n\pi}{\lambda}}\right) - \int 2x\frac{\sin\frac{n\pi x}{\lambda}}{\frac{n\pi}{\lambda}}dx\right\}_{-\lambda/2}^{\lambda/2} + \left(\frac{\lambda}{4}\frac{\sin\frac{n\pi x}{\lambda}}{\frac{n\pi}{\lambda}}\right)_{\lambda/2}^{\lambda}\right]$$

$$= \frac{1}{\lambda}\left[\frac{-\lambda^2}{4n\pi}\cdot\sin\frac{n\pi}{2} + \frac{1}{\lambda}\left\{\left(\frac{\lambda x^2}{n\pi}\sin\frac{n\pi x}{\lambda}\right) - 2x\left(\frac{-\cos\frac{n\pi x}{\lambda}}{\frac{n^2\pi^2}{\lambda^2}}\right) - \left[\frac{2\sin\frac{n\pi x}{\lambda}}{\frac{n^3\pi^3}{\lambda^3}}\right]\right\}_{-\lambda/2}^{\lambda/2} - \frac{\lambda^2}{4n\pi}\sin\frac{n\pi}{2}\right]$$

$$= \frac{1}{\lambda}\left[-\frac{2\lambda^2}{4n\pi}\sin\frac{n\pi}{2} + \frac{1}{\lambda}\left\{\frac{\lambda\lambda^2}{4n\pi}\sin\frac{n\pi}{2} + \frac{\lambda\lambda^2}{4n\pi}\sin\frac{n\pi}{2} + \frac{2\lambda\lambda^2}{2n^2\pi^2}\cos\frac{n\pi}{2} + \frac{2\lambda\lambda^2}{2n^2\pi^2}\cos\frac{n\pi}{2}\right.\right.$$

$$\left.\left.\frac{2\lambda^3}{n^3\pi^3}\sin\frac{n\pi}{2} - \frac{2\lambda^3}{n^3\pi^3}\sin\frac{n\pi}{2}\right\}\right]$$

$$= -\frac{\lambda}{2n\pi}\sin\frac{n\pi}{2} + \frac{1}{\lambda^2}\left[\frac{2\lambda^3}{4n\pi}\sin\frac{n\pi}{2} + \frac{2\lambda^3}{n^2\pi^2}\cos\frac{n\pi}{2} - \frac{4\lambda^3}{n^3\pi^3}\sin\frac{n\pi}{2}\right]$$

$$\Rightarrow \quad a_n = \frac{2\lambda}{n^2\pi^2}\left[\cos\frac{n\pi}{2} - \frac{2}{n\pi}\sin\frac{n\pi}{2}\right] \qquad \text{...(iii)}$$

and $b_n = \frac{1}{\lambda}\int_{-\lambda}^{\lambda} f(x)\sin\frac{n\pi x}{\lambda}dx$

$$= \frac{1}{\lambda}\left[\int_{-\lambda}^{-\lambda/2} f(x)\sin\frac{n\pi x}{\lambda}dx + \int_{-\lambda/2}^{\lambda/2} f(x)\sin\frac{n\pi x}{\lambda}dx + \int_{\lambda/2}^{\lambda} f(x)\sin\frac{n\pi x}{\lambda}dx\right]$$

$$= \frac{1}{\lambda}\left[\int_{-\lambda}^{-\lambda/2}\frac{\lambda}{4}\sin\frac{n\pi x}{\lambda}dx + \int_{-\lambda/2}^{\lambda/2}\frac{x^2}{\lambda}\sin\frac{n\pi x}{\lambda}dx + \int_{\lambda/2}^{\lambda}\frac{\lambda}{4}\sin\frac{n\pi x}{\lambda}dx\right]$$

$$= \frac{1}{\lambda}\left[\left\{\frac{\lambda}{4}\left(\frac{-\cos\frac{\pi x}{\lambda}}{\frac{n\pi}{\lambda}}\right)_{-\lambda}^{-\frac{1}{2}} + \frac{1}{\lambda}\left\{\left(-x^2\frac{\cos\frac{n\pi x}{\lambda}}{\frac{n\pi}{\lambda}}\right) + \int 2x\frac{\cos\frac{n\pi x}{\lambda}dx}{\frac{n\pi}{\lambda}}\right\}_{-\lambda/2}^{\lambda/2} + \left(-\frac{\lambda}{4}\frac{\cos\frac{n\pi x}{\lambda}}{\frac{n\pi}{\lambda}}\right)_{\lambda/2}^{\lambda}\right\}\right]$$

$$= \frac{1}{\lambda}\left[\left(-\frac{\lambda^2}{4n\pi}\cos\frac{nx\pi}{\lambda}\right)_{-\lambda}^{-\lambda/2} + \frac{1}{\lambda}\left\{\left(-\frac{\lambda x^2}{n\pi}\cos\frac{n\pi x}{\lambda}\right) + \left(2x\frac{\sin\frac{n\pi x}{\lambda}}{\frac{n^2\pi^2}{\lambda^2}}\right) + \left(2\frac{\cos\frac{n\pi x}{\lambda}}{\frac{n^3\pi^3}{\lambda^3}}\right)\right\}_{-\lambda/2}^{\lambda/2}\right.$$

$$\left. = \left(\frac{\lambda^2}{4n\pi}\cos\frac{n\pi x}{\lambda}\right)_{\lambda/2}^{\lambda}\right]$$

$$= \frac{1}{\lambda}\left[-\frac{\lambda^2}{4n\pi}\cos\frac{n\pi}{2} + \frac{\lambda^2}{4n\pi}\cos n\pi + \frac{1}{\lambda}\left\{-\frac{\lambda\lambda^2}{4n\pi}\cos\frac{n\pi}{2} + \frac{\lambda\lambda^2}{4n\pi}\cos\frac{n\pi}{2} + \frac{2\lambda}{2}\cdot\frac{\lambda^2}{n^2\pi^2}\right.\right.$$

$$\left.\left.\times\sin\frac{n\pi}{2} - \frac{2\lambda}{2}\cdot\frac{\lambda^2}{n^2\pi^2}\sin\frac{n\pi}{2} + \frac{2\lambda^3}{n^3\pi^3}\cos\frac{n\pi}{2} - \frac{2\lambda^3}{n^3\pi^3}\cos\frac{n\pi}{2}\right) - \frac{\lambda^3}{4n\pi}\cos n\pi + \frac{\lambda^2}{4n\pi}\cos\frac{n\pi}{2}\right]$$

$$\Rightarrow \qquad b_n = 0 \qquad \text{...(iv)}$$

Substituting the value of a_0, a_n and b_n from equations (ii), (iii) and (iv) respectively in equations (i), we have

$$f(x) = \frac{\lambda}{6} + \sum_{n=1}^{\infty}\frac{2\lambda}{n^2\pi^2}\left(\cos\frac{n\pi}{2} - \frac{2}{n\pi}\sin\frac{n\pi}{2}\right)\cos\frac{n\pi x}{\lambda} \qquad \textbf{\textit{Ans.}}$$

Example 16 Find half range cosine series for the function

$$f(t) = (t - 1)^2,\ 0 < t < 1$$

hence, show that

$$\pi^2 = 8\left(\frac{1}{1^2} + \frac{1}{3^2} + \frac{1}{5^2} + \ldots\right)$$

Solution: As given in the example, half range is $0 < t < 1$, therefore, full range can be defined as $-1 < t < 1$.

$$\Rightarrow \qquad c = 1$$

Hence, half range cosine series for $f(t)$ is given by

$$f(t) = a_0 + \sum_{n=1}^{\infty} a_n \cos \frac{n\pi x}{c},\ 0 < t < c \qquad \ldots\text{(i)}$$

$$= a_0 + \sum_{n=1}^{\infty} a_n \cos (n\pi t),\ 0 < t < 1 \qquad \ldots\text{(ii)}$$

Here,

$$a_0 = \frac{1}{c}\int_0^c f(t)\, dt$$

$$= \frac{1}{1}\int_0^1 (t - 1)^2\, dt$$

$$= \left[\frac{(t-1)^3}{3}\right]_0^1 = \frac{1}{3}$$

$$\Rightarrow \qquad a_0 = \frac{1}{3} \qquad \ldots\text{(iii)}$$

$$a_n = \frac{2}{c}\int_0^c f(t) \cos\left(\frac{n\pi t}{c}\right) dt$$

$$= \frac{2}{1}\int_0^1 (t - 1)^2\, dt$$

$$= 2\left[(t-1)^2 \frac{\sin n\pi t}{n\pi} + 2(t \quad 1)\frac{\cos n\pi t}{(n\pi)^2} - \frac{2\sin n\pi t}{(n\pi)^3}\right]_0^1$$

$$= 2\left[\frac{2}{(n\pi)^2}\right]$$

$$= 4 \cdot \frac{1}{(n\pi)^2} \quad \text{...(iv)}$$

$$= \frac{4}{n^2\pi^2}$$

Using a_0, a_n, we get

$$(t-1)^2 = \frac{1}{3} + \sum_{n=1}^{\infty} \frac{4}{n^2\pi^2} \cos n\pi t$$

$$= \frac{1}{3} + \frac{4}{\pi^2} \sum_{n=1}^{\infty} \frac{1}{n^2} \cos n\pi t \quad \text{(A)}$$

Now, putting $t = 0$, in equation (A), we get

$$1 = \frac{1}{3} + \frac{4}{\pi^2}\left(\frac{1}{1^2} + \frac{1}{2^2} + \frac{1}{3^2} + \frac{1}{4^2} + \frac{1}{5^2} + \ldots\right)$$

$$\Rightarrow \quad \frac{2}{3} = \frac{4}{\pi^2}\left(\frac{1}{1^2} + \frac{1}{2^2} + \frac{1}{3^2} + \frac{1}{4^2} + \frac{1}{5^2} + \ldots\right)$$

$$\frac{\pi^2}{6} = \frac{1}{1^2} + \frac{1}{2^2} + \frac{1}{3^2} + \frac{1}{4^2} + \frac{1}{5^2} + \ldots \quad \text{...(v)}$$

On putting $t = 1$, in equation (A)

$$0 = \frac{1}{3} + \frac{4}{\pi^2}\left(-\frac{1}{1^2} + \frac{1}{2^2} - \frac{1}{3^2} + \frac{1}{4^2} - \frac{1}{5^2} + \ldots\right)$$

$$\Rightarrow \quad \frac{\pi^2}{12} = \frac{1}{1^2} - \frac{1}{2^2} + \frac{1}{3^2} - \frac{1}{4^2} + \frac{1}{5^2} \ldots \quad \text{...(vi)}$$

On adding (v) and (vi), we get

$$\frac{\pi^2}{6} + \frac{\pi^2}{12} = 2\left(\frac{1}{1^2} + \frac{1}{3^2} + \frac{1}{5^2} + \ldots\right)$$

$$\pi^2\left(\frac{2+1}{1^2}\right) = 2\left(\frac{1}{1^2} + \frac{1}{3^2} + \frac{1}{5^2} + \ldots\right)$$

$$\pi^2 = 8\left(\frac{1}{1^2} + \frac{1}{3^2} + \frac{1}{5^2} + \ldots\right)$$

Example 17 Find the half range sine series for the function $f(t) = t(\pi - t)$ for $0 \le t \le \pi$ and hence find the sum of the series $\frac{1}{1^3} - \frac{1}{3^3} + \frac{1}{5^3} \dots$.

Solution: The half range sine series for $f(t)$ is

$$f(t) = x(\pi - t) = \sum_{n=1}^{\infty} b_n \sin nt; \quad 0 \le t \le \pi \qquad \text{...(i)}$$

Then
$$b_n = \frac{2}{\pi}\int_0^{\pi} f(t) \sin nt\, dt$$

$$\Rightarrow \qquad b_n = \frac{2}{\pi}\int_0^{\pi} t(\pi - t) \sin nt\, dt$$

On integrating by parts, we get

$$b_n = \frac{2}{\pi}\left[\left(\frac{t(\pi - t)\cos nt}{n}\right) + \int \left\{(\pi - 2t)\frac{\cos nt}{n}\right\} dt\right]_0^x$$

$$= \frac{2}{\pi}\left[\frac{-t(\pi - t)}{n}\cos nt + (\pi - 2t)\frac{\sin nt}{n^2} + 2\left(-\frac{\cos nt}{n^3}\right)\right]_0^{\pi}$$

$$= \frac{2}{\pi}\left[\frac{2}{n^3}(-\cos n\pi + 1)\right]$$

$$= -\frac{4}{\pi n^3}(\cos n\pi - 1)$$

$$\Rightarrow \qquad b_n = \frac{4}{n\pi^3}[(-1)^n - 1] \qquad \text{...(ii)}$$

Using equation (ii) in equation (i), we get

$$f(t) = \sum_{n=1}^{\infty} \frac{-4}{\pi n^3}[(-1)^n - 1] \sin nt$$

$$\Rightarrow \qquad t(\pi - t) = f(t) = \frac{8}{\pi}\left[\frac{\sin t}{1^3} + \frac{\sin 3t}{3^3} + \frac{\sin 5t}{5^3} + \dots\right] \qquad \text{...(iii)}$$

Put $t = \frac{\pi}{2}$ in equation (iii), we get

$$\frac{\pi}{2}\left(\pi - \frac{\pi}{2}\right) = \frac{8}{\pi}\left[\frac{1}{1^3} - \frac{1}{3^3} + \frac{1}{5^3} - \dots\right]$$

$$\Rightarrow \qquad \frac{\pi^2}{2} = \frac{8}{\pi}\left[\frac{1}{1^3} - \frac{1}{3^3} + \frac{1}{5^3} - \dots\right]$$

$$\Rightarrow \qquad \frac{1}{1^3} - \frac{1}{3^3} + \frac{1}{5^3} + \dots = \frac{\pi^3}{32}$$

Example 18 Find the half range cosine series for the function $f(x) = \dfrac{\sin \pi x}{L}$; $0 < x < L$.

Solution: Draw the graph of the function $f(x) = \dfrac{\sin \pi x}{L}$ in the interval $0 < x < L$ is the line OAB.

Let us extend the function in the interval $-L < x < 0$ shown by the line $OA'B'$ such that new function is symmetrical about the y-axis and therefore represents an even function in the interval $- L < x < L$.

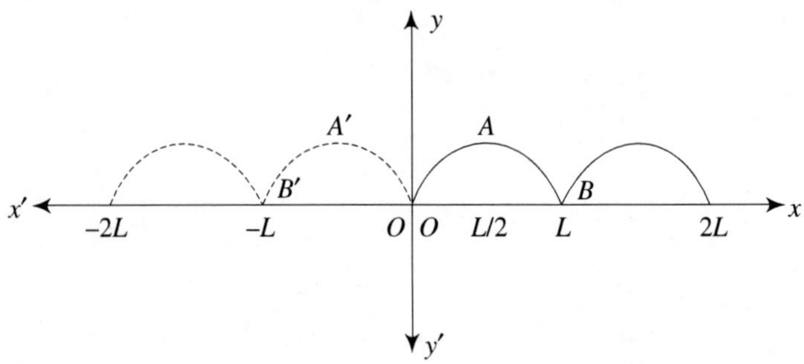

Fig. 7.5

$$g(x) = \frac{\sin \pi x}{L}, \quad 0 < x < L$$

$$g(x) = \frac{\sin \pi x}{L}, \quad -L < x < L$$

Hence, $g(x)$ is an even periodic function with period $2L$.

Now the Fourier series expansion of $g(x)$ in the interval $(-L, L)$ is

$$g(x) = a_0 + \sum_{n=1}^{\infty} a_n \cos\frac{n\pi x}{L} \quad \text{...(i)}$$

where,

$$a_0 = \frac{1}{L}\int_0^L f(x)dx$$

$$= \frac{1}{L}\int_0^L \sin\left(\frac{\pi x}{L}\right)dx$$

$$= \frac{1}{L}\cdot\frac{L}{\pi}\left(-\cos\frac{\pi x}{L}\right)_0^L$$

$\Rightarrow$

$$a_0 = \frac{1}{\pi}(1+1) = \frac{2}{\pi} \quad \text{...(ii)}$$

and

$$a_n = \frac{2}{L}\int_0^L f(x)\cos\left(\frac{n\pi x}{L}\right)dx$$

$$= \frac{2}{L}\int_0^L \sin\left(\frac{\pi x}{L}\right)\cos\left(\frac{n\pi x}{L}\right)dx$$

$$= \frac{1}{L}\int_0^L\left[\sin(1+n)\frac{\pi x}{L} + \sin(1-n)\frac{\pi x}{L}\right]dx$$

$$= \frac{1}{L}\left[\left(-\frac{\cos(1+n)\frac{\pi x}{L}}{\frac{(1+n)\pi}{L}}\right) + \left(-\frac{\cos(1+n)\frac{\pi x}{L}}{\frac{(1-n)\pi}{L}}\right)\right]_0^L$$

$$= -\left[\frac{\cos\frac{\pi x}{L}\cos\frac{n\pi x}{L} - \sin\frac{\pi x}{L}\sin\frac{n\pi x}{L}}{(1+n)\pi} + \frac{\cos\frac{\pi x}{L}\cos\frac{n\pi x}{L} + \sin\frac{\pi x}{L}\sin\frac{n\pi x}{L}}{(1-n)\pi}\right]_0^L ; n \neq 1$$

$$= -\left[-\frac{\cos n\pi}{(1+n)\pi} - \frac{\cos n\pi}{(1-n)\pi}\right] + \left[\frac{1}{(1+n)\pi} + \frac{1}{(1-n)\pi}\right]; n \neq 1$$

$$= \frac{1}{\pi}\left[\frac{1}{(n+1)}[(-1)^n + 1] - \frac{1}{(n-1)}[(-1)^n + 1]\right]; n \neq 1$$

$$= \frac{1}{\pi}\left[\frac{1}{(n+1)} - \frac{1}{(n-1)}\right][(-1)^n + 1]; n \neq 1$$

$$\Rightarrow \quad a_n = \frac{-2[(-1)^n + 1]}{\pi(n^2 - 1)}; n \neq 1 \qquad \text{...(iii)}$$

Now, $a_1 = \frac{2}{L}\int_0^L f(x)\cos\left(\frac{\pi x}{L}\right)dx$

$$= \frac{2}{L}\int_0^L \sin\left(\frac{\pi x}{L}\right)\cos\left(\frac{\pi x}{L}\right)dx$$

$$= \frac{1}{L}\int_0^L \sin\left(\frac{2\pi x}{L}\right)dx$$

$$= \frac{1}{L}\left(-\frac{\cos\left(\frac{2\pi x}{L}\right)}{\frac{2\pi}{L}}\right)_0^L$$

$$= -\frac{1}{L}\frac{L}{2\pi}(1-1) = 0$$

$$\Rightarrow \quad a_1 = 0 \qquad \text{...(iv)}$$

Using equations (ii), (iii) and (iv) in equation (i), we have

$$g(x) = \frac{2}{\pi} + 0 - \sum_{n=2}^{\infty} \frac{2[(-1)^n + 1]}{\pi(n^2 - 1)} \cos\left(\frac{n\pi x}{L}\right)$$

$$g(x) = \frac{2}{\pi} - \frac{4}{\pi}\left[\frac{1}{(2^2 - 1)}\cos\frac{2\pi x}{L} + \frac{1}{(4^2 - 1)}\cos\frac{4\pi x}{L} + \dots\right]$$

$$g(x) = \frac{2}{\pi} - \frac{4}{\pi}\left[\frac{1}{1.3}\cos\frac{2\pi x}{L} + \frac{1}{3.5}\cos\frac{4\pi x}{L} + \frac{1}{5.7}\cos\frac{6\pi x}{L} + \dots\right]$$

Hence, the half range cosine series of $f(x)$ in $(0, L)$ is

$$f(x) = \frac{2}{\pi} - \frac{4}{\pi}\left[\frac{1}{1.3}\cos\frac{2\pi x}{L} + \frac{1}{3.5}\cos\frac{4\pi x}{L} + \frac{1}{5.7}\cos\frac{6\pi x}{L} + \dots\right]$$ ***Ans.***

Example 19 Find the Fourier sine series for the function $f(x) = 1 - \frac{x}{L}$ in the interval $0 < x < L$.

Solution: Draw the graph of the function $f(x) = 1 - \frac{x}{L}$ in the interval $0 < x < L$ is the line AB.

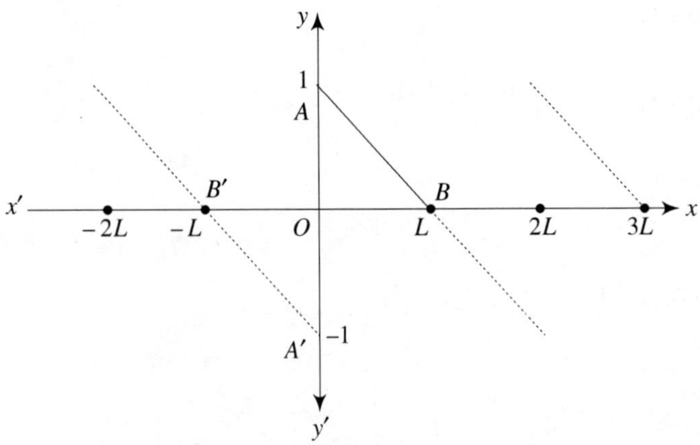

Fig. 7.6

Now let us extend the function $f(x)$ in the interval $-L < x < 0$ shown by AB, such that new function is symmetrical about the origin, i.e., in opposite quadrant and therefore represents an odd functon in the interval $-L < x < L$

$$g(x) = 1 - \frac{x}{L}, \; 0 < x < L$$

$$g(x) = -\left(1 + \frac{x}{L}\right) \text{ in } -L < x < L$$

and $$g(x + 2L) = g(x)$$

Thus, $g(x)$ is an odd periodic function in the interval $(-L, L)$ with period $2L$.

Now the Fourier series for $g(x)$ in the interval $(-L, L)$ is

$$g(x) = \sum_{n=1}^{\infty} b_n \sin\frac{n\pi x}{L} \qquad \text{...(i)}$$

where,
$$b_n = \frac{2}{L}\int_0^L f(x)\sin\frac{n\pi x}{L}dx$$

$$\Rightarrow \qquad b_n = \frac{2}{L}\int_0^L\left(1-\frac{x}{L}\right)\sin\frac{n\pi x}{L}dx$$

On integrating by parts, we get

$$b_n = \frac{2}{L}\left[\left\{-\left(1-\frac{x}{L}\right)\frac{\cos\frac{n\pi x}{L}}{\frac{n\pi}{L}}\right\} - \frac{1}{L}\frac{\sin\frac{n\pi x}{L}}{\frac{n^2\pi^2}{L^2}}\right]_0^L$$

$$\Rightarrow \qquad b_n = \frac{2}{L}\cdot\frac{L}{n\pi}$$

$$\Rightarrow \qquad b_n = \frac{2}{n\pi} \qquad \text{...(ii)}$$

Using equation (ii) and in equation (i), we have

$$g(x) = \sum_{n=1}^{\infty}\frac{2}{n\pi}\sin\frac{n\pi x}{L}$$

$$g(x) = \frac{2}{\pi}\left[\sin\frac{\pi x}{L}+\frac{1}{2}\sin\frac{2\pi x}{L}+\frac{1}{3}\sin\frac{3\pi x}{L}+\ldots\right]$$

Thus, the Fourier sine series of $f(x)$ in $(0, L)$ is

$$f(x) = g(x) = \frac{2}{\pi}\left[\sin\frac{\pi x}{L}+\frac{1}{2}\sin\frac{2\pi x}{L}+\frac{1}{3}\sin\frac{3\pi x}{L}+\ldots\right]$$

[Since $f(x)$ equals of $g(x)$ in the interval $(0, L)$] ***Ans.***

Example 20 If $f(x) = 1 - \dfrac{x}{L}$ in the interval $0 < x < L$, find the Fourier cosine series.

Solution: Draw the graph of the funcion $f(x) = 1 - \dfrac{x}{L}$ in $0 < x < L$ is the line AB. Now let us extend the funcion $f(x)$ in the interval $-L < x < 0$ shown by the line AB', such that the new function is symmetrical about the y-axis, therefore it represents an even function in the interval $-L < x < L$

$$g(x) = 1 - \frac{x}{L} \quad 0 < x < 2$$
$$= 1 + \frac{x}{L} \quad -2 < x < 0$$

Thus, $g(x)$ is an even periodic function in the interval $(-L, L)$ with period $2L$. Hence, the Fourier series for $g(x)$ over the interval $(-L, L)$ is given by

$$g(x) = a_0 + \sum_{n=1}^{\infty} a_n \cos\frac{n\pi x}{L} \qquad \text{...(i)}$$

where, $$a_0 = \frac{1}{L}\int_0^L f(x)dx = \frac{1}{L}\int_0^L \left(1 - \frac{x}{L}\right)dx$$

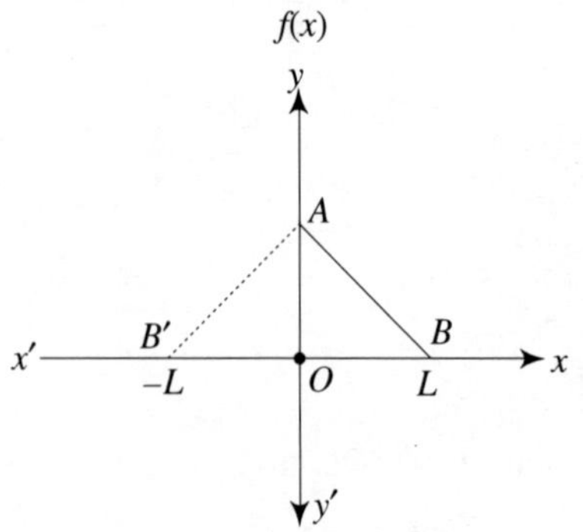

Fig. 7.7

$$\Rightarrow \qquad a_0 = \frac{1}{L}\left(x - \frac{x^2}{2L}\right)_0^L = \frac{1}{L}\left(L - \frac{L}{2}\right)$$

$$\Rightarrow \qquad a_0 = \frac{1}{2} \qquad \text{...(ii)}$$

and $$a_0 = \frac{2}{L}\int_0^L f(x)\cos\frac{n\pi x}{L}dx$$
$$= \frac{2}{L}\int_0^L \left(1 - \frac{x}{L}\right)\cos\frac{n\pi x}{L}dx$$

On integrating by parts,

$$a_n = \frac{2}{L}\left[\left(1 - \frac{x}{L}\right)\frac{\sin\frac{n\pi x}{L}}{\frac{n\pi}{L}} - \frac{1}{L}\frac{\cos\frac{n\pi x}{L}}{\frac{n^2\pi^2}{L^2}}\right]_0^L$$

$$\Rightarrow \qquad a_n = \frac{2}{L^2}\cdot\frac{L^2}{n^2\pi^2}(1-\cos n\pi)$$

$$\Rightarrow \qquad a_n = \frac{2}{n^2\pi^2}[1-(-1)^n] \qquad \text{...(iii)}$$

Using equations (ii) and (iii) in equation (i), we have

$$g(x) = \frac{1}{2}+\sum_{n=1}^{\infty}\frac{2}{n^2\pi^2}[1-(-1)^n]\cos\frac{n\pi x}{L}$$

$$g(x) = \frac{1}{2}+\frac{4}{\pi^2}\left[\frac{1}{1^2}\cos\frac{\pi x}{L}+\frac{1}{3^2}\cos\frac{3\pi x}{L}+\frac{1}{5^2}\cos\frac{5\pi x}{L}+\ldots\right]$$

Then the Fourier cosine series of $f(x)$ in (0, 2) is

$$f(x) = \frac{1}{2}+\frac{4}{\pi^2}\left[\frac{1}{1^2}\cos\frac{\pi x}{L}+\frac{1}{3^2}\cos\frac{3\pi x}{L}+\frac{1}{5^2}\cos\frac{5\pi x}{L}+\ldots\right]$$

[since $f(x)$ is equal to $g(x)$ in the interval $(0, L)$] ***Ans.***

Example 21 Express $f(x) = x$ as a half range sine series in the interval $0 < x < 2$.

Solution: Draw the graph of $f(x) = x$ in $0 < x < 2$, then we have the line OA. Now we extend the function $f(x)$ in the interval $-2 < x < 0$, given the line OA', such that the new function is symmetrical about the origin, i.e., in opposite quadrant, therefore represents an odd function in the interval (– 2, 2)

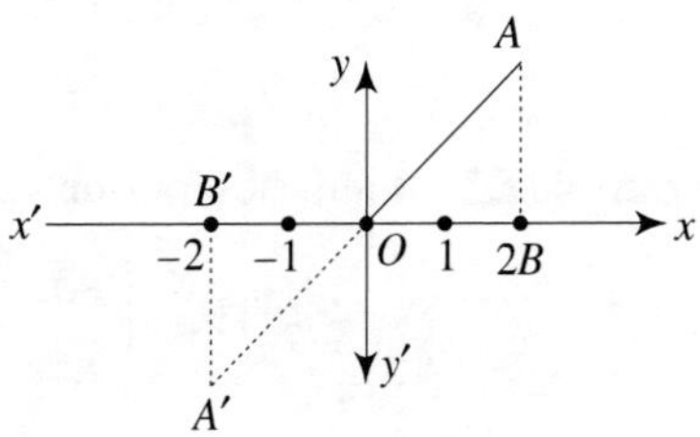

Fig. 7.8

$$g(x) = x \qquad 0 < x < 2$$
$$= x \qquad -2 < x < 0$$

Thus, $g(x)$ is an odd periodic function in the interval (– 2, 2) with period 4.

Then the Fourier series for the function $g(x)$ in the interval (– 2, 2) is given by

$$g(x) \neq \sum_{n=1}^{\infty} b_n \sin\frac{n\pi x}{2} \qquad [\because \ 2c = 4, c = 2] \qquad \text{...(i)}$$

where

$$b_n = \int_0^2 f(x)\sin\frac{n\pi x}{2}dx$$

$$\Rightarrow \qquad b_n = \int_0^2 x\sin\frac{n\pi x}{2}dx$$

On integrating by parts, we get

$$b_n = \left[x\left(\frac{-\cos\frac{n\pi x}{2}}{\frac{n\pi}{2}}\right)+\left(\frac{\sin\frac{n\pi x}{2}}{\frac{n^2\pi^2}{2^2}}\right)\right]_0^2$$

$$\Rightarrow \qquad b_n = \frac{2.2}{n\pi}(0 - \cos n\pi)$$

$$\Rightarrow \qquad b_n = \frac{-4}{n\pi}(-1)^n \qquad \text{...(ii)}$$

Substituting the value of b_n from equation (ii) in equation (i), we have

$$g(x) = \sum_{n=1}^{\infty} \frac{-4}{n\pi}(-1)^n \sin\frac{n\pi x}{2}$$

$$g(x) = \frac{-4}{\pi}\left[-\sin\frac{n\pi}{2} + \frac{\sin\frac{2\pi x}{2}}{2} - \frac{\sin\frac{3\pi x}{2}}{3} + \frac{\sin\frac{4\pi x}{2}}{4} - \cdots\right]$$

$$g(x) = \frac{-4}{\pi}\left[\sin\frac{n\pi}{2} - \frac{1}{2}\sin\frac{2\pi x}{2} + \frac{1}{3}\sin\frac{3\pi x}{2} - \frac{1}{4}\sin\frac{4\pi x}{2} + \cdots\right]$$

Then the half range sine series of $f(x)$ in (0, 2) is

$$f(x) = \frac{4}{\pi}\left[\sin\frac{n\pi}{2} - \frac{1}{2}\sin\frac{2\pi x}{2} + \frac{1}{3}\sin\frac{3\pi x}{2} - \frac{1}{4}\sin\frac{4\pi x}{2} + \cdots\right]$$

[since $f(x)$ is equal to $g(x)$ in the interval (0, 2)] ***Ans.***

Example 22 Find the Fourier series of the function

$$F(x) = \begin{cases} -a, & -l \le x < 0 \\ a, & 0 < x \le l \end{cases}$$

Solution: Here the length of interval, $2c = 2l$

i.e., $c = l$

Hence, the Fourier series is

$$f(x) = a_0 + \sum_{n=1}^{\infty}\left(a_n \cos\frac{n\pi x}{l} + b_n \sin\frac{n\pi x}{l}\right) \qquad \text{...(i)}$$

where,

$$a_0 = \frac{1}{2l}\int_{-1}^{l} f(x)dx$$

$$= \frac{1}{2l}\left[\int_{-1}^{0} f(x)dx + \int_{0}^{l} f(x)dx\right]$$

$$= \frac{1}{2l}\left[\int_{-1}^{0} -a dx + \int_{0}^{l} a dx\right]$$

$$= \frac{a}{2l}\left[(-x)_{-l}^{0} + (x)_0^l\right]$$

$$= \frac{a}{2l}[-0 - l + l - 0]$$

$$\Rightarrow \qquad a_0 = 0 \qquad \text{...(ii)}$$

and
$$a_n = \frac{1}{l}\int_{-l}^{l} f(x)\cos\frac{n\pi x}{l}\,dx$$

$$= \frac{1}{l}\left[\int_{-l}^{0} f(x)\cos\frac{n\pi x}{l}\,dx + \int_{0}^{l} f(x)\cos\frac{n\pi x}{l}\,dx\right]$$

$$= \frac{1}{l}\left[\int_{-l}^{0} -a\cos\frac{n\pi x}{l}\,dx + \int_{0}^{l} a\cos\frac{n\pi x}{l}\,dx\right]$$

$$= \frac{a}{l}\left[\left(-\frac{\sin\frac{n\pi x}{l}}{\frac{n\pi}{l}}\right)_{-l}^{0} + \left(\frac{\sin\frac{n\pi x}{l}}{\frac{n\pi}{l}}\right)_{0}^{l}\right]$$

$$= \frac{a}{l}[0+0]$$

$$\therefore \qquad a_0 = 0 \qquad \text{...(iii)}$$

Also
$$b_n = \frac{1}{l}\int_{-l}^{l} f(x)\sin\frac{n\pi x}{l}\,dx$$

$$= \frac{1}{l}\left[\int_{-1}^{0} -a\sin\frac{n\pi x}{l}\,dx + \int_{0}^{l} a\sin\frac{n\pi x}{l}\,dx\right]$$

$$= \frac{1}{l}\left[\left(\frac{a\sin\frac{n\pi x}{l}}{\frac{n\pi}{l}}\right)_{-l}^{0} + \left(\frac{-a\cos\frac{n\pi x}{l}}{\frac{n\pi}{l}}\right)_{0}^{l}\right]$$

$$= \frac{1}{l}\left[\frac{la}{n\pi}\{1-\cos\pi-\cos n\pi+1\}\right]$$

$$= \frac{2a}{n\pi}[1-\cos n\pi]$$

$$\therefore \qquad b_0 = \frac{2a}{n\pi}[1-(-1)^n] \qquad \text{...(iv)}$$

Substituting the value of a_0, a_n and b_n from equations (ii), (iii) and (iv) respectively in equation (i), we have

$$f(x) = 0 + 0 + \sum_{n=1}^{\infty} \frac{2a}{n\pi}[1-(-1)^n]\sin\frac{n\pi x}{l}$$

$$= \frac{2a}{\pi}\left[\frac{2\sin\frac{\pi x}{l}}{1} + \frac{2\sin\frac{3\pi x}{l}}{3} + \frac{2\sin\frac{5\pi x}{l}}{5} + \cdots\right]$$

$$\therefore \qquad f(x) = \frac{4a}{\pi}\left[\sin\frac{\pi x}{l} + \frac{1}{3}\sin\frac{3\pi x}{l} + \frac{1}{5}\sin\frac{5\pi x}{l} + \cdots\right] \qquad \textbf{\textit{Ans.}}$$

Example 23 Obtain the Fourier series expansion for the function $f(x) = \dfrac{(\pi - x)}{2}$ in $0 < x < 2$.

Solution: Here the length of interval, $2c = 2$

i.e., $c = 1$

Hence, the required Fourier series is

$$f(x) = a_0 + \sum_{n=1}^{\infty}\left(a_n \cos\frac{n\pi x}{1} + b_n \sin\frac{n\pi x}{1}\right); \; 0 < x < 2 \qquad \text{...(i)}$$

where,
$$a_0 = \frac{1}{2.1}\int_0^2 f(x)dx$$

$$= \frac{1}{2}\int_0^2 \left(\frac{n - x}{2}\right)dx$$

$$\Rightarrow \qquad a_0 = \frac{1}{4}\left[n\pi - \frac{x^2}{2}\right]_0^2 = \frac{1}{4}[2\pi - 2] = \frac{1}{2}(\pi - 1) \qquad \text{...(ii)}$$

Also
$$a_n = \frac{1}{l}\int_0^2 f(x)\cos\frac{n\pi x}{1}dx$$

$$= \int_0^2 \frac{(\pi - x)}{2}\cos n\pi\, x\, dx$$

$$= \frac{1}{2}\int_0^2 (\pi - x)\cos n\pi\, x\, dx$$

On integrating by parts, we get

$$a_n = \frac{1}{2}\left[(\pi - x)\frac{\sin n\pi x}{nx} + \int\frac{\sin n\pi x}{n\pi}dx\right]_0^2$$

$$= \frac{1}{2}\left[(\pi - x)\frac{\sin n\pi x}{nx} - \frac{\cos n\pi x}{n^2\pi^2}\right]_0^2$$

$$= \frac{1}{2}\left[-\frac{\cos 2n\pi}{n^2x^2} + \frac{\cos n\pi 0}{n^2\pi^2}\right]$$

$$\Rightarrow \qquad a_n = \frac{1}{2n^2\pi^2}[1-1] = 0 \qquad \text{...(iii)}$$

Again $$b_n = \frac{1}{1}\int_0^2 f(x)\sin\frac{n\pi x}{1}dx$$

$$= \int_0^2 \frac{(\pi - x)}{2}\sin n\pi x\, dx$$

On integrating by parts, we get

$$b_n = \frac{1}{2}\left[(\pi - x)\left(-\frac{\cos n\pi x}{n\pi}\right) - \int \frac{\cos n\pi x}{n\pi}dx\right]_0^2$$

$$= \frac{1}{2}\left[(\pi - x)\left(-\frac{\cos n\pi x}{n\pi}\right) - \frac{\sin n\pi x}{n^2\pi^2}\right]_0^2$$

$$= \frac{1}{2}\left[(\pi - x)\left(-\frac{\cos 2\, n\pi}{n\pi}\right) + \frac{\pi \cos n\pi 0}{n\pi}\right]$$

$$\Rightarrow \qquad b_n = \frac{1}{n\pi} \qquad \text{...(iv)}$$

Substituting the value of a_0, a_n and b_n from equations (ii), (iii) and (iv) respectively in equation (i), then

$$f(x) = \frac{\pi - x}{2} = \frac{(\pi - 1)}{2} + 0 + \sum_{n=1}^{\infty} \frac{1}{n\pi}\sin n\pi x$$

$$\Rightarrow \qquad \frac{\pi - x}{2} = \frac{(\pi - 1)}{2} + \frac{1}{\pi}\left[\frac{\sin \pi x}{1} + \frac{\sin 2\pi x}{2} + \frac{\sin 3\pi x}{3} + \cdots\right] \qquad \textit{Ans.}$$

Example 24 Find the Fourier series to represent the function:

$$f(x) = \begin{cases} x, & 0 < x < 1 \\ 0, & 1 < x < 2 \end{cases}$$

Solution: Here the length of interval $2c = c$

i.e., $$c = 1$$

Hence, the required Fourier series is

$$f(x) = a_0 + \sum_{n=1}^{\infty}\left(a_n \cos\frac{n\pi x}{1} + b_n \sin\frac{n\pi x}{1}\right);\ 0 < x < 2 \qquad \text{...(i)}$$

where, $$a_0 = \frac{1}{2.1}\int_0^2 f(x)dx = \frac{1}{2}\left[\int_0^1 f(x)dx + \int_1^2 f(x)dx\right]$$

$$= \frac{1}{2}\left[\int_0^1 x dx \int_1^2 0 dx\right] = \frac{1}{2}\left(\frac{x^2}{2}\right)_0^1$$

$\therefore$ $$a_0 = \frac{1}{4} \qquad \text{...(ii)}$$

Also $$a_n = \frac{1}{1}\int_0^2 f(x)\cos\frac{n\pi x}{1}dx$$

$$= \left[\int_0^1 f(x)\cos\frac{n\pi x}{1}dx + \int_1^2 f(x)\cos\frac{n\pi x}{1}dx\right]$$

$$= \left[\int_0^1 x\cos n\pi x\, dx + \int_0^2 0\cos n\pi x\, dx\right]$$

On integrating by parts, we get

$$a_n = \left[\left(x\frac{\sin n\pi x}{n\pi}\right)_0^1 - \left(-\frac{\cos n\pi x}{n^2\pi^2}\right)_0^1 + 0\right]$$

$$= \frac{\cos n\pi - \cos 0}{n^2\pi^2}$$

$\Rightarrow$ $$a_n = \frac{1}{n^2\pi^2}[(-1)^n - 1] \qquad \text{...(iii)}$$

and $$b_n = \frac{1}{1}\int_0^2 f(x)\sin\frac{n\pi x}{1}dx$$

$$= \left[\int_0^1 f(x)\sin\frac{n\pi x}{1}dx + \int_1^2 f(x)\sin\frac{n\pi x}{1}dx\right]$$

$$= \left[\int_0^1 x\sin n\pi x\, dx + \int_0^2 0\sin n\pi x\, dx\right]$$

On interating by parts, we get

$$b_n = \left[x\left(-\frac{\cos n\pi x}{n\pi}\right) + \frac{\sin n\pi x}{n^2\pi^2}\right]_0^1$$

$$= -\frac{\cos n\pi}{n\pi} = \frac{-(-1)^n}{n\pi}$$

$\Rightarrow$ $$b_n = \frac{(-1)^{n+1}}{n\pi} \qquad \text{...(iv)}$$

Substituting the values of a_0, a_n and b_n from equations (ii), (iii) and (iv) respectively in equation (i), we get

$$f(x) = \frac{1}{4} + \sum_{n=1}^{\infty} \frac{1}{n^2\pi^2}[(-1)^n - 1]\cos n\pi x + \sum_{n=1}^{\infty} \frac{(-1)^{n+1}}{n\pi}\sin n\pi x$$

$$f(x) = \frac{1}{4} - \frac{2}{\pi^2}\left[\frac{\cos \pi x}{1^2} + \frac{\cos 3\pi x}{3^2} + \frac{\cos 5\pi x}{5^2} + \cdots\right] + \frac{1}{\pi}\left[\frac{\sin \pi x}{1} - \frac{\sin 2\pi x}{2} + \frac{\sin 3\pi x}{3} - \cdots\right]$$

Ans.

Example 25 Express $f(x) = x$ as half range cosine series in the interval $0 < x < 2$.

Solution: Draw the graph of $f(x) = x$ in the interval $0 < x < 2$. Then we have the line OA.

Now extend the function $f(x)$ in the interval $(-2, 0)$ shown by OA', then the new function is symmetrical about the y-axis, therefore represents an even function in $(-2, 2)$

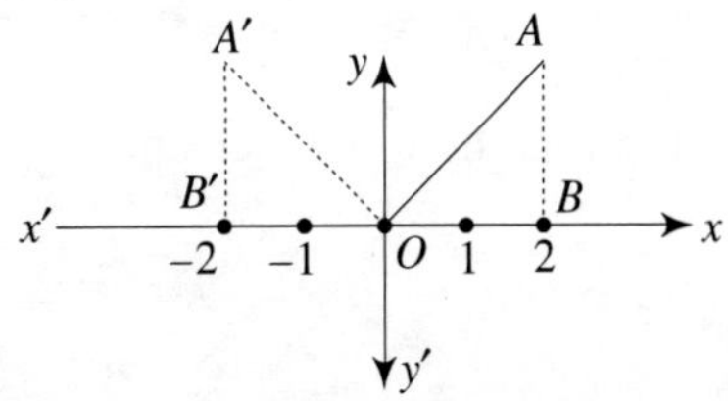

Fig. 7.9

$$g(x) = x \quad 0 < x < 2$$
$$= x \quad -2 < x < 0$$

Thus, $g(x)$ is an even periodic function in the interval $(-2, 2)$ with period 4.

Then the Fourier series for $g(x)$ over the interval $(-2, 2)$ is given by

$$g(x) = a_0 + \sum_{n=1}^{\infty} a_n \cos\frac{n\pi x}{2} \qquad [\because \; 2c = 4, c = 2] \qquad \text{...(i)}$$

where,
$$a_0 = \frac{1}{2}\int_0^2 f(x)dx = \frac{1}{2}\int_0^2 x\,dx$$

$$a_0 = 1 \qquad \text{...(ii)}$$

Also
$$a_n = \int_0^2 f(x)\cos\frac{n\pi x}{2}dx$$

$\Rightarrow$
$$a_n = \int_0^2 x\cos\frac{n\pi x}{2}dx$$

On integrating by parts, we get

$$a_n = \left[\left(\frac{x\sin\frac{n\pi x}{2}}{\frac{n\pi}{2}}\right) + \left(\frac{\cos\frac{n\pi x}{2}}{\frac{n^2\pi^2}{2^2}}\right)\right]_0^2$$

$\Rightarrow$
$$a_n = \frac{4}{n^2\pi^2}(\cos n\pi - 1)$$

$$\Rightarrow \qquad a_n = \frac{4}{n^2\pi^2}[(-1)^n - 1)] \qquad ...(iii)$$

Substituting the value of a_0 and a_n from equations (ii) and (iii), respectively in equation (i), then

$$g(x) = 1+\sum_{n=1}^{\infty} \frac{4}{n^2\pi^2}[(-1)^n - 1]\cos\frac{n\pi x}{2}$$

$$g(x) = 1+\frac{4}{\pi^2}\left[\frac{2}{1^2}\cos\frac{\pi x}{2}+\frac{2}{3^2}\cos\frac{3\pi x}{2}+\frac{2}{5^2}\cos\frac{5\pi x}{2}+\cdots\right]$$

$$g(x) = 1+\frac{8}{\pi^2}\left[\cos\frac{\pi x}{2}+\frac{1}{3^2}\cos\frac{3\pi x}{2}+\frac{1}{5^2}\cos\frac{5\pi x}{2}+\cdots\right]$$

Then the half range cosine series of $f(x)$ in (0, 2) is

$$f(x) = 2 - \frac{8}{\pi^2}\left[\cos\frac{\pi x}{2}+\frac{1}{3^2}\cos\frac{3\pi x}{2}+\frac{1}{5^2}\cos\frac{5\pi x}{2}+\cdots\right]$$

[since $f(x)$ equals to $g(x)$ in the interval (0, 2)] ***Ans.***

EXERCISE

1. $f(x) = \begin{cases} 0, & 0 \le x < l \\ a, & l \le x < 2l \end{cases}$

$$\left[\textbf{\textit{Ans.}}\ f(x) = \frac{a}{2} - \frac{2a}{\pi}\left[\sin\frac{\pi x}{l}+\frac{1}{3}\sin\frac{3\pi x}{l}+\frac{1}{5}\sin\frac{5\pi x}{l}+\cdots\right]\right]$$

2. $f(x) = x\cos\left(\frac{\pi x}{L}\right), -L \le x \le L.$

$$\left[\textbf{\textit{Ans.}}\ f(x) = -\frac{L}{2\pi}\sin\left(\frac{\pi x}{L}\right)+\frac{2L}{\pi}\sum_{n=2}^{\infty}\frac{(-1)^n n}{n^2-1}\sin\left(\frac{n\pi x}{L}\right)\right]$$

3. $f(x) = \pi x, -c \le x \le c.$

$$\left[\textbf{\textit{Ans.}}\ f(x) = 2c\left[\sin\left(\frac{\pi x}{c}\right)-\frac{1}{2}\sin\left(\frac{2\pi x}{c}\right)+\frac{1}{3}\sin\left(\frac{3\pi x}{c}\right)-\frac{1}{4}\sin\left(\frac{4\pi x}{c}\right)+\cdots\right]\right]$$

4. (i) $f(x) = \begin{cases} x, & -1 < x < 0 \\ x+2, & 0 < x < 1 \end{cases}$

Deduce that $\frac{\pi}{4} = 1-\frac{1}{3}+\frac{1}{5}-\frac{1}{7}+...$

[***Hint:*** Put $x = \frac{1}{2}$ on both sides to deduce the result]

$$\left[\textbf{Ans. } f(x) = 1 + \frac{2}{\pi} \sum_{n=1}^{\infty} \frac{1}{n} [1 - 2(-1)^n] \sin n\pi x \right]$$

(ii) $f(x) = \dfrac{x(\pi^2 - x^2)}{12}, (-\pi, \pi)$

$$\left[\textbf{Ans. } \frac{\sin x}{1^3} - \frac{\sin 2x}{2^3} + \frac{\sin 3x}{3^3} + \cdots \right]$$

5. $f(x) = x^2 - 2, \quad -2 \le x \le 2.$

$$\left[\textbf{Ans. } x^2 - 2 = -\frac{2}{3} - \frac{16}{\pi^2} \left(\cos \frac{\pi x}{2} - \frac{\cos \frac{2\pi x}{2}}{2^2} + \frac{\cos \frac{3\pi x}{2}}{3^2} - \cdots \right) \right]$$

6. Obtain the Fourier series of the periodic function

$$f(x) = \begin{cases} 1, & 0 < x < 1 \\ 2, & 1 < x < 2 \end{cases}$$

$$\left[\textbf{Ans. } f(x) = 3 - \sum_{n=1}^{\infty} \frac{2}{(2n-1)\pi} \sin (2n - 1)\pi x \right]$$

7. (i) In the interval (0, 2), $f(x)$ is defined as

$$f(x) = \begin{cases} 0, & 0 \le x < l \\ a, & l \le x < 2l \end{cases}$$

Expand $f(x)$ in a Fourier series.

$$\left[\textbf{Ans. } f(x) = \frac{a}{2} - \frac{2a}{\pi} \left[\sin \frac{\pi x}{l} + \frac{1}{3} \sin \frac{3\pi x}{l} + \frac{1}{5} \sin \frac{5\pi x}{l} + \cdots \right] \right]$$

(ii) Find the Fourier series for the function $f(x) = x + x^2$ in the interval $-\pi < x < x$. Hence, show that

$$\frac{\pi^2}{6} = 1 + \frac{1}{2^2} + \frac{1}{3^2} + \cdots$$

$$\left[\textbf{Ans. } \frac{\pi^2}{3} - 4\left(\cos x - \frac{1}{2^2} \cos 2x + \frac{1}{3^2} \cos 3x - \cdots \right) + 2\left(\sin x - \frac{1}{2} \sin 2x + \frac{1}{3} \sin 3x - \cdots \right) \right]$$

8. $f(x) = (x - 1)^2$, $0 < x < 1$, hence show that $\dfrac{\pi^2}{8} = \dfrac{1}{1^2} + \dfrac{1}{3^2} + \dfrac{1}{5^2} + \cdots$.

$$\left[\textbf{Ans. } (x - 1)^2 = \frac{1}{3} + \frac{4}{\pi^2} \sum_{n=1}^{\infty} \frac{1}{n^2} \cos n\pi x \right]$$

9. $f(x) = x$, $0 < x < \pi$.

$$\left[\textbf{Ans. } x = \frac{\pi}{2} - \frac{4}{\pi} \left[\cos x + \frac{\cos 3x}{3^2} + \frac{\cos 5x}{5^2} + \cdots \right] \right]$$

10. $f(x) = 2x - 1,\ 0 < x < 1.$ $\left[\textbf{Ans. } 2x - 1 = -\frac{8}{\pi^2}\left[\cos \pi x + \frac{\cos 3\pi x}{3^2} + \frac{\cos 5\pi x}{5^2} + \cdots\right]\right]$

11. $f(x) = \sin x,\ 0 < x < \pi.$ $\left[\textbf{Ans. } \sin x = \frac{4}{\pi}\left[\frac{1}{2} - \frac{\cos 2x}{3} - \frac{\cos 4x}{15} - \cdots\right]\right]$

12. $f(x) = \begin{cases} x & \text{when } 0 < x < \dfrac{\pi}{2} \\ 0 & \text{when } \dfrac{\pi}{2} < x < \pi \end{cases}$

$$\left[\textbf{Ans. } f(x) = \frac{\pi}{8} + \sum_{n=1}^{\infty}\left[\frac{1}{n}\sin\frac{n\pi}{2} + \frac{2}{\pi n^2}\left(\cos\frac{n\pi}{2} - 1\right)\right]\cos n\pi\right]$$

13. $f(x) = \begin{cases} kx, & 0 \le x \le \dfrac{l}{2} \\ k(l - x), & \dfrac{l}{2} < x \le l \end{cases}$

$$\left[\textbf{Ans. } f(x) = \frac{kl}{4} - \frac{8kl}{\pi^2}\left[\frac{1}{2^2}\cos\frac{2\pi x}{l} + \frac{1}{6^2}\cos\frac{6\pi x}{l} + \frac{1}{10^2}\cos\frac{10\pi x}{l} + \cdots\right]\right]$$

14. $f(x) = \begin{cases} x & \text{for } 0 < x < \dfrac{\pi}{2} \\ \pi - x & \text{for } \dfrac{\pi}{2} < x < \pi \end{cases}$

$$\left[\textbf{Ans. } f(x) = \frac{\pi}{4} - \frac{2}{\pi}\left(\frac{\cos 2x}{1^2} + \frac{\cos 6x}{3^2} + \frac{\cos 10x}{5^2} + \cdots\right)\right]$$

15. $f(x) = \begin{cases} 1 & \text{if } 0 < x < \dfrac{1}{2} \\ 0 & \text{if } \dfrac{1}{2} < x < 1 \end{cases}$ $\left[\textbf{Ans. } f(x) = \frac{1}{2} + \frac{2}{\pi}\sum_{n=1}^{\infty}\frac{1}{n}\sin\frac{n\pi}{2}\cos n\pi x\right]$

16. $f(x) = \begin{cases} 1 & \text{if } 0 < x < \dfrac{1}{2} \\ 0 & \text{if } \dfrac{1}{2} < x < 1 \end{cases}$ $\left[\textbf{Ans. } f(x) = \frac{2}{\pi}\sum_{n=1}^{\infty}\left(1 - \cos\frac{n\pi}{2}\right)\frac{1}{n}\sin n\pi x\right]$

17. Find the half range sine series for the following functions:

$f(x) = 2x - 1,\ 0 < x < 1.$

$$\left[\textbf{Ans. } 2x - 1 = -\frac{2}{\pi}\left[\sin 2\pi x + \frac{1}{2}\sin 4\pi x + \frac{1}{3}\sin 6\pi x + \cdots\right]\right]$$

18. $f(x) = \begin{cases} \frac{1}{4} - x, & 0 < x < \frac{1}{2} \\ x - \frac{3}{4}, & \frac{1}{2} < x < 1 \end{cases}$

$$\left[\textbf{Ans. } f(x) = \left(\frac{1}{\pi} - \frac{4}{\pi^2}\right)\sin \pi x + \left(\frac{1}{3\pi} + \frac{4}{3^2\pi^2}\right)\sin 3x + \left(\frac{1}{5\pi} - \frac{4}{5^2\pi^2}\right)\sin 5x + \cdots\right]$$

19. $f(x) = C$, where, C is constant; for $0 < x < \pi$.

$$\left[\textbf{Ans. } C = \frac{4C}{\pi}\left[\sin x + \frac{\sin 3x}{3} + \frac{\sin 5x}{5} + \cdots\right]\right]$$

20. $f(x) = \begin{cases} x, & 0 < x \le \frac{\pi}{2} \\ \frac{\pi}{2}, & \frac{\pi}{2} < x \le \pi \end{cases}$

$$\left[\textbf{Ans. } f(x) = \left(\frac{2}{\pi} + 1\right)\sin x - \frac{1}{2}\sin 2x + \left(-\frac{2}{9\pi} + \frac{1}{3}\right)\sin 3x + \cdots\right]$$

21. $f(x) = \frac{l}{2} - x,\ 0 < x < 1.$ $\left[\textbf{Ans. } \frac{l}{2} - x = \frac{l}{\pi}\sum_{n=1}^{\infty}\frac{1}{n}\sin\frac{2\pi x}{l}\right]$

22. $f(x) = \begin{cases} x & \text{for } 0 < x \le \frac{\pi}{2} \\ \pi - x & \text{for } \frac{\pi}{2} < x < \pi \end{cases}$ $\left[\textbf{Ans. } f(x) = \frac{4}{\pi}\left[\frac{\sin x}{1^2} - \frac{\sin 3x}{3^2} + \frac{\sin 5x}{5^2} - \cdots\right]\right]$

23. $f(x) = x \sin x,\ 0 \le x \le 1.$ $\left[\textbf{Ans. } x\sin x = \frac{\pi}{2}\sin x - \frac{8}{\pi}\left[2\frac{\sin 2x}{3^2} + 4\frac{\sin 4x}{15^2} + \cdots\right]\right]$

24. $f(x) = x,\ 0 < x < \pi.$ $\left[\textbf{Ans. } x = 2\left[\sin x - \frac{1}{2}\sin 2x + \frac{1}{3}\sin 3x + \cdots\right]\right]$

25. Find the Fourier series for the given functions:

(i) $f(t) = |\sin t|, -\pi < t < \pi$ $\left[\textbf{Ans.}\ \frac{2}{\pi}-\frac{4}{\pi}\left(\frac{\cos 2t}{2^2-1}+\frac{\cos 4t}{4^2-1}+\ldots+\frac{\cos 4nt}{4n^2-1}+\ldots\right)\right]$

(ii) $f(t) = \begin{cases} 0, & 0 \le t \le \pi \\ 2\pi - t & \pi \le t \le 2\pi \end{cases}$ $\left[\textbf{Ans.}\ \frac{\pi}{2}-\frac{4}{\pi}\left[\cos t+\frac{\cos 3t}{3^2}+\frac{\cos 5t}{5^2}+\ldots\right]\right]$

(iii) $f(t) = \begin{cases} 0, & -\pi < t < 0 \\ \frac{\pi t}{4}, & 0, < t < \pi \end{cases}$ $\left[\textbf{Ans.}\ f(t) = \frac{\pi^2}{16}-\frac{1}{2}\left(\frac{\cos t}{1^2}+\frac{\cos 3t}{3^2}+\frac{\cos 5t}{5^2}+\ldots\right)\right.$

$$\left. +\frac{\pi}{4}\left(\sin t-\frac{\sin 2t}{2}+\frac{\sin 3t}{3}+\ldots\right)\right]$$

(iv) $f(t) = \frac{1}{4}(\pi - t)^2, 0 < t < 2\pi$ $\left[\textbf{Ans.}\ f(t) = \frac{\pi^2}{1^2}+\frac{\cos t}{1^2}+\frac{\cos^2 t}{2^2}+\frac{\cos^3 t}{3^2}+\ldots\right]$

(v) $f(t) = \cos h\ at, -\pi < t < \pi$

$$\left[\textbf{Ans.}\ f(t) = \frac{2a^2}{\pi}\sin h\ a\pi\left[\frac{1}{2a^2}+\sum_{n=1}^{\infty}\frac{(-1)^n}{n^2+a^2}\cos nt\right]\right]$$

(vi) $f(t) = \sin pt; -\pi < t < \pi$

$$\left[\textbf{Ans.}\ \sin pt = \frac{2}{\pi}\sin p\pi\left[\frac{\sin t}{1^2-p^2}-\frac{2\sin 2t}{2^2-p^2}+\frac{3\sin 3t}{3^2-p^2}+\ldots\right]\right]$$

26. Find the Fourier series for the function

$$f(t) = \begin{cases} 1+\frac{2t}{\pi} & -\pi \le t \le 0 \\ 1-\frac{2t}{\pi} & 0 \le t \le \pi \end{cases}$$

hence, deduce that

$$\frac{\pi^2}{8} = 1+\frac{1}{3^2}+\frac{1}{5^2}+\frac{1}{7^2}+\ldots$$

$$\left[\textbf{Ans.}\ f(t) = \frac{8}{\pi^2}\sum_{\substack{n=\text{odd}\\ \text{number}}}\cos nt\right]$$

27. Find the Fourier series to represent the function

$f(t) = \sin$ at, $-\pi < t < \pi$; a being a fraction

$$\left[\textbf{Ans.}\ \sin at = \frac{2\sin a\pi}{\pi}\left[\frac{\sin t}{1^2-a^2}-\frac{2\sin t}{2^2-a^2}+\ldots\right]\right]$$

28. Find Fourier series for the function

$$f(t) = \begin{cases} \pi + t, & -\pi < t < 0 \\ \pi - t, & 0 < t < \pi \end{cases}$$

$$\left[\textbf{Ans. } f(t) = \frac{\pi}{2} + \frac{4}{\pi}\left[\frac{\cos t}{1^2} + \frac{\cos 3t}{3^2} + \frac{\cos 5t}{5^2} + ... \right] \right]$$

29. Suppose that $\alpha > 0$ is not integer. Show that

$$\cos \alpha t = \frac{2a \sin \alpha\pi}{\pi}\left[\frac{1}{2\alpha^2} - \frac{\cos t}{\alpha^2 - 1^2} + \frac{\cos 2t}{\alpha^2 - 2^2} - \frac{\cos 3t}{\alpha^2 - 3^2} + ... \right] \quad \forall\, t \in [-\pi, \pi]$$

30. Find the Fourier series for $f(t) = x^2 - 2, -2 \le t \le 2$.

$$\left[\textbf{Ans. } f(t) = -\frac{2}{3} - \frac{16}{\pi^2}\left[\frac{1}{2\alpha^2} - \frac{\cos t}{\alpha^2 - 2^2} + \frac{\cos 2t}{\alpha^2 - 2^2} - \frac{\cos 3t}{a^2 - 3^2} + ... \right] \right]$$

31. Find half range cosine series for the function

$$f(t) = t, 0 < t < 2$$

$$\left[\textbf{Ans. } f(t) = 1 - \frac{8}{\pi^2}\left[\cos\left(\frac{\pi t}{2}\right) + \frac{1}{3^2}\cos\left(\frac{3\pi t}{2}\right) + \frac{1}{5^2}\cos\left(\frac{5\pi t}{2}\right) + ... \right] \right]$$

32. Find the Fourier series to the function

$$f(t) = \pi t, -c \le t \le c$$

$$\left[\textbf{Ans. } f(t) = 2c\left[\sin\frac{\pi t}{c} - \frac{1}{2}\sin\frac{2\pi t}{c} + ... \right] \right]$$

33. Find the half range sine series for the function

$$f(t) = 2t - 1, 0 < x \le 1$$

$$\left[\textbf{Ans. } f(t) = -\frac{2}{\pi}\left[\sin 2\pi t + \frac{1}{2}\sin 4\pi t + \frac{1}{3}\sin 6\pi t + ... \right] \right]$$

34. Find half Fourier series for the function

$$f(t) = \begin{matrix} 1, & 0 < t < 1 \\ 2, & 1 < t < 2 \end{matrix}$$

$$\left[\textbf{Ans. } 3 - \sum_{n=1}^{\infty} \frac{2}{(2n-1)\pi}\sin(2n-1)\pi t \right]$$

35. Find Fourier series for the function:

$$f(t) = \begin{cases} t, & -1 < t < 0 \\ t+2, & 0 < t < 1 \end{cases}$$

and deduce $\dfrac{\pi}{4} = 1 - \dfrac{1}{3} + \dfrac{1}{5} + \dfrac{1}{7} + \dots$

$$\left[\textbf{Ans. } f(t) = 1 + \frac{2}{\pi} \sum_{n=1}^{\infty} \frac{1}{n} [1 - 2(-1)^n] \sin n\pi t\right]$$

36. Find the Fourier series of the function:

$$f(t) = \sin t,\ 0 < t < \pi$$

$$\left[\textbf{Ans. } f(t) = \frac{2}{\pi} - \frac{4}{\pi}\left[\frac{\cos 2t}{2^2 - 1^2} + \frac{\cos 4t}{4^2 - 1^2} + \frac{\cos 6t}{6^2 - 1^2} + \dots\right]\right]$$

37. Find the half range sine series for the function:

$$f(t) = 2t - 1,\ 0 < x \le 1$$

$$\left[\textbf{Ans. } \sin 2\pi t + \frac{1}{2} \sin 4\pi t + \frac{1}{3} \sin 6\pi t + \dots\right]$$

38. Find the half range sine series for the function:

$$f(t) = \begin{cases} \dfrac{1}{4} - t & 0 < t < \dfrac{1}{2} \\ t - \dfrac{3}{4} & \dfrac{1}{2} < t < 1. \end{cases}$$

$$\left[\textbf{Ans. } f(t) = \left(\frac{1}{\pi} - \frac{4}{\pi^2}\right)\sin \pi t + \left(\frac{1}{3\pi} + \frac{4}{3^2\pi^2}\right)\sin 3t + \left(\frac{1}{5\pi} - \frac{4}{5^2\pi^2}\right)\sin 5t + \dots\right]$$

7.5 HARMONIC ANALYSIS

We have discussed, the problem for expanding $y = f(x)$ as Fourier series with period 2π in the interval $(0, 2\pi)$ such that

$$f(x) = a_0 + \sum_{n=1}^{\infty} (a_n \cos nx + b_n \sin nx) \qquad \dots(1)$$

where,

$$a_0 = \frac{1}{2\pi} \int_0^{2\pi} f(x)\, dx \qquad \dots(2)$$

$$a_n = \frac{1}{\pi} \int_0^{2\pi} f(x) \cos nx\, dx \qquad \dots(3)$$

$$b_n = \frac{1}{\pi}\int_0^{2\pi} f(x) \sin nx \, dx \qquad ...(4)$$

When $f(x)$ is given in analytical form, i.e., an explicit function of an independent variable, the integrals in the R.H.S. of equations (2), (3), (4) can be evaluated and then the Fourier coefficients (a_0, a_n and b_n) can be found. However, in many problems the function is often given not by a formula but by a graph or by a table of the corresponding values. In such cases, the integral equations of (2), (3) and equation (4) cannot be evaluated, then we find out the Fourier coefficients, i.e., a_0, a_n and b_n by ordinary summation instead of integration. This process of finding the Fourier series is known as **harmonic analysis.**

Mean value of $f(x)$ over the interval (a, b)

The mean value of $f(x)$ over the interval (a, b) is given by $\frac{1}{(b-a)}\int_a^b f(x)dx$

Now from equation (2)

$$a_0 = \frac{2}{2\pi}\int_0^{2\pi} f(x)dx; \text{ over the interval } (0, 2\pi), \text{ i.e., } a = 0, b = 2\pi$$

$$= \frac{1}{(2\pi - 0)}\int_0^{2\pi} (x)dx$$

$\Rightarrow$ a_0 = Mean value of $f(x)$ over the interval $(0, 2\pi)$

$$\Rightarrow \quad a_0 = \frac{\Sigma f(x)}{N} \qquad ...(5)$$

where, $\Sigma f(x)$ = Sum of the values of $f(x)$ at the equally spaced points

N = Total number of equally spaced points at which the values of $f(x)$ are given over the interval

and

$$a_n = \frac{1}{\pi}\int_0^{2\pi} f(x) \cos nx \, dx$$

$$= 2\frac{1}{2\pi}\int_0^{2\pi} f(x) \cos nx \, dx$$

$$= 2\left[\frac{1}{(2\pi - 0)}\int_0^{2\pi} f(x) \cos nx \, dx\right]$$

$= 2 \times$ mean value of $f(x) \cos nx$ over the interval $(0, 2\pi)$

$$\Rightarrow \quad a_n = \frac{2}{N}\Sigma f(x) \cos nx \qquad ...(6)$$

Again

$$b_n = \frac{1}{\pi}\int_0^{2\pi} f(x) \sin nx \, dx$$

$$= 2.\frac{1}{2\pi}\int_0^{2\pi} f(x) \sin nx \, dx$$

$$= 2\left[\frac{1}{(2\pi - 0)}\int_0^{2\pi} f(x) \sin nx \, dx\right]$$

= 2 × Mean value of $f(x) \sin nx$ over the interval $(0, 2\pi)$

$$\Rightarrow \qquad b_n = \frac{2}{N}\Sigma f(x) \sin nx \qquad ...(7)$$

Remark 4

1. In equation (1), the term $(a_1 \cos x + b_1 \sin x)$ is known as **fundamental or first harmonic;** the term $(a_2 \cos 2x + b_2 \sin 2x)$ is called the **second harmonic** and so on.
2. Fourier series for the function $f(x)$ over the interval $(-c, c)$ is given by

$$f(x) = a_0 + \sum_{n=1}^{\infty}\left[a_n \cos\frac{n\pi x}{c} + b_n \sin\frac{n\pi x}{c}\right] \qquad ...(8)$$

where,
$$a_0 = \frac{1}{2c}\int_{-c}^{c} f(x) \, dx, \qquad ...(9)$$

$$a_n = \frac{1}{c}\int_{-c}^{c} f(x) \cos\frac{n\pi x}{c} \, dx, \qquad ...(10)$$

$$b_n = \frac{1}{c}\int_{-c}^{c} f(x) \sin\frac{n\pi x}{c} \, dx \qquad ...(11)$$

Now,
$$a_0 = \frac{1}{2c}\int_{-c}^{c} f(x) \, dx = \frac{1}{\{c-(-c)\}}\int_{-c}^{c} f(x) dx$$

= Mean value of $f(x)$ over the interval $(-c, c)$

$$\Rightarrow \qquad a_0 = \frac{\Sigma f(x)}{N}$$

and
$$a_n = \frac{1}{c}\int_{-c}^{c} f(x) \cos\frac{n\pi x}{c} dx$$

$$= 2.\frac{1}{2c}\int_{-c}^{c} f(x) \cos\frac{n\pi x}{c} dx$$

$$= 2\frac{1}{\{c-(-c)\}}\int_{-c}^{c} f(x) \cos\frac{n\pi x}{c} dx$$

= 2 × [Mean value of $f(x) \cos \dfrac{n\pi x}{c}$ over the inverval $(-c, c)$]

$$\Rightarrow \qquad a_n = \frac{2}{N}\Sigma f(x)\cos\frac{n\pi x}{c}$$

Again $\qquad b_n = \frac{1}{c}\int_{-c}^{c} f(x)\sin\frac{n\pi x}{c}dx$

$$= 2.\frac{1}{2c}\int_{-c}^{c} f(x)\sin\frac{n\pi x}{c}dx$$

$$= 2\frac{1}{\{c-(-c)\}}\int_{-c}^{c} f(x)\sin\frac{n\pi x}{c}dx$$

$$= 2 \times \left[\text{Mean value of } f(x)\sin\frac{n\pi x}{c} \text{ over the interval } (-c, c)\right]$$

$$\Rightarrow \qquad b_n = \frac{2}{N}\Sigma f(x)\sin\frac{n\pi x}{c}$$

3. The half range sine series over the interval $(0, c)$ is given by

$$f(x) = \sum_{n=1}^{\infty} b_n \sin\frac{n\pi x}{c}$$

where, $\qquad b_n = \frac{2}{c}\int_{0}^{c} f(x)\sin\frac{n\pi x}{c}dx$

$$= 2 \times \left[\frac{1}{c-0}\int_{0}^{c} f(x)\sin\frac{n\pi x}{c}dx\right]$$

$$= 2 \times \left[\text{Mean value of } f(x)\sin\frac{n\pi x}{c} \text{ over the interval } (0, c)\right]$$

$$\Rightarrow \qquad b_n = \frac{2}{N}\Sigma f(x)\sin\frac{n\pi x}{c}$$

4. The half range cosine series in the interval $(0, c)$ is given by

$$f(x) = a_0 + \sum_{n=1}^{\infty} a_n \cos\frac{n\pi x}{c}$$

where, $\qquad a_0 = \frac{1}{c}\int_{0}^{c} f(x)dx$

$$= 1 \times \left[\frac{1}{c-0}\int_{0}^{c} f(x)\,dx\right]$$

$$= 1 \times \left[\text{Mean value of } f(x) \text{ over the interval } (0, c)\right]$$

$$\Rightarrow \qquad a_0 = \frac{\Sigma f(x)}{N}$$

and $$a_n = \frac{2}{c}\int_0^c f(x) \cos\frac{n\pi x}{c}\,dx$$

$$= \left[\frac{1}{c-0}\int_0^c f(x) \cos\frac{n\pi x}{c}\,dx\right]$$

$$= 2 \times [\text{Mean value of } f(x) \cos\frac{n\pi x}{c} \text{ over the interval } (0, c)]$$

$$= \frac{2}{N}\Sigma f(x) \cos\frac{n\pi x}{c}$$

Example 26 Obtain the expansion for $y = f(x)$ from the given table up to the first harmonic. Also find the amplitude of the first harmonic.

x	:	0	1	2	3	4	5
y	:	9	18	24	28	26	20

Solution: Here the total number of equally spaced points, $N = 6$

The interval $(0, 2\pi)$ is divided into 6 equally spaced points of size $\frac{(2\pi - 0)}{6} = \frac{\pi}{3}$, i.e., 60°

Hence, the period is 6, i.e., $\Rightarrow 2c = 6 \Rightarrow c = 3$.

∴ The Fourier series is given by

$$f(x) = a_0 + \sum_{n=1}^{\infty}\left(a_n \cos\frac{n\pi x}{3} + b_n \sin\frac{n\pi x}{3}\right)$$

$$\Rightarrow \qquad f(x) = a_0 + a_1 \cos\frac{\pi x}{3} + b_1 \sin\frac{\pi x}{3} \text{ (Up to first harmonic)} \qquad \text{...(i)}$$

where, $$a_0 = \frac{\Sigma f(x)}{N},$$

$$a_1 = \frac{2}{N}\Sigma f(x) \cos\frac{\pi x}{3},$$

$$b_1 = \frac{2}{N}\Sigma f(x) \sin\frac{\pi x}{3},$$

Put $a_1 = A$, $\sin\phi$, $b_1 = A$, $\cos\phi$ in equation (i), then

$$f(x) = a_0 + A_1 \sin\left(\frac{\pi x}{3} + \phi\right)$$

where, $A_1 = \sqrt{a_1^2 + b_1^2}$ = amplitude of the first harmonic

We have to find only a_0, a_1 and b_1 only.

Table 7.1

x	θ	$y = f(x)$	$\cos\frac{\pi x}{3}$	$\sin\frac{\pi x}{3}$	$f(x)$ cos	$f(x)\sin\frac{\pi x}{3}$
0	0, i.e., 0°	9	1	0	9	0
1	$\frac{\pi}{3}$, i.e., 60°	18	$+\frac{1}{2}$	$\frac{\sqrt{3}}{2}$	9	$9\sqrt{3}$
2	$\frac{2\pi}{3}$, i.e., 120°	24	$-\frac{1}{2}$	$\frac{\sqrt{3}}{2}$	−12	$12\sqrt{3}$
3	$\frac{3\pi}{3}$, i.e., 180°	28	−1	0	−28	0
4	$\frac{4\pi}{3}$, i.e., 240°	26	$-\frac{1}{2}$	$-\frac{\sqrt{3}}{2}$	−13	$-13\sqrt{3}$
5	$\frac{5\pi}{3}$, i.e., 300°	20	$\frac{1}{2}$	$-\frac{\sqrt{3}}{2}$	10	$-10\sqrt{3}$
		$\Sigma f(x) = 125$			$\Sigma f(x)\cos\frac{\pi x}{3} = -25$	$\Sigma f(x)\sin\frac{\pi x}{3} = -2\sqrt{3}$

Now, $$a_0 = \frac{1}{N}\Sigma f(x) = \frac{1}{6}(125)$$

$$\Rightarrow \quad a_0 = 20.833 \qquad \text{...(ii)}$$

and $$a_1 = \frac{2}{N}\Sigma f(x)\cos\frac{\pi x}{3} \qquad \text{(Put } n = 1)$$

$$\Rightarrow \quad a_1 = \frac{2}{6}(-25) = -8.333 \qquad \text{...(iii)}$$

and $$b_1 = \frac{2}{N}\Sigma f(x)\sin\frac{\pi x}{3} \qquad \text{(Put } n = 1)$$

$$= \frac{2}{6}\left(-2\sqrt{3}\right)$$

$$\Rightarrow \quad b_1 = -1.1547 \qquad \text{...(iv)}$$

Using equations (ii), (iii) and equation (iv) in equation (i), we have

$$y = f(x) = 20.833 - 8.333\cos\frac{\pi x}{3} - 1.1547\sin\frac{\pi x}{3}$$

amplitude of the first harmonic $= \sqrt{a_1^2 + b_1^2}$

$$= \sqrt{(-8.333)^2 + (-1.1547)^2}$$

$$= 8.4126$$ ***Ans.***

Example 27 In a machine the displacement y of a given point is given for a certain angle θ as follows:

θ°	0	30	60	90	120	150	180	210	240	270	300	330
y	0.9	8	7.2	5.6	3.6	1.7	9.5	0.2	0.9	2.5	4.7	6.8

Find the coefficient of sin 2θ in the Fourier series representing y.

Solution: In this case $N = 12$

Now the Fourier series is

$$y = f(x) = a_0 + \sum_{n=1}^{\infty} (a_n \cos n\theta + b_n \sin n\theta);$$

$$\Rightarrow \qquad y = f(x) = a_0 + (a_1 \cos \theta + b_1 \sin \theta) + (a_2 \cos 2\theta + b_2 \sin 2\theta) + \ldots\ldots$$

We have to calculate the coefficient of sin 2θ or b_2

and $$b_2 = \frac{2}{N} \Sigma f(x) \sin 2\theta = \frac{2}{N} \Sigma y \sin 2\theta$$

Table 7.2

θ°	0	30	60	90	120	150	180	210	240	270	300	300	
y	0.9	8	7.2	5.6	3.6	1.7	0.5	0.2	0.9	2.5	4.7	6.8	
sin 2θ	0	0.866	0.866	0	−0.866	−0.866	0	0.866	0.866	0	−0.866	−0.866	
y sin 2θ	0	6.928	6.235	0	−3.117	−1.472	0	0.173	0.779	0	−4.070	−5.888	Σy sin 2θ = −0.432

Hence, $$b_2 = \frac{2}{12} \times (-0.432)$$

$$b_2 = -0.072 \text{ coefficient of } \sin 2\theta$$ ***Ans.***

Example 28 Variation of periodic current A (amperes) with time t (seconds) is given in the following table

t**(secs):**	**0**	$\frac{T}{6}$	$\frac{T}{3}$	$\frac{T}{2}$	$\frac{2T}{3}$	$\frac{5T}{6}$	$\boldsymbol{T}$
A(amps):	**1.98**	**1.30**	**1.05**	**1.30**	**−0.88**	**−0.25**	**1.98**

Show that there is a direct current part of 0.175 ampere in the varibale current. Also find the amplitude of the first harmonic.

Solution: Here the period is T, i.e., $2C = T$

$$\Rightarrow \qquad C = \frac{T}{2}$$

and $N = 6$ [Seventh value is the repetition of the first value]

Hence, the required Fourier series is

$$f(x) = A = a_0 + \sum_{n=1}^{\infty} \left(a_n \cos \frac{n\pi t}{T/2} + b_n \sin \frac{n\pi t}{T/2} \right) \quad \text{...(i)}$$

$$\Rightarrow \quad f(x) = A = a_0 + a_1 \cos \frac{2\pi t}{T} + b_1 \sin \frac{2\pi t}{T} + \ldots \quad \text{...(ii)}$$

Put $\quad a_1 = A_1 \sin \phi,\ b_1 = A_1 \cos \phi$

then $\quad A = a_0 + A_1 \sin \left(\frac{2\pi t}{T} + \phi \right)$

where, $A_1 = \sqrt{a_1^2 + b_1^2}$ = Amplitude of the first harmonic and a_0 is the direct current part.

Table 7.3

t	A	$\cos \frac{2\pi t}{T}$	$\sin \frac{2\pi t}{T}$	$A \cos \frac{2\pi t}{T}$	$A \sin \frac{2\pi t}{T}$
0	1.98	1	0	1.98	0
$\frac{T}{6}$	1.30	0.5	0.866	0.65	1.1258
$\frac{T}{3}$	1.05	−0.5	0.866	−0.525	0.9093
$\frac{T}{2}$	1.30	−1	0	−1.30	0
$\frac{2T}{3}$	−0.88	−0.5	−0.866	0.44	0.76208
$\frac{5T}{6}$	−0.25	0.5	−0.866	−0.125	0.2165
				$\Sigma A \cos \frac{2\pi t}{T} = 1.12$	$\Sigma A \sin \frac{2\pi t}{T} = 3.01368$

Now, $\quad a_0 = \frac{1}{N} \Sigma A = \frac{1}{6} \times 4.5 = 0.75 \quad$ ...(iii)

and $\quad a_1 = \frac{2}{N} \Sigma A \cos \frac{2\pi t}{T} = \frac{2}{6} \times 1.12$

$\Rightarrow \quad a_1 = 0.373 \quad$ ...(iv)

Also $\quad b_1 = \frac{2}{N} \Sigma A \sin \frac{2\pi t}{T} = \frac{2}{6} \times 3.01368$

$\Rightarrow \quad b_1 = 1.005 \quad$ ...(v)

Using equations (iii), (iv) and (v) in equation (ii), we get

$$A = 0.75 + 0.373 \cos \frac{2\pi t}{T} + 1.005 \sin \frac{2\pi t}{T} + \ldots$$

∴ The direct current part in the variable current, $a_0 = 0.75$

Amplitude of first harmonic

$$= \sqrt{a_1^2 + b_1^2}$$

$$= \sqrt{(0.373)^3 + (1.005)^2}$$

$$= 1.072$$ ***Ans.***

Example 29 The turning moment (T) units of the crankshaft of a steam engine is given for a series of values of the crank angle (θ) in degrees:

$\theta°$	0	30	60	90	120	150	180
T	0	5224	8097	7850	5499	2626	0

Find the first four terms in a series of sine to represent T and find T when $\theta = 75°$.

Solution: Here $N = 6$ [seventh value is the repetition of the first value]

The half range sine series in the interval $(0, \pi)$ is given by

$$f(x) = T = \sum_{n=1}^{\infty} b_n \sin n\theta$$

$$T = b_1 \sin \theta + b_2 \sin 2\theta + b_3 \sin 3\theta + b_4 \sin 4\theta + \ldots \qquad \ldots\text{(i)}$$

$$b_1 = \frac{2}{N} \Sigma A \sin \theta = \frac{2}{6}(23549) \qquad \ldots\text{(ii)}$$

Table 7.4

$\theta°$	0	30	60	90	120	150	
T	0	5224	8097	7850	5499	2626	
sin θ	0	0.5	0.866	1	0.866	0.5	
sin 2θ	0	0.866	0.866	0	−0.866	−0.866	
sin 3θ	0	0	0	−1	0	1	
sin 4θ	0	0.866	−0.866	0	0.866	−0.866	
T sin θ	0	2612	7012	7850	+4762	1313	$\Sigma T \sin \theta = 23549$
T sin 2θ	0	4524	7012	0	−4762	−2274	$\Sigma T \sin 2\theta = 4500$
T sin 3θ	0	5224	0	−7850	0	2626	$\Sigma T \sin 3\theta = 0$
T sin 4θ	0	4524	−7012	0	4762	−2274	$\Sigma T \sin 4\theta = 0$

$$\Rightarrow \qquad b_1 = 7850$$

$$b_2 = \frac{2}{N}\Sigma A \sin 2\theta = \frac{2}{6} \times 4500$$

$$\Rightarrow \qquad b_2 = 1500 \qquad \text{...(iii)}$$

$$b_3 = \frac{2}{N}\Sigma A \sin 3\theta = \frac{2}{6} \times 0$$

$$\Rightarrow \qquad b_3 = 0 \qquad \text{...(iv)}$$

and
$$b_4 = \frac{2}{N}\Sigma A \sin 4\theta = \frac{2}{6} \times 0$$

$$\Rightarrow \qquad b_4 = 0 \qquad \text{...(v)}$$

Using equations (ii), (iii), (iv) and (v) in equation (i), we get

$$T = 7850 \sin \theta + 1500 \sin 2\theta \qquad \text{...(vi)}$$

when $\theta = 75°$, then

$$T = 7850 \sin 75 + 1500 \sin (2 \times 75)$$

$$= 7582 + 750$$

$$\Rightarrow \qquad T = 8332$$ ***Ans.***

Example 30 Obtain the first three coefficients in the Fourier cosine series, for the function $y = f(x)$, given by the table as follows:

x	0	1	2	3	4	5
y	4	8	15	7	6	6

Solution: Here the total number of equally spaced points, $6 = N$

The interval $(0, 2\pi)$ is divided into 6 equally spaced point of size $\frac{(2\pi - 0)}{6} = \frac{\pi}{3}$, i.e., 60°

Here the period is 6, i.e., $2c = 6 \Rightarrow c = 3$.

Therefore the Fourier cosine series is given by

$$f(x) = a_0 + \sum_{n=1}^{\infty}\left(a_n \cos \frac{n\pi x}{3}\right)$$

$$\Rightarrow \qquad f(x) = a_0 + a_1 \cos \frac{\pi x}{3} + a_2 \cos \frac{2\pi x}{3} + a_3 \cos \frac{3\pi x}{3} + \ldots \qquad \text{...(i)}$$

We have to find out a_0, a_1, a_2 and a_3 only.

Table 7.5

x	0	1	2	3	4	5	
$y = f(x)$	4	8	15	7	6	2	$\Sigma y = 42$
$\cos \frac{\pi x}{3}$	1	0.5	−0.5	−1	−0.5	0.5	
$\cos \frac{2\pi x}{3}$	1	−0.5	−0.5	1	−0.5	−0.5	
$\cos \frac{3\pi x}{3}$	1	−1	1	−1	1	−1	
$y \cos \frac{\pi x}{3}$	4	4	−7.5	−7	−3	1	$\Sigma y \cos \frac{\pi x}{3} = -8.5$
$y \cos \frac{2\pi x}{3}$	4	−4	−7.5	7	−3	−1	$\Sigma y \cos \frac{2\pi x}{3} = -4.5$
$y \cos \frac{3\pi x}{3}$	4	−8	15	−7	6	−2	$\Sigma y \cos \frac{3\pi x}{3} = 8$

Now, $$a_0 = \frac{1}{N} \Sigma f(x) = \frac{1}{6} \times 42 = 7 \quad \text{...(ii)}$$

$$a_1 = \frac{2}{N} \Sigma f(x) \cos \frac{\pi x}{3} = \frac{2}{6} \times (-8.5)$$

$\Rightarrow$ $$a_1 = -2.8 \quad \text{...(iii)}$$

$$a_1 = \frac{2}{N} \Sigma y \cos \frac{2\pi x}{3} = \frac{2}{6} \times (-4.5)$$

$$a_2 = -1.5 \quad \text{...(iv)}$$

and $$a_3 = \frac{2}{N} \Sigma y \cos \frac{3\pi x}{3}$$

$$= \frac{2}{6} \times (8)$$

$$a_3 = 2.67 \quad \text{...(v)}$$

Using equations (ii), (iii), (iv) and (v) in equation (i), then

$$y = f(x) = 7 - 2.8 \cos \frac{\pi x}{3} - 1.5 \cos \frac{2\pi x}{3} + 2.67 \cos \frac{3\pi x}{3} \quad \textbf{\textit{Ans.}}$$

$\therefore$ The direct current part in the variable current $= a_0 = 0.75$

EXERCISE

1. The following values of y give the displacement of a certain machine part for the rotation of the flywheel,

x	0	$\frac{\pi}{3}$	$\frac{2\pi}{3}$	π	$\frac{4\pi}{3}$	$\frac{5\pi}{3}$	2π
y	1.98	2.15	2.77	−0.22	−0.31	1.43	1.98

Expand y as a Fourier series up to the third harmonic.

[***Ans.*** $y = 1.30 + (0.92 \cos x + 1.097 \sin x) - (0.42 \cos 2x + 0.681 \sin 2x) + 0.36 \cos 3x + ...$]

2. The part of a machine has an oscillatory motion. The displacement y at any time t is given by the following table

t	0.02	0.04	0.06	0.08	0.10	0.12	0.14	0.16	0.18	0.20
y	0.64	1.13	1.34	0.95	0	−0.92	−1.33	−1.666	−0.66	0

Find the constants in the following form of the Fourier series

$y = A \sin (10 \pi t + \alpha_1) + B \sin (20 \pi + \alpha_2)$

$$\left[\textbf{\textit{Ans.}}\ y = -0.0016 - 0.0109 \cos 10 \pi t + 1.317 \sin 10 \pi t + 0.0048 \cos 20 \pi t - 0.1524 \sin 20 \pi t,\ A = 1.317,\ B = -0.1524,\ \alpha_1 = \tan \alpha_1 = \frac{-0.0109}{1.317} = -0.0083,\ \alpha_2 = \tan \alpha_2 = \frac{-0.0048}{0.1524} = -0.0315\right]$$

3. Analyse the data given in the table harmonically and express y in a Fourier series up to the third harmonic.

$x°$	0	30	60	90	120	150
y	2.34	3.01	3.69	4.15	3.69	2.20
$x°$	180	210	240	270	300	330
y	0.83	0.51	0.88	1.09	1.19	1.64

[***Ans.*** $y = 2.102 + 0.558 \cos x + 1.531 \sin x - 0.519 \cos 2x - 0.091 \sin 2x + 0.20 \cos 3x + 0.135 \sin 3x$]

4. Analyse the current i given by the table below into its constitutent harmonics as far as third one.

θ(deg.)	0	30	60	90	120	150
i(amp.)	0	24.0	33.5	27.5	18.2	13.0
θ(deg.)	180	210	240	270	300	330
i(amp.)	0	−24.0	−33.5	−27.5	−18.2	−13.0

[***Ans.*** $i = 5.73 \cos \theta + 30.26 \sin \theta - 5.10 \cos 3\theta + 3.17 \sin 3\theta + ...$]

5. Harmonically analyse the data given and express y is Fourier series.

x	0	2	4	6	8	10	12
y	9.0	18.2	24.4	27.8	27.5	22.0	9.0

$$\left[\textbf{Ans. } y = 21.45 - 8.22 \cos \frac{\pi x}{6} - 3.08 \cos \frac{2\pi x}{6} - 2.36 \cos \frac{3\pi x}{6} - 1.99 \sin \frac{\pi x}{6} - 0.202 \sin \frac{2\pi x}{6}\right]$$

6. The following values of x and y are taken from a periodic curve which has a period 2π. Expand y as a Fourier series up to the third harmonic.

x°	0	30	60	90	120	150
y	0.50	0.60	0.83	1.00	0.80	0.42
x°	180	210	240	270	300	330
y	0.00	−0.34	−0.50	−0.20	0.67	0.70

[**Ans.** $y = 0.373 + 0.359 \cos x + 0.466 \sin x - 0.085 \cos 2x - 0.289 \sin 2x - 0.117 \cos 3x - 0.090 \sin 3x$.]

Chapter 8

The Fourier Transform

Some Useful Results

(i) $\int_{-\infty}^{\infty} e^{-x^2} dx = \sqrt{\pi}$

(ii) $\int_{0}^{\infty} e^{-x^2} dx = \frac{\sqrt{\pi}}{2}$

(iii) $\int e^{ax} \sin bx \, dx = \frac{e^{ax}}{a^2+b^2}(a \sin bx - b \cos bx)$

(iv) $\int e^{ax} \cos bx \, dx = \frac{e^{ax}}{a^2+b^2}(a \cos bx + b \sin bx)$

(v) $\int_{0}^{\infty} e^{-ax} \sin bx \, dx = \frac{b}{a^2+b^2}$

(vi) $\int_{0}^{\infty} e^{-ax} \cos bx \, dx = \frac{a}{a^2+b^2}$

(vii) $\int_{0}^{\infty} \frac{\sin px}{x} dx = \begin{cases} -\pi/2, & p<0 \\ \pi/2, & p=1 \\ \pi/2, & p>0 \end{cases}$

(viii) $\int_{0}^{\infty} \frac{f(ax)-f(bx)}{x} dx = f(0) \log (b/a)$ for, $a, b > 0$

(ix) $\int_{-\infty}^{\infty} \frac{\sin px}{(x-b)^2+a^2} dx = \frac{\pi}{4} e^{-ap} \sin bp$ for, $p > 0$

(x) $\frac{d^n}{dx^n}\left(\frac{x}{a^2+x^2}\right) = \frac{(-1)^n n!}{(a^2+x^2)^{\frac{n+1}{2}}} \cos[(n+1)\tan^{-1}(a/x.)]$

(xi) $\frac{d^n}{dx^n}\left(\frac{a}{a^2+x^2}\right) = \frac{(-1)^n n!}{(a^2+x^2)^{(n+1)/2}}.\ \sin[(n+1)\tan^{-1}(a/x)]$

(xii) $\int_0^\infty \frac{e^{ax}-e^{-ax}}{e^{\pi x}-e^{-\pi x}}dx = \frac{1}{2}\tan^{-1}(a/2)$

(xiii) $\int_0^\infty \frac{e^{ax}+e^{-ax}}{e^{\pi x}-e^{-\pi x}}dx = \frac{1}{2}\sec(a/2)$

(xiv) $\int_0^\infty \frac{\cos bx}{a^2+x^2}dx = \frac{\pi}{2a}e^{-ab}, (a, b > 0)$

(xv) $\int_0^\infty \frac{x\sin bx}{a^2+x^2}dx = \frac{\pi}{2}e^{-ab}, (a, b > 0)$

8.1 IMPORTANT TERMS AND THEIR DEFINITIONS

8.1.1 Dirichlet Conditions

A function $f(x)$ is said to satisfy Dirichlet condition in the open interval (a, b), if

(i) $f(x)$ is defined and single-valued function except at finite number of points in the interval (a, b) (if possible), or

(ii) $f(x)$ and $f'(x)$ are piecewise continuous in the open interval (a, b).

These conditions play a vital role in the study of Fourier transforms.

8.1.2 Fourier Integral Formula

Consider a function $f(x)$, satisfying Dirichlet conditions in every finite interval $[-l, l]$ and defined as $\frac{1}{2}$ $[f(x+0)+f(x+0)]$ at every point of discontinuity.

Further, $\int_{-\infty}^{\infty} |f(x)|dx$ converges or $f(x)$ is absolutely integrable in $(-\infty, \infty)$ then by Fourier integral formula,

$$f(x) = \frac{1}{2\pi}\int_{-\infty}^{\infty} f(v)\left[\int_{-\infty}^{\infty}\cos w(x-v)dw\right]dv \quad ...(i)$$

$$\Rightarrow \quad f(x) = \frac{1}{2\pi}\int_{-\infty}^{\infty} dw \int_{-\infty}^{\infty} f(v)\cos w(x-v)dv$$

and we know that

$$\frac{1}{2\pi}\int_{-\infty}^{\infty} f(w)\left[\int_{-\infty}^{\infty} \sin w(x-v)dw\right]dv = 0 \qquad \text{...(ii)}$$

Multiplying equation (ii) by i and adding equation (i), we have

$$f(x) = \frac{1}{2\pi}\int_{-\infty}^{\infty} f(v)\left[\int_{-\infty}^{\infty} e^{iw(x-v)}dw\right]dv$$

$$\Rightarrow \qquad f(x) = \frac{1}{2\pi}\int_{-\infty}^{\infty} e^{iwx}\, dw \int_{-\infty}^{\infty} f(v)e^{iwx}\, dv.$$

Fourier Transform or Complex Fourier Transform:

Let $f(x)$ be a function defined on open interval $(-\infty, \infty)$, piecewise continuously differentiable and absolutely integrable in an open interval $(-\infty, +\infty)$ then the Fourier transform of $f(x)$ is defined as

$$F(f(x)) = \frac{1}{2\pi}\int_{-\infty}^{\infty} e^{ipx} f(x)dx,\ p > 0 \qquad \text{...(i)}$$

Fourier transform denoted by $F(f(x))$ and kernel of the Fourier transform is $\frac{1}{2\pi}e^{ipx}$.

Fourier Cosine Transform: Let function $f(x)$ be piecewise continuously differentiable, absolutely in open interval $(0, \infty)$ and defined for $x > 0$, then the infinite Fourier cosine transform or Fourier transform is defined as

$$F_c(f(x)) = F_c(p) = \sqrt{\frac{2}{\pi}}\int_0^{\infty} f(x)\cos px\, dx$$

Fourier Sine Transform: Let function $f(x)$ be piecewise continuously differentiable, absolutely integrable in an open interval $(0, \infty)$ and defined for $x > 0$, then the infinite Fourier sine transform or Fourier sine transform is defined as

$$F_s(f(x)) = F_s(p) = \sqrt{\frac{2}{\pi}}\int_0^{\infty} f(x)\sin px\, dx,\ p > 0.$$

where, $\sqrt{\frac{2}{\pi}}\cos px$ and $\sqrt{\frac{2}{\pi}}\sin px$ known as kernel of the Fourier cosine transform and Fourrier sine transform, respectively.

Inversion Theorem: Let $F(p) = Lf(x)$ or $F(p)$ be the Fourier transform of $f(x)$, then

$$F(p) = \frac{1}{2\pi}\int_{-\infty}^{\infty} e^{ipx} f(x)dx \qquad \text{...(i)}$$

Further assume that $f(x)$ is piecewise continuously differentiable and absolutely integrable is an open interval $(-\infty, \infty)$, then at every point of continuity x of function $f(x)$,

$$f(x) = \frac{1}{\sqrt{2\pi}} \int_{-\infty}^{\infty} e^{-ipx} F(p)\, dp \qquad \text{...(ii)}$$

where, $f(x)$ is called inverse Fourier transform.

Solution: We know that

$$f(x) = \frac{1}{2\pi} \int_{-\infty}^{\infty} e^{ivx}\, dv \int_{-\infty}^{\infty} e^{ivx} f(u)\, du \qquad \text{Substitute} \Rightarrow \begin{Bmatrix} v = -p \\ dv = -dp \end{Bmatrix}$$

$$= \frac{1}{\sqrt{2\pi}} \int_{-\infty}^{\infty} e^{-ipx}\, dp \frac{1}{\sqrt{2\pi}} \int_{-\infty}^{\infty} e^{ipu} F(u)\, du$$

$$= \frac{1}{\sqrt{2\pi}} \int_{-\infty}^{\infty} e^{-ipx} F(p) dp$$

In this chapter we are using kernel of Fourier transform as $\frac{1}{\sqrt{2\pi}} e^{ipx}$. Some authors assume kernel or Fourier transform as e^{ipx}, in this case Fourier transform and inverse Fourier transform can be defined as

$$F(p) = \int_{-\infty}^{\infty} e^{ipx} F(x) dx$$

$$f(x) = \frac{1}{2\pi} \int_{-\infty}^{\infty} e^{-ipx} F(p) dp$$

Fourier Cosine Transform: Let $F_c f(x) = F_c(p)$ be the Fourier cosine transform of $f(x)$ as given below

$$F_c(p) = \sqrt{\frac{2}{\pi}} \int_0^{\infty} f(x) \cos px\, dx$$

and if $f(x)$ piecewise continuously differentiable, absolutely integrable in an open interval $(0, \infty)$ then at every point of continuity of the function $f(x)$,

$$f(x) = \sqrt{\frac{2}{\pi}} \int_0^{\infty} F_c(p) \cos px\, dx.$$

Solution: We know that

$$f(x) = \frac{1}{\pi} \int_0^{\infty} dw \int_{-\infty}^{\infty} f(v) \cos w\,(x - v)\, dv$$

$$= \frac{1}{\pi} \int_0^{\infty} dp \int_{-\infty}^{\infty} f(v) \cos p(x - v)\, dv \text{ for } p = w$$

$$= \frac{1}{\pi}\int_0^\infty dp \int_{-\infty}^\infty f(v)\,[\cos px \cos pv + \sin px \sin pv]\,dv$$

$$= \frac{1}{\pi}\int_0^\infty dp \int_{-\infty}^\infty f(v) \cos px \cos pv\,dv + \frac{1}{\pi}\int_0^\infty dp \int_{-\infty}^\infty f(v) \sin px \sin pv\,dv$$

$$= \frac{1}{\pi}\int_0^\infty \cos px\,dp \int_{-\infty}^\infty f(v) \cos pv\,dv + \frac{1}{\pi}\int_0^\infty \sin px\,dp \int_{-\infty}^\infty f(v) \sin pv\,dv$$

$$\Rightarrow \qquad f(x) = \frac{1}{\pi}\int_0^\infty \cos px\,dp \int_{-\infty}^\infty f(x) \cos px\,dx + \frac{1}{\pi}\int_0^\infty \sin px\,dx \int_{-\infty}^\infty f(x) \sin px\,dx$$

(For $v \equiv x$)

Let us define $f(x)$ is the open interval $(-\infty, 0)$, such that $f(-x) = f(x)$. Then $f(x)$ is an even function in the open interval $(-\infty, \infty)$, we have

$$\int_{-\infty}^\infty f(x) \sin (px)\,dx = 0 \qquad \text{(as } f(x) \sin px \text{ is an odd function)}$$

$$\int_{-\infty}^\infty f(x) \cos px\,dx = 2\int_0^\infty f(x) \cos px\,dx \qquad \text{(as } f(x) \cos px \text{ is an even function)}$$

Then,
$$f(x) = \frac{2}{\pi}\int_0^\infty \cos px\,dp \int_{-\infty}^\infty f(x) \cos px\,dx$$

$$f(x) = \frac{2}{\pi}\int_0^\infty F_c(p) \cos px\,dp$$

If the kernel of the Fourier cosine transform taken as cos px, then

$$F_c(p) = \int_0^\infty f(x) \cos px\,dx$$

$$f(x) = \frac{2}{\pi}\int_0^\infty F_c(p) \cos px\,dx$$

Fourier Cosine Transform from Complex Fourier Transform

Let us assume a function $f(x)$ is defined for positive value of x only. Further assume a function $f_+(x)$, known as extension of $f(x)$ or even extension of $f(x)$ which is defined on an open interval $(-\infty, \infty)$ as below

$$f_+(x) = \begin{cases} f(-x) & , \quad x < 0 \\ f(x) & , \quad x \ge 0 \end{cases}$$

Then, $\int_{-\infty}^{\infty} f_+(x)e^{ipx}\,dx = \int_{-\infty}^{0} e^{ipx} f(-x)dx + \int_{0}^{\infty} e^{ipx} f(x)dx$

$$= \int_{0}^{\infty} (e^{ipx} + e^{-ipx}) f(x)dx$$

$$= 2\int_{0}^{\infty} f(x) \cos px\, dx$$

$$\left[\begin{array}{l} I = \int_{-\infty}^{0} e^{ipx} f(-x)dx, \text{ if } -x = t \\ \text{Then, } dx = -dt \\ \text{Thus, } I = \int_{-\infty}^{0} e^{-ipt} f(t)dt \\ \quad = \int_{0}^{\infty} e^{-ipt} f(t)dt \\ \quad = \int_{0}^{\infty} e^{-ipx} + f(x)dx \end{array}\right]$$

Thus, $$F[f_+(x)] = \frac{2}{\sqrt{2\pi}} \int_{-\infty}^{\infty} e^{ipx} f_+(x)dx$$

$$= \frac{2}{\sqrt{2\pi}} \left[2 \cdot \int_{0}^{\infty} f(x)\cos px\, dx\right]$$

$$= \sqrt{\frac{2}{\pi}} \int_{0}^{\infty} f(x) \cos px\, dx$$

Now, using inversion formula

$$f(x) = \frac{2}{\sqrt{2\pi}} \int_{-\infty}^{\infty} e^{-ipx} F(p)dp$$

Thus, $$f_+(x) = \frac{2}{\sqrt{2\pi}} \int_{-\infty}^{\infty} e^{-ipx} F_c(p)dp$$

$$= \frac{2}{\sqrt{2\pi}} \left[\int_{-\infty}^{0} e^{-ipx} F_c(p)dp + \int_{0}^{\infty} e^{-ipx} F_c(p)dp\right]$$

$$= \frac{2}{\sqrt{2\pi}} \left[\int_{0}^{\infty} e^{-ipx} F_c(-p)dp + \int_{0}^{\infty} e^{-ipx} F_c(p)dp\right]$$

$\Rightarrow$ $$f_+(x) = \frac{2}{\sqrt{2\pi}} \int_{0}^{\infty} (e^{ipx} + e^{-ipx}) F_c(p)\, dp$$

(For $x \geq 0, f_+(x) = f(x)$, then)

$$\Rightarrow \qquad f(x) = \sqrt{\frac{2}{\pi}} \int_0^{\infty} F_c(p) \cos px \, dx$$

and
$$F(f(x)) = \frac{2}{\sqrt{2\pi}} \int_{-\infty}^{\infty} e^{ipx} f(x) dx$$

$$= \frac{2}{\sqrt{2\pi}} \left[\int_{-\infty}^{0} e^{ipx} f(x) dx + \int_0^{\infty} e^{ipx} f(x) dx \right]$$

$$= \frac{2}{\sqrt{2\pi}} \left[\int_0^{\infty} e^{-ipx} f(-x) dx + \int_0^{\infty} e^{ipx} f(x) dx \right]$$

Let the function $f(x)$ be even, then

$$f\_(-x) = f(x)$$

Therefore,
$$F\{f(x)\} = \frac{2}{\sqrt{2\pi}} \int_0^{\infty} (e^{-ipx} + e^{ipx}) f(x) dx$$

$$= \sqrt{\frac{2}{\pi}} \int_0^{\infty} f(x) \cos px \, dx$$

$\Rightarrow$ If the function $f(x)$ is an even function

Then, $\qquad F(x) = F_c(f(x)) = F_c(p)$

Fourier sine transform: Let $Fs[f(x)]$ $(= F_s(p))$ be the Fourier sine transform of $f(x)$ as given below:

$$Fs(p) = \sqrt{\frac{2}{\pi}} \int_0^{\infty} f(x) \sin px \, dx \qquad ...(i)$$

and if, $f(x)$ be piecewise continuously differentiable, absolutely integrable in an open interval $(-\infty, \infty)$ at every point of continuity of the function $f(x)$, then

$$f(x) = \sqrt{\frac{2}{\pi}} \int_0^{\infty} F_s(p) \sin px \, dx \qquad ...(ii)$$

Solution: We know that

$$f(x) = \int_0^{\infty} dp \int_{-\infty}^{\infty} f(u) \cos [p(x-u)] du$$

$$= \frac{1}{\pi} \int_0^{\infty} dp \int_{-\infty}^{\infty} f(u) [\cos px \cos pu + \sin px \sin pu)] du$$

$$\Rightarrow \qquad f(x) = \frac{1}{\pi}\int_0^\infty \cos px\, dp \int_{-\infty}^\infty f(u)\cos pu\, du$$

$$+ \frac{1}{\pi}\int_0^\infty \sin px\, dp \int_{-\infty}^\infty f(u)\sin pu\, du \qquad \text{...(iii)}$$

Now, let us define $f(x)$ is the open interval $(-\infty, 0)$ such that $f(x) = -f(-x)$, in this case $(-\infty, \infty)$ hence, $f(x)\cos px$, product of an odd and an even function, will be an odd function. therefore, equation (iii) reduces to

$$f(x) = \frac{2}{\pi}\int_0^\infty \sin px\, dp \int_0^\infty f(x)\sin px\, dx$$

$$= \sqrt{\frac{2}{\pi}}\int_0^\infty \sin px\, dp\ \frac{\sqrt{2}}{\pi}\int_0^\infty f(x)\sin px\, dx$$

$$= \sqrt{\frac{2}{\pi}}\int_0^\infty F_s(p)\sin px\, dp, \qquad \text{(Using equation (i))}$$

If the kernel of the Fourier sine transform taken as $\sin px$ in place of $\sqrt{\frac{2}{\pi}}\sin px$, then

$$F_s(p) = \int_0^\infty f(x)\sin px\, dx$$

$$f(x) = \frac{2}{\pi}\int_0^\infty F_s(p)\sin px\, dp$$

Fourier Sine Transform from Complex Fourier Transform

Let us assume a function $f(x)$, defined for positive value of x only, further assume a function $f(x)$, known as odd extension of $f(x)$ of x which is defined on an open interval $(-\infty, \infty)$ as given below.

$$f\_(x) = \begin{cases} -f(-x) & , \quad x < 0 \\ f(x) & , \quad x \geq 0 \end{cases}$$

Then, $$\int_{-\infty}^\infty f(x)\, e^{ipx}\, dx = \int_{-\infty}^0 f\_(x)\, e^{ipx}\, dx + \int_0^\infty f\_(x)\, e^{ipx}\, dx$$

$$= -\int_{-\infty}^0 f(-x)\, e^{ipx}\, dx + \int_0^\infty f(x)\, e^{ipx} f(x)\, dx$$

$$= -\int_0^\infty e^{-ipx} f(x)\, dx + \int_0^\infty e^{ipx} f(x)\, dx$$

$$= \int_0^\infty f(x)(e^{ipx} - e^{-ipx})\, dx$$

$$= \int_0^\infty f(x)(2i \sin px)\, dx$$

$$= 2i \int_0^\infty f(x) \sin px\, dx$$

Then, $$F(f\_(x)) = 2i \frac{2}{\sqrt{2\pi}} \int_0^\infty f(x) \sin px\, dx$$

$$= i \sqrt{\frac{2}{\pi}} \int_0^\infty f(x) \sin px\, dx$$

$$= i\, F_s(p)$$

Now, using inversion formula, we have

$$f\_(x) = \frac{2}{\sqrt{2\pi}} \int_{-\infty}^{\infty} e^{-ipx}\, i\, F_s(p)dp$$

$$= \frac{i}{\sqrt{2\pi}} \left[\int_{-\infty}^{0} e^{-ipx} F_s(p)dp + \int_0^\infty e^{-ipx} F_s(p)dp \right]$$

$$= \frac{i}{\sqrt{2\pi}} \left[\int_0^\infty e^{ipx} F_s(-p)dp + \int_0^\infty e^{-ipx} F_s(p)dp \right]$$

$$= \frac{i}{\sqrt{2\pi}} \left[-\int_0^\infty e^{ipx} F_s(p)dp + \int_0^\infty e^{-ipx} F_s(p)dp \right]$$

$$= \frac{i}{\sqrt{2\pi}} \left[-2i \int_0^\infty \sin px\, F_s(p)dp \right]$$

$$= \sqrt{\frac{2}{\pi}} \int_0^\infty \sin px\, F_s(p)\, dp$$

$$\Rightarrow \qquad f(x) = \sqrt{\frac{2}{\pi}} \int_0^\infty F_s(p) \sin px\, dp,\ x > 0$$

We know that

$$F(f(x)) = \frac{1}{\sqrt{2\pi}} \left[\int_0^\infty e^{-ipx} f(-x)dx + \int_0^\infty e^{ipx} f(x)dx \right]$$

For an odd function $f(-x) = -f(x)$, then

$$F(f(x)) = \frac{1}{\sqrt{2\pi}}\left[\int_0^\infty (e^{ipx} - e^{-ipx}) f(x)dx\right]$$

$$= i\sqrt{\frac{2}{\pi}}\int_0^\infty \sin px\, f(x)\, dx$$

$$\Rightarrow \qquad F(f(x)) = i\, F_s\,(f(x))$$

Relation between Fourier and Laplace Transform

Let us assume a function $f(x)$ is defined as

$$f(x) = \begin{matrix} e^{-kx}\psi(x) & , & x > 0 \\ 0 & , & x < 0 \end{matrix}$$

Then, $\qquad F(f(x)) = \int_{-\infty}^{\infty} e^{ipx} f(x)dx \qquad$ (Using non-symmetrical form)

$$= \int_{-\infty}^{0} e^{ipx} f(x)dx + \int_0^\infty e^{ipx} f(x)dx$$

$$= \int_{-\infty}^{0} e^{ipx} \cdot 0 \cdot dx + \int_0^\infty e^{-ipx} \cdot e^{-kx}\, \psi(x)dx$$

$$= \int_0^\infty e^{-(k-ip)x}\, \psi(x)dx$$

$$= \int_0^\infty e^{-\lambda x}\, \psi(x)dx,\ \lambda = k - ip$$

$$= L\,(\psi(x))$$

$\Rightarrow \quad F(f(x)) = L\,(\psi(x))$, where, p is a parameter for Fourier transform and λ is the parameter for Laplace transform.

Illustrative Examples

Example 1 Find the Fourier transform of $f(x)$, if

$$f(x) = \begin{cases} e^{iwx} & , \quad a < x < b \\ 0 & , \quad b < x < a \end{cases}$$

Solution: $\qquad F(f(x)) = \frac{2}{\sqrt{2\pi}}\int_{-\infty}^{\infty} e^{ipx} f(x)dx$

$$= \frac{2}{\sqrt{2\pi}} \int_a^b e^{ibx} \cdot e^{iwx}\, dx$$

$$= \frac{2}{\sqrt{2\pi}} \int_a^b e^{i(b+w)bx} \cdot dx$$

$$= \frac{1}{\sqrt{2\pi}} \left[\frac{e^{i(p+w)x}}{i(p+w)} \right]_a^a$$

$$= \frac{1}{\sqrt{2\pi}} \left[\frac{e^{i(p+w)b} - e^{i(p+w)a}}{i(p+w)} \right]$$

$$= \frac{i}{\sqrt{2\pi}} \left[\frac{e^{i(p+w)b} - e^{i(b+w)a}}{p+w} \right]$$

Example 2 Find the Fourier transform of $f(x)$, if

$$f(x) = \begin{cases} \dfrac{\sqrt{2\pi}}{2a} &, \quad |x| \le a \\ 0 &, \quad |x| > a \end{cases}$$

Solution: We know that

$$F(f(x)) = \frac{2}{\sqrt{2\pi}} \int_{-\infty}^{\infty} f(x) e^{ipx}\, dx$$

$$= \frac{1}{\sqrt{2\pi}} \int_{-a}^{a} \frac{\sqrt{2\pi}}{2a} \cdot e^{ipx} dx$$

$$= \int_{-a}^{a} \frac{1}{2a} e^{ipx} dx$$

$$= \frac{1}{2a} \int_{-a}^{a} e^{ipx} dx$$

$$= \frac{1}{2aip} \left[e^{ipx} \right]_{-a}^{a}$$

$$= \frac{1}{2aip} \, [e^{ipa} - e^{-ipa}]$$

$$= \frac{1}{2aip} \, [2i \sin pa]$$

$$= \frac{1}{ap} \sin ap$$

Example 3 Find the Fourier transform of

$$f(x) = \begin{cases} 1 - x^2 & , \ |x| \le 1 \\ 0 & , \ |x| > 1 \end{cases}$$

and hence deduce $\int\limits_0^\infty \frac{x\cos x - \sin x}{x^3} \cos x / 2dx$.

Solution: We know that

$$F(f(x)) = \frac{2}{\sqrt{2\pi}} \int\limits_{-\infty}^{\infty} e^{ipx} f(x)dx$$

$$= \frac{2}{\sqrt{2\pi}} \int\limits_{-1}^{1} (1 - x^2)e^{ipx}\, dx$$

$$= \frac{1}{\sqrt{2\pi}}\left[\frac{(1-x^2)(e^{ipx})}{ip}\right]_{-1}^{1} + \frac{2}{\sqrt{2\pi}} \int\limits_{-1}^{1} x\frac{e^{ipx}}{ip} dx$$

$$= 0 + \frac{\sqrt{2}}{ip\sqrt{\pi}}\left[\left(\frac{x\, e^{ipx}}{ip}\right)_{-1}^{1} - \int\limits_{-1}^{1} \frac{e^{ipx}}{ip} dx\right]$$

$$= \frac{\sqrt{2}}{ip\sqrt{\pi}}\left[\frac{e^{ip} + e^{-ip}}{ip} - \frac{1}{(ip)^2}\left(e^{ipx}\right)_{-1}^{1}\right]$$

$$= \frac{\sqrt{2}}{\sqrt{\pi}}\left[\frac{2\cos p}{i^2 p^2} + \frac{e^{ip} - e^{-ip}}{ip^3}\right]$$

$$= \sqrt{\frac{2}{\pi}}\left[-\frac{2\cos p}{p^2} + \frac{2i\sin p}{i\, p^3}\right]$$

$$= -2\sqrt{\frac{2}{\pi}}\left[\frac{p\cos p - \sin p}{p^3}\right]$$

We have

$$f(x) = \frac{2}{\sqrt{2\pi}} \int\limits_{-\infty}^{\infty} e^{-ipx} F(p)dp$$

$$\Rightarrow \qquad -\frac{1}{\sqrt{2\pi}} \int\limits_{-\infty}^{\infty} \frac{2\sqrt{(2/\pi)}(p\cos p - \sin p)}{p^3} e^{-ipx} dp$$

$$= \begin{cases} 1 - x^2 & , \ |x| \le 1 \\ 0 & , \ |x| > 1 \end{cases}$$

$$\Rightarrow \qquad -\int_{-\infty}^{\infty} \frac{p\cos p - \sin p}{p^3} \cos pxdp + i \int_{-\infty}^{\infty} \frac{p\cos p - \sin p}{p^3} \sin pxdp$$

$$= \begin{cases} \frac{\pi}{2}(1 - x^2) & , \ |x| \le 1 \\ 0 & , \ |x| > 1 \end{cases}$$

$$\Rightarrow \qquad -\int_{-\infty}^{\infty} \frac{p\cos p - \sin p}{p^3} \cos pxdp = \begin{cases} \frac{\pi}{2}(1 - x^2) & , \ |x| \le 1 \\ 0 & , \ |x| > 1 \end{cases}$$

Substituting $x = 1/2$, we get

$$-\int_{-\infty}^{\infty} \frac{p\cos p - \sin p}{p^3} \cos p/2dp = \frac{\pi}{2}\left(1 - \frac{1}{4}\right)$$

$$= 3\pi/8$$

$$\Rightarrow \qquad 2\int_{0}^{\infty} \frac{p\cos p - \sin p}{p^3} \cos p/2\,dp = -\frac{3\pi}{8}$$

$$\Rightarrow \qquad \int_{0}^{\infty} \frac{p\cos p - \sin p}{p^3} \cos p/2\,dp = -\frac{3\pi}{8}$$

or
$$\int_{0}^{\infty} \frac{x\cos x - \sin x}{x^3} \cos(x/2)\,dx = -\frac{3\pi}{16}$$

Example 4 Find the Fourier transform of $f(x)$ such that

$$f(x) = \begin{cases} 1 & , \ |x| \le a \\ 0 & , \ |x| > a \end{cases}$$

Hence, deduce

(i) $\displaystyle\int_{-\infty}^{\infty} \frac{\sin ap \cos px}{p} dp$

(ii) $\displaystyle\int_{0}^{\infty} \frac{\sin p}{p} dp$

Solution: We know that

$$F(p) = \frac{2}{\sqrt{2\pi}} \int_{-\infty}^{\infty} e^{ipx} f(x)dx$$

$$= \frac{2}{\sqrt{2\pi}} \int_{-a}^{a} e^{ipx} dx$$

$$= \frac{2}{\sqrt{2\pi}} \left[\frac{e^{ipx}}{ip}\right]_{-a}^{a}$$

$$= \frac{1}{\sqrt{2\pi}} \cdot \frac{1}{ip} (e^{ipa} - e^{-ipa})$$

$$= \frac{1}{ip\sqrt{2\pi}} 2i \sin ap$$

$$= \frac{2}{p\sqrt{2\pi}} \sin ap, \; p \neq 0$$

For $p = 0, F(p) = \dfrac{2a}{\sqrt{2\pi}}$

(i) We know that

$$f(x) = \frac{2}{\sqrt{2\pi}} \int_{-\infty}^{\infty} -e^{ipx} F(p)dp$$

$$\Rightarrow \frac{2}{\sqrt{2\pi}} \int_{-\infty}^{\infty} \frac{2\sin ap}{p\sqrt{2\pi}} \cdot e^{-ipx} \, dp = \begin{bmatrix} 1 & , & |x| \leq a \\ 0 & , & |x| > a \end{bmatrix}$$

L.H.S.
$$\frac{1}{\left(\sqrt{2\pi}\right)^2} \int_{-\infty}^{\infty} \frac{2\sin ap}{p} \cdot (\cos px - i \sin px)dp$$

$$= \frac{2}{\sqrt{2\pi}} \int_{-\infty}^{\infty} \frac{\sin ap}{p} (\cos px - i \sin px)dp$$

$$= \frac{1}{\pi} \int_{-\infty}^{\infty} \frac{\sin ap \cos px}{p} dp - \frac{i}{\pi} \int_{-\infty}^{\infty} \frac{\sin ap \sin px}{p} dp$$

$$= \frac{1}{\pi} \int_{-\infty}^{\infty} \frac{\sin ap \cos px}{p} dp + 0$$

$$= \int_{-\infty}^{\infty} \frac{\sin ap \cos px}{p}\, dp = \begin{cases} \pi, & |x| \leq a \\ 0, & |x| > a \end{cases}$$

(ii) Let us assume, $a = 1$, $x = 0$ then,

$$\int_{-\infty}^{\infty} \frac{\sin p}{p}\, dp = \pi$$

$$\Rightarrow \qquad 2\int_{0}^{\infty} \frac{\sin p}{p}\, dp = \pi$$

$$\Rightarrow \qquad \int_{0}^{\infty} \frac{\sin p}{p}\, dp = \pi/2$$

Elementary Properties of the Fourier Transform

(i) *Linear property:* For any two functions $f_1(x)$ and $f_2(x)$ with their Fourier transforms $F_1(p)$ and $F_2(p)$, we have

$$F\{c_1 f_1(x) + c_2 f_2(x)\} = c_1 F_1(p) + c_2 F_2(p)$$

where c_1 and c_2 are any constants.

Proof: We know that

$$F\{f_1(x)\} = F_1(p) = \frac{2}{\sqrt{2\pi}} \int_{-\infty}^{\infty} e^{ipx} f_1(x)dx$$

and $$F\{f_2(x)\} = F_2(p) = \frac{2}{\sqrt{2\pi}} \int_{-\infty}^{\infty} e^{ipx} f_2(x)dx$$

Then, $$F\{c_1 f_1(x) + c_2 f_2(x)\}$$

$$= \frac{2}{\sqrt{2\pi}} \int_{-\infty}^{\infty} e^{ipx} \{c_1 f_1(x) + c_2 f_2(x)\}dx$$

$$= \frac{1}{\sqrt{2\pi}} \int_{-\infty}^{\infty} c_1 f_1(x) e^{ipx} dx + \frac{1}{\sqrt{2\pi}} \int_{-\infty}^{\infty} c_2 f_2(x) e^{ipx} dx$$

$$= \frac{c_1}{\sqrt{2\pi}} \int_{-\infty}^{\infty} f_1(x) e^{ipx} dx + \frac{c_2}{\sqrt{2\pi}} \int_{-\infty}^{\infty} f_2(x) e^{ipx} dx$$

$$= c_1 \left[\frac{1}{\sqrt{2\pi}} \int_{-\infty}^{\infty} f_1(x) e^{ipx} dx\right] + c_2 \left[\frac{1}{\sqrt{2\pi}} \int_{-\infty}^{\infty} f_2(x) e^{ipx} dx\right]$$

$$= c_1 F_1(p) + c_2 F_2(p)$$

$\Rightarrow \qquad F\{c_1 f_1(x) + c_2 f_2(x)\} = c_1 F_1(p) + c_2 F_2(p)$

This result can be generalised for any number of finite functions, then

$F\{c_1 f_1(x) + c_2 f_2(x) + ... + c_n f_n(x) = c_1 F_1(p) + c_2 F_2(p) + ... + c_n F_n(x)$

Change of Scale Property

(i) *Complex Fourier transform:* If $F(p)$ is the complex Fourier transform of $f(x)$ then complex Fourier transform of $f(ax)$ is

$$\frac{1}{a} F(p/a)$$

Proof: We know that

$$F(p) = \frac{2}{\sqrt{2\pi}} \int_{-\infty}^{\infty} e^{ipr} f(x)dx \qquad ...(i)$$

For any number, $a > 0$

$$\mathbf{F}[f(ax)] = \int_{-\infty}^{\infty} e^{ipx} f(ax)dx$$

Let $\qquad ax = u$

$\Rightarrow \qquad a\,dx = du$

or $\qquad dx = \frac{1}{a} du$

$$\mathbf{F}\{f(ax)\} = \frac{1}{a} \int_{-\infty}^{\infty} e^{ipu/a} f(u)du$$

$$= \frac{1}{a} \int_{-\infty}^{\infty} e^{i\left(\frac{p}{a}\right)u} f(u)du$$

$$= \frac{1}{a} \int_{-\infty}^{\infty} e^{-i\frac{p}{a}u} f(x)dx$$

$$= \frac{1}{a} F(p/a)$$

(ii) *Fourier sine transform:* If $F_s(p)$ is the Fourier sine transform of the function $f(x)$, then, the Fourier sine transform of the function $f(ax)$ is $\frac{1}{a} Fs(p/a)$

Proof: We know that

$$\mathbf{F}_s\{f(x)\} = F_s(p) = \sqrt{\frac{2}{\pi}} \int_0^{\infty} \sin px\, F(x)dx$$

Then, $$F_s\{f(ax)\} = \sqrt{\frac{2}{\pi}}\int_0^\infty \sin px\, f(ax)dx$$

$$= \sqrt{\frac{2}{\pi}}\int_0^\infty \sin\frac{u}{a}p\, f(u)\frac{dx}{a} \qquad (ax = u \Rightarrow a\,dx = dt)$$

$$= \frac{1}{a}\sqrt{\frac{2}{\pi}}\int_0^\infty \sin\frac{u}{a}p\, f(u)dx/a$$

$$= \frac{1}{a}\left[\sqrt{\frac{2}{\pi}}\int_0^\infty \sin\frac{p}{a}xf(x)dx\right]$$

$$= \frac{1}{a}F_s(p/a)$$

(iii) *Fourier cosine transform:* If $F_c(p)$ is the Fourier cosine transform of the function $f(x)$, then the Fourier cosine transform of the function $f(ax)$ is $\frac{1}{a}F_c(p/a)$.

$$F_c\{f(x)\} = F_c(p) = \sqrt{\frac{2}{\pi}}\int_0^\infty \cos px\, f(x)dx$$

Then, $$F_c\{f(ax)\} = \sqrt{\frac{2}{\pi}}\int_0^\infty \cos px\, f(ax)dx$$

$$= \sqrt{\frac{2}{\pi}}\int_0^\infty \cos p\cdot\frac{u}{a}f(u)\frac{du}{a}$$

$$= \frac{1}{a}\sqrt{\frac{2}{\pi}}\int_0^\infty \cos\frac{p}{a}u\, f(u)du$$

$$= \frac{1}{a}\sqrt{\frac{2}{\pi}}\int_0^\infty \cos\frac{p}{a}x\, f(x)dx$$

$$= \frac{1}{a}\left[\sqrt{\frac{2}{\pi}}\int_0^\infty \cos\frac{p}{a}x\, f(x)dx\right]$$

$$= \frac{1}{a}F_c(p/a)$$

(iv) *Shifting property:* If $F(p)$ is the complex Fourier transform of $f(x)$, then the complex Fourier transform of $f(x-a)$ is $e^{ipa}\,F(p)$.

Proof: We know that

$$\mathbf{F}\{f(x)\} = F(p) = \frac{2}{\sqrt{2\pi}} \int_{-\infty}^{\infty} e^{ipx} f(x)dx$$

Then, $$\mathbf{F}\,[f(x-a)] = \frac{2}{\sqrt{2\pi}} \int_{-\infty}^{\infty} e^{ipx} f(x-a)dx \qquad [\text{Let } x - a = u,\ dx = du,\ x = u + a]$$

$$= \frac{2}{\sqrt{2\pi}} \int_{-\infty}^{\infty} e^{ip(u+a)} f(u)du$$

$$= \frac{2}{\sqrt{2\pi}} e^{ipa} \int_{-\infty}^{\infty} e^{ipu} f(u)du$$

$$= \frac{2}{\sqrt{2\pi}} \cdot e^{ipa} \int_{-\infty}^{\infty} e^{ipx} f(x)dx$$

$$= e^{ipa}\left[\frac{1}{\sqrt{2\pi}} \int_{-\infty}^{\infty} e^{ipx} f(x)dx\right]$$

$$= e^{ipa} F(p)$$

(v) *Modulation theorem:* If $F(p)$ is the Fourier transform of $f(x)$ then $\frac{1}{2}[F(p - a) + F(p + a)]$ will be the Fourier transform of $f(x)$ cos ax.

Proof: We know that

$$\mathbf{F}[f(x)] = \frac{2}{\sqrt{2\pi}} \int_{-\infty}^{\infty} e^{ipx} f(x)dx$$

Then, $$\mathbf{F}[f(x)\cos ax] = \frac{2}{\sqrt{2\pi}} \int_{-\infty}^{\infty} e^{ipx} f(x) \cos ax\, dx$$

$$= \frac{2}{\sqrt{2\pi}} \int_{-\infty}^{\infty} e^{ipx} \cdot f(x) \frac{e^{iax} + e^{-iax}}{2} dx$$

$$= \frac{1}{2}\left[\frac{1}{\sqrt{2\pi}} \int_{-\infty}^{\infty} e^{i(p+a)x} f(x)dx + \frac{1}{\sqrt{2\pi}} \int_{-\infty}^{\infty} e^{i(p-a)x} f(x)dx\right]$$

$$= \frac{1}{2}[F(p + a) + F(p - a)]$$

Some Results:

(i) $\mathbf{F_s}[f(x)\cos ax] = \frac{1}{2}[F_s(p+a) + F_s(p-a)]$

(ii) $\mathbf{F_c}[f(x)\sin ax] = \frac{1}{2}[F_s(p+a) - F_s(p-a)]$

(iii) $\mathbf{F_s}[f(x)\sin ax] = \frac{1}{2}[F_c(p-a) - F_s(p+a)]$

Illustrative Examples

Example 1 If $\psi(p)$ is the Fourier sine transform of $f(x)$ $p > 0$, then for $p < 0$,

$$F_s\{f(x)\} = -\psi(-p)$$

Solution: We know that

$$F_s\{f(x)\} = \sqrt{\frac{2}{\pi}}\int_0^\infty f(x)\sin px\,dx$$

$$= \psi(p),\ p > 0$$

Let $p < 0 \Rightarrow p = -m$ for $m > 0$

Then,

$$F_s\{f(x)\} = \sqrt{\frac{2}{\pi}}\int_0^\infty f(x)\sin(-mx)dx$$

$$= -\sqrt{\frac{2}{\pi}}\int_0^\infty f(x)\sin mx\,dx$$

$$= -\psi(-p)$$

Hence,

$$F_s\{f(x)\} = \begin{matrix}\psi(|p|) & , & p>0 \\ -\psi(|p|) & , & p<0\end{matrix}$$

Example 2 Find the complex Fourier transform of $f(x) = e^{-a|x|}$, where $a > 0$ and $x \in (-\infty, \infty)$.

Solution:

$$F(f(x)) = \frac{2}{\sqrt{2\pi}}\int_{-\infty}^{\infty} e^{ipx} f(x)dx$$

$$= \frac{2}{\sqrt{2\pi}}\int_{-\infty}^{\infty} e^{ipx}\, e^{-a|x|}\, dx$$

$$= \frac{1}{\sqrt{2\pi}}\int_{-\infty}^{0} e^{ipx}e^{ax}dx + \frac{1}{\sqrt{2\pi}}\int_0^\infty e^{ipx}e^{-ax}dx$$

$$= \frac{1}{\sqrt{2\pi}}\left[\frac{e^{(a+ip)x}}{a+ip}\right]_{-\infty}^{0} + \frac{1}{\sqrt{2\pi}}\left[\frac{e^{-i(a-ip)x}}{-(a-ip)}\right]_{0}^{\infty}$$

$$= \frac{1}{\sqrt{2\pi}}\left[\frac{1}{a+ip}\right] + \frac{1}{\sqrt{2\pi}}\left[\frac{1}{a-ip}\right]$$

$$= \frac{2}{\sqrt{2\pi}}\,\frac{2a}{a^2+p^2}$$

$$= \sqrt{\frac{2}{\pi}}\,\frac{a}{a^2+p^2}$$

Thus, $$\mathbf{F}\{f(x)\} = \sqrt{\frac{2}{\pi}}\cdot\frac{a}{a^2+p^2}.$$

Example 3 Find the Fourier sine and cosine transforms of $f(x)$, if

$$f(x) = \begin{cases} x &, \quad 0<x<1 \\ 2-x &, \quad 1<x<2 \\ 0 &, \quad x>2 \end{cases}$$

Solution: We know that

$$F_s[f(x)] = \sqrt{\frac{2}{\pi}}\int_0^{\infty} f(x)\sin px\, dx$$

$$= \sqrt{\frac{2}{\pi}}\left[\int_0^1 f(x)\sin px\, dx + \int_1^2 f(x)\sin px\, dx + \int_2^{\infty} f(x)\sin px\, dx\right]$$

$$= \sqrt{\frac{2}{\pi}}\left[\int_0^1 x\sin px\, dx + \int_1^2 (2-x)\sin px\, dx + \int_0^{\infty} 0\cdot\sin px\, dx\right]$$

$$= \sqrt{\frac{2}{\pi}}\left[\left(\frac{-x\cos px}{p} + \frac{\sin px}{p^2}\right)_0^1 + \left(\frac{-(2-x)\cos px}{p}\right)_1^2 - \left(\frac{\sin px}{p^2}\right)_1^2\right]$$

$$= \sqrt{\frac{2}{\pi}}\left[-\frac{\cos p}{2} + 0 + \frac{\sin p}{p^2} - 0\right] + \sqrt{\frac{2}{\pi}}\left[0 + \frac{\cos p}{p} - \frac{\sin 2p}{p^2} + \frac{\sin p}{p^2}\right]$$

$$= \sqrt{\frac{2}{\pi}}\left[\frac{2\sin p}{p^2} - \frac{\sin 2p}{p^2}\right]$$

$$= \sqrt{\frac{2}{\pi}}\left[\frac{1}{p^2}(2\sin p - 2\sin p\cos p)\right]$$

$$= \sqrt{\frac{2}{\pi}} \cdot \frac{1}{p^2}\ 2\sin p(1 - \cos p)$$

and

$$F_c(f(x)) = \sqrt{\frac{2}{\pi}}\left[\int_0^1 x\cos x\,dx + \int_1^2 (2-x)\cos px\,dx\right]$$

$$= \sqrt{\frac{2}{\pi}}\left[\left(x \cdot \frac{\sin px}{p}\right)_0^1 - \frac{1}{p}\int_0^1 \sin px\,dx\right]$$

$$+\sqrt{\frac{2}{\pi}}\left[\left((2-x)\frac{\sin px}{p}\right)_1^2 + \frac{1}{p}\int_1^2 \sin px\,dx\right]$$

$$= \sqrt{\frac{2}{\pi}}\left[x\frac{\sin px}{p} - \left(\frac{-\cos px}{p^2}\right)\right]_0^1 + \sqrt{\frac{2}{\pi}}\left[(2-x)\frac{\sin px}{p} + \left(\frac{-\cos px}{p^2}\right)_1^2\right]$$

$$= \sqrt{\frac{2}{\pi}}\left[\frac{\sin p}{p} - 0 + \frac{\cos p}{p^2} - \frac{1}{p^2}\right] + \sqrt{\frac{2}{\pi}}\left[0 - \frac{\sin p}{p} - \frac{\cos 2p}{p^2} + \frac{\cos p}{p^2}\right]$$

$$\Rightarrow \qquad F_c(f(x)) = \sqrt{\frac{2}{\pi}}\left[\frac{2}{p^2}\cos p(1 - \cos p)\right]$$

Multiple Variable Fourier Transforms: Let us consider a function (f) of two variables x and y, i.e., $f(x, y)$. For a function of single variable x, the Fourier transform of the function can be defined as

$$F(p, y) = \frac{1}{2\pi}\int_{-\infty}^{\infty} f(x, y)e^{ipx}\,dx \qquad \text{...(i)}$$

Now, consider the function $F(p, y)$ a function of y, its Fourier transform can be defined as

$$\phi(p, q) = \frac{2}{\sqrt{2\pi}}\int_{-\infty}^{\infty} F(p, y)e^{iqy}\,dy$$

$$\Rightarrow \qquad \phi(p, q) = \frac{2}{\sqrt{2\pi}}\int_{-\infty}^{\infty}\int_{-\infty}^{\infty} f(x, y)\,e^{i(px+qy)}dx\,dy$$

and inversion formula can be defined on two variables as

$$f(x, y) = \frac{1}{2\pi}\int_{-\infty}^{\infty} \phi(p, q)\, e^{-i(px+qy)} dp\, dq$$

Convolution: Convolution or falting of two integrable functions, F and G, in the open interval $(-\infty, \infty)$ can be defined as

$$H(x) = F * G = \frac{2}{\sqrt{2\pi}}\int_{-\infty}^{\infty} F(u)\, G(x-u)du$$

Some authors defined convolution as

$$F * G = \int_{-\infty}^{\infty} F(u)\cdot G(x-u)du$$

Convolution or Falting Theorem for Fourier Transform: If $F\{f_1(x)\}$ and $F\{f_2(x)\}$ are the Fourier transforms of the functions $f_1(x)$ and $f_2(x)$ respectively, and

(i) Functions $f_1(x)$ and $f_2(x)$ have first order derivatives which are continuous in open interval $(-\infty, \infty)$.

(ii) Functions $f_1(x)$ and $f_2(x)$ are absolutely integrable in open interval $(-\infty, \infty)$.

Then the Fourier transform of the convolution of $f_1(x)$ and $f_2(x)$ exists and is the product of the Fourier transform of $f_1(x)$ and $f_2(x)$ such that

$$\begin{aligned} F\{f_1(x) * f_2(x)\} &= F\{f_1(x)\}\, F\{f_2(x)\} \\ &= F_1(p)\, F_2(p) \end{aligned}$$

or $$\int_{-\infty}^{\infty} f_1(u)\, f_2(x-u)du = \int_{-\infty}^{\infty} e^{-ipx}\, F_1(p)\, F_2(p)dp$$

Proof: We know that

$$\begin{aligned} F\{f_1(x)\} &= \frac{2}{\sqrt{2\pi}}\int_{-\infty}^{\infty} e^{ipx} f_1(x)dx \\ &= F_1(p) \end{aligned}$$

$$\begin{aligned} F\{f_2(x)\} &= \frac{2}{\sqrt{2\pi}}\int_{-\infty}^{\infty} e^{ipx} f_2(x)dx \\ &= F_2(p) \end{aligned}$$

Using definition of convolution, we have

$$F\{f_1 * f_2\} = F\left[\frac{1}{\sqrt{2\pi}}\int_{-\infty}^{\infty} f_1(u) f_2(x-u)du\right]$$

$$= \frac{2}{\sqrt{2\pi}} \int_{-\infty}^{\infty} e^{ipx} \left[\frac{1}{\sqrt{2\pi}} \int_{-\infty}^{\infty} f_1(u) f_2(x-u) du \right] dx$$

$$= \frac{1}{2\pi} \int_{-\infty}^{\infty} f_1(u) \left[\int_{-\infty}^{\infty} e^{ipx} f_2(x-u)\, dx \right] du$$

(by changing of order of integration)

$$= \frac{1}{2\pi} \int_{-\infty}^{\infty} f_1(u) \left[\int_{-\infty}^{\infty} e^{ip(u+t)} f_2(t) dt \right] du \qquad \text{Assume } x - u = t \Rightarrow dx = dt$$

$$= \frac{1}{2\pi} \int_{-\infty}^{\infty} e^{ipu} f_1(u) \left[\frac{1}{\sqrt{2\pi}} \int_{-\infty}^{\infty} e^{ipt} f_2(t) dt \right] du$$

$$= \frac{2}{\sqrt{2\pi}} \int_{-\infty}^{\infty} e^{ipu} f_1(u)\, F_2(p) du\; F_2(p) \frac{1}{\sqrt{2\pi}} \int_{-\infty}^{\infty} e^{ipu} f_1(u) du$$

where, $F_2(p) = \dfrac{i}{\sqrt{2\pi}} \int e^{ipt} f_2(t) dt$

$$= F_2(p) \cdot F_1(p)$$

Applying inverse Fourier transform of above the equation, we find

$$\int_{-\infty}^{\infty} F_1(p) F_2(p)\, e^{-ipx}\, dp$$

$$= \int_{-\infty}^{\infty} F_1(u) F_2(x-u) du$$

Fourier Transform of Derivatives

(i) The Fourier transform of $f'(x)$, the derivative of $f(x)$ is $-\, ip\, F(p)$, where, $F(p)$ is the Fourier transform of the function $f(x)$.

Proof: We know that

$$F(p) = F\{f'(x)\} = \frac{2}{\sqrt{2\pi}} \int_{-\infty}^{\infty} f'(x) e^{ipx}\, dx \qquad \text{...(i)}$$

$$= \frac{2}{\sqrt{2\pi}} \int_{-\infty}^{\infty} e^{ipx} f'(x)\, dx$$

$$\Rightarrow \qquad F\{f'(x)\} = \frac{1}{\sqrt{2\pi}} \left[e^{ipx} f(x) \right]_{-\infty}^{\infty} - \frac{ip}{\sqrt{2\pi}} \int_{-\infty}^{\infty} e^{ipx} f(x)\, dx \qquad \text{...(ii)}$$

It is given that $f(x)$ is integrable in open interval $(-\infty, \infty)$. Therefore, $\lim_{|x|\to 0} f(x) = 0$ hence, first integral in the equation (ii) vanishes

$$\Rightarrow \qquad F\{f'(x)\} = -ip\,F(p)$$

(ii) The Fourier transform of $f^n(x)$, i.e., nth derivative of the function $f(x)$, is $(-ip)^n F(p)$ provided that all $(n-1)^{\text{th}}$ derivatives of the function $f(x)$ are continuous, n^{th} derivative is piecewise continuous and differentiable in the open interval $(-\infty, \infty)$, and each derivative is absolutely integrable in the open integrable $(-\infty, \infty)$.

Then,

$$F\{f^n(x)\} = (-ip)^n F\{f(x)\}$$

Proof: We know that

$$F\{f(x)\} = \frac{2}{\sqrt{2\pi}} \int_{-\infty}^{\infty} e^{ipx} f(x)\,dx$$

Then,

$$F\{f^n(x)\} = \frac{2}{\sqrt{2\pi}} \int_{-\infty}^{\infty} e^{ipx} f^n(x)\,dx$$

$$= \frac{1}{\sqrt{2\pi}}\left[\left(e^{ipx}\cdot f^{n-1}(x)\right)_{-\infty}^{\infty} - ip\int_{-\infty}^{\infty} e^{ipx} f^{n-1}(x)dx\right]$$

$$= \frac{1}{\sqrt{2\pi}}\left(e^{ipx} f^{n-1}(x)\right)_{-\infty}^{\infty} - \frac{ip}{\sqrt{2\pi}}\int_{-\infty}^{\infty} e^{ipx} f^{n-1}(x)dx$$

$$= -\frac{2}{\sqrt{2\pi}}(ip)\int_{-\infty}^{\infty} e^{ipx} f^{n-1}(x)dx$$

$$\left[\because \lim_{x\to\infty} f^{n-1}(x) = 0\right]$$

Continuing in the same manner of integration for $(n-1)$ times, we get

$$F\{f^n(x)\} = \frac{1}{2\pi}(-ip)^n \int_{-\infty}^{\infty} e^{ipx} f(x)dx$$

$$F\{f^n(x)\} = (-ip)^n F\{f(x)\}.$$

Theorem: Let $F(p) = F\{f(x)\}$ and if

(a) (i) $f(x)$ is continuous and absolutely integrable in open interval $(-\infty, \infty)$, and

(ii) $x\,f(x)$ is absolutely integrable in open interval $(-\infty, \infty)$

Then, $\quad F'(p) = i\,F\{x\,f(x)\}$

Proof: We know that

$$F(p) = \frac{2}{\sqrt{2\pi}} \int_{-\infty}^{\infty} e^{ipx} f(x)dx$$

Differentiating both sides with respect to p, we have

$$F'(p) = \frac{1}{\sqrt{2\pi}} \int_{-\infty}^{\infty} \left(\frac{\partial}{\partial p} e^{ipx} \right) f(x) dx$$

(b) (i) $f(x)$ is continuous in open interval $(-\infty, \infty)$

(ii) every $x^m f(x)$, $m = 0, \ldots n$, is absolutely integrable in open interval $(-\infty, \infty)$

then $F^m(p) = i^m F\{x^m f(x)\}$, $m = 0, 1, \ldots n$ and using the results,

$$F\{x^k f^r(x)\} = (-1)^k \frac{d^k}{dp^k} F[f^r(x)]$$

$$= (-1)^{r+k} \frac{d^k}{dp^k} p^r F(p)$$

$$\Rightarrow \qquad F^{(r)}(p) = i^r F(x^r f(x))$$

$$\therefore \qquad p^n F^{(r)}(p) = i^{r+n} F\left\{ \frac{d^k}{dp^k} x^r f(x) \right\}$$

Applying Leibnitz's theorem

$$\frac{d^n}{dx^n}[x^r f(x)] = \sum_{m=0}^{q} \frac{r!n!}{(n-m)!(r-m)!} F[r^{r-m} f^{(n-m)}(x)]$$

Parseval's Identity for Fourier Transform

If $F_1(p)$ and $F_2(p)$ are the complex transform of $f_1(x)$ and $f_2(x)$ respectively, then

(i) $\int_{-\infty}^{\infty} F_1(p)\bar{F}_2(p)dp = \int_{-\infty}^{\infty} f_1(x) \cdot \bar{f}_2(x)dx$

(ii) $\int_{-\infty}^{\infty} |F_1(p)|^2 dp = \int_{-\infty}^{\infty} |f_1(x)|^2 dx$

where, $\bar{F}_2(p)$ and $\bar{f}_2(x)$ represents the complex conjugate of the functions $F_2(p)$ and $f_2(x)$, respectively.

Proof: (i) Inversion of the Fourier transform can be defined as

$$f_2(x) = \frac{2}{\sqrt{2\pi}} \int_{-\infty}^{\infty} F_2(p)e^{-ipx}\, dp \qquad \text{...(i)}$$

Applying complex conjugate to the both sides of the equation, we have

$$\bar{f}_2(x) = \frac{2}{\sqrt{2\pi}} \int_{-\infty}^{\infty} \bar{F}_2(p)\, e^{ipx}\, dp \qquad \text{...(ii)}$$

Then, $\int_{-\infty}^{\infty} f_1(x)\bar{f}_2(x)dx = \int_{-\infty}^{\infty} f_1(x)\left[\frac{1}{\sqrt{2\pi}} \int_{-\infty}^{\infty} \bar{F}_2(p)e^{ipx} dx \right] dp$

$$= \int_{-\infty}^{\infty} \bar{F}_2(p)F(p)dp$$

$$\Rightarrow \quad \int_{-\infty}^{\infty} f_1(x)\bar{f}_2(x)dx = \int_{-\infty}^{\infty} F_2(p)F_1(p)dp \qquad \text{...(iii)}$$

(ii) Let us assume $f_1(x) = f_2(x)$ in equation (iii), we get

$$\int_{-\infty}^{\infty} |F_1(p)|^2\, dp = \int_{-\infty}^{\infty} |f_1(x)|^2 dx$$

Example 4 Find Fourier transform of $f(x)$ defined by function

$$f(x) = \begin{cases} 1 & , \quad |x| < a \\ 0 & , \quad |x| > 0 \end{cases}$$

and, hence deduce

$$\int_{0}^{\infty} \frac{\sin^2 ax}{x^2} dx = \frac{\pi a}{2}.$$

Solution: (i) We know that

$$F\{f(x)\} = \frac{2}{\sqrt{2\pi}} \int_{-\infty}^{\infty} e^{ipx} f(x)dx \qquad \text{...(i)}$$

$$= \frac{2}{\sqrt{2\pi}} \int_{-a}^{a} e^{ipx} \cdot 1 \cdot dx$$

$$= \frac{1}{\sqrt{2\pi}} \left[\frac{e^{ipx}}{ip} \right]_{-a}^{a}$$

$$= \frac{1}{\sqrt{2\pi}} \left[\frac{e^{ipa} - e^{-ipa}}{ip} \right]$$

$$= \frac{1}{\sqrt{2\pi}} \frac{i\, 2 \sin ap}{ip}$$

$$= \frac{1}{\sqrt{2\pi}} \frac{2 \sin ap}{p}, \; p \neq 0 \qquad \text{...(ii)}$$

We have Parseval's identity

$$\int_{-\infty}^{\infty} |f(x)|^2 dx = \int_{-\infty}^{\infty} |F(p)|^2 dp$$

by equation (ii), we get

$$\int_{-a}^{a} 1^2 dx = \frac{1}{2\pi}\int_{-\infty}^{\infty} \frac{4\sin^2 ap}{p^2}\, dp$$

$$\Rightarrow \qquad \frac{2a\cdot 2\pi}{4} = \int_{-\infty}^{\infty} \frac{\sin^2 ap}{p^2}\, dp$$

or

$$\pi a = 2\int_{0}^{\infty} \frac{\sin^2 ap}{p^2}\, dp$$

$$\Rightarrow \qquad \frac{\pi a}{2} = \int_{0}^{\infty} \frac{\sin^2 ap}{p^2}\, dp$$

On changing the variable, we get

$$\int_{0}^{\infty} \frac{\sin^2 ax}{x^2}\, dx = \frac{\pi a}{2}.$$

Example 5 Use Parseval's identity to prove that

(i) $\displaystyle\int_{0}^{\infty} \frac{dt}{(a^2+t^2)(b^2+t^2)} = \frac{\pi}{2ab\,(a+b)}$, $a, b > 0$

(ii) $\displaystyle\int_{0}^{\infty} \frac{\sin at}{t(a^2+t^2)}\, dt = \frac{\pi}{2}\left[\frac{1-e^{-a^2}}{a^2}\right].$

Solution: (i) Let us assume

$$f_1(x) = e^{-ax} \text{ and } f_2(x) = e^{-px}$$

Then,

$$F_{1c}(p) = \sqrt{\frac{2}{\pi}}\frac{a}{a^2+b^2}$$

and

$$F_{2c}(p) = \sqrt{\frac{2}{\pi}}\frac{b}{a^2+b^2}$$

Parseval's identity for Fourier cosine transform can be defined as

$$\int_{0}^{\infty} F_{1c}(p)\, F_{2c}(p) dp = \int_{0}^{\infty} f_1(x)\, f_2(x) dx$$

or

$$\frac{2}{\pi}\int_{0}^{\infty} \frac{dp}{(a^2+p^2)(b^2+p^2)}\, dp = \int_{0}^{\infty} e^{-(a+b)x}\, dx$$

$$\Rightarrow \quad \int_0^\infty \frac{dp}{(a^2+p^2)(b^2+p^2)} = \frac{\pi}{ab}\left[\frac{e^{-(a+b)x}}{-(a+b)}\right]_0^\infty$$

$$= \frac{\pi}{2ab(a+b)}$$

On changing variable as per property of definite integral, we have

$$\int_0^\infty \frac{dt}{(a^2+t^2)(b^2+t^2)} = \frac{\pi}{2ab(a+b)}$$

(ii) Let us assume,

$$f_1(x) = e^{-ax}$$

$$\Rightarrow \quad F_{1c}(p) = \sqrt{\frac{2}{\pi}}\frac{a}{a^2+p^2}$$

and

$$f_2(x) = \begin{cases} 1 \ , & 0 < x < a \\ 0 \ , & x > a \end{cases}$$

Then,

$$F_{2c}(p) = \sqrt{\frac{2}{\pi}}\int_0^\infty f_i(x)\cos px\, dx$$

$$= \sqrt{\frac{2}{\pi}}\int_0^a \cos px\, dx$$

$$= \sqrt{\frac{2}{\pi}}\frac{\sin ap}{p}$$

We know that

$$\int_0^\infty f_1(x)\, f_2(x)dx = \int_0^\infty F_{1c}(p)\, F_{2c}(p)dp$$

$$\int_0^\infty e^{ax} f_2(x)dx = \frac{2}{\pi}\int_0^\infty \frac{a}{a^2+p^2}\frac{\sin ap}{p}dp$$

or

$$\int_0^\infty 1\cdot e^{-ax}dx = \frac{2a}{\pi}\int_0^\infty \frac{\sin ap}{p(a^2+p^2)}dp$$

or

$$\left[\frac{e^{-ax}}{-a}\right]_0^a = \frac{2a}{\pi}\int_0^\infty \frac{\sin ap}{p(a^2+p^2)}dp$$

or $$\frac{1}{a}\left(1-e^{-a^2}\right) = \frac{2a}{\pi}\int_0^\infty \frac{\sin ap}{p(a^2+p^2)}\,dp$$

$$\Rightarrow \quad \int_0^\infty \frac{\sin ap}{p(a^2+p^2)}\,dp = \frac{\pi}{2a^2}\left(1-e^{-a^2}\right)$$

Using property of definite integral to change the variable from p to t, we get

$$\int_0^\infty \frac{\sin at}{t(a^2+t^2)}\,dt = \frac{\pi}{2a^2}\left(1-e^{-a^2}\right).$$

Example 6 Solve the integral equation

$$\int_0^\infty f(x)\sin px\,dx = \begin{cases} 1 \;, & 0 \le p < 1 \\ 2 \;, & 1 \le p < 2 \\ 0 \;, & p > 2 \end{cases}$$

Solution: We know that

$$F_s(p) = F_s\{f(x)\} = \sqrt{\frac{2}{\pi}}\int_0^\infty f(x)\sin px\,dx$$

and $$F_s(p) = \sqrt{\frac{2}{\pi}}\int_0^\infty f(x)\sin px\,dx$$

$$= \sqrt{\frac{2}{\pi}}\left[\sqrt{\frac{2}{\pi}}\int_0^1 1\cdot\sin px\,dp + \frac{\sqrt{2}}{\pi}\int_1^2 2\sin px\,dp\right]$$

$$= \frac{2}{\pi}\left[\frac{-\cos px}{x}\right]_0^1 + \frac{4}{\pi}\left[\frac{-\cos px}{x}\right]_1^2$$

$$= \frac{2}{x\pi}[-\cos p + 1] + \frac{4}{x\pi}[-\cos 2x + \cos x]$$

$$= \frac{2}{\pi x}[1 - \cos 2x + \cos x].$$

Example 7 Solve the integral equation

$$\int_0^\infty f(x)\cos px\,dx = \begin{cases} 1-p \;, & 0 \le p \le 1 \\ 0 \;, & p > 1 \end{cases}$$

Hence, show that $\int_0^\infty \frac{\sin^2 t}{t^2}\,dt = \frac{\pi}{2}$.

Solution: Given that

$$\int_0^\infty f(x)\cos px\,dx = \begin{cases} 1-p & , \quad 0 \le p \le 1 \\ 0 & , \quad p > 1 \end{cases}$$

Then, $$\sqrt{\frac{2}{\pi}}\int_0^\infty f(x)\cos px\,dx = \sqrt{\frac{2}{\pi}}\begin{cases} 1-p & , \quad 0 \le p \le 1 \\ 0 & , \quad p > 1 \end{cases}$$

Hence, $$F_c(p) = \sqrt{\frac{2}{\pi}}\int_0^\infty f(x)\cos px\,dp = \sqrt{\frac{2}{\pi}}\begin{cases} 1-p & , \quad 0 \le p \le 1 \\ 0 & , \quad p > 1 \end{cases}$$

Then, by using Fourier cosine inversion formula

$$f(x) = \sqrt{\frac{2}{\pi}}\int_0^\infty F_c(p)\cos px\,dp$$

$$= \sqrt{\frac{2}{\pi}}\left[\int_0^1 \sqrt{\frac{2}{\pi}}(1-p)\cos px\,dp\right]$$

$$= \frac{2}{\pi}\int_0^1 (1-p)\cos px\,dp$$

$$= \frac{2}{\pi}\left[\left[(1-p)\frac{\sin px}{x}\right]_0^1 + \int_0^1 \frac{\sin px}{x}\,dp\right]$$

$$= \frac{2}{\pi x^2}[1 - \cos x]$$

$$\Rightarrow \qquad f(x) = \frac{2}{\pi x^2}(1 - \cos x)$$

We have

$$\int_0^\infty f(x)\cos px\,dx = \begin{cases} 1-p & , \quad 0 \le p \le 1 \\ 0 & , \quad p > 1 \end{cases}$$

$$\Rightarrow \int_0^\infty \frac{2}{\pi x^2}(1 - \cos x)\cos px\,dx = \begin{cases} 1-p & , \quad 0 \le p \le 1 \\ 0 & , \quad p > 1 \end{cases}$$

Applying limit $p \to 0$, we get

$$\lim_{p\to 0}\left[\int_0^\infty \frac{2}{\pi x^2}(1 - \cos x)\,dx = 1\right.$$

$$\Rightarrow \qquad \frac{2}{\pi}\int_0^\infty \frac{(1-\cos x)}{x^2}\,dx = 1$$

$$\int_0^\infty \frac{(1-\cos x)}{x^2}\,dx = \frac{\pi}{2}$$

$$\Rightarrow \qquad \int_0^\infty \frac{2\sin^2 x/2}{x^2}\,dx = \frac{\pi}{2}$$

Let us assume $x/2 = t$

$$\Rightarrow \qquad dx = 2\,dt$$

Then, $$\int_0^\infty \frac{\sin^2 t}{t^2} = \frac{\pi}{2}$$

Example 8 Prove that

$$\int_0^\infty \frac{x^2 dx}{(x^2+a^2)^4} = \frac{\pi}{(2a)^5},\ a > 0.$$

Solution: Given that

$$\int_0^\infty \frac{x^2 dx}{(x^2+a^2)^4} = \frac{\pi}{(2a)^5}$$

Let us assume $f(x) = \dfrac{1}{2(x^2+a^2)}$

$$\Rightarrow \qquad f'(x) = -\frac{x}{(x^2+a^2)}$$

and we have

$$F[e^{-a|x|}] = \sqrt{\frac{2}{\pi}}\frac{a}{a^2+p^2}$$

or $$F^{-1}\left\{\frac{1}{a^2+p^2}\right\} = \frac{1}{a}\sqrt{\frac{\pi}{2}}\,e^{-a|x|}$$

or $$\frac{1}{\sqrt{2\pi}}\int_{-\infty}^{\infty} \frac{e^{-ipx}}{a^2+p^2}\,dp = \frac{1}{a}\sqrt{\frac{\pi}{2}}\,e^{-a|x|}$$

On putting $-p$ in place of p, we get

$$\frac{1}{\sqrt{2\pi}}\int_{-\infty}^{\infty}\frac{e^{ipx}}{a^2+p^2}\,dp = \left(\sqrt{\frac{\pi}{2}}\right)\frac{1}{a}\cdot e^{-a|x|}$$

$$\Rightarrow \qquad F(p) = \sqrt{\frac{\pi}{2}}\left(\frac{1}{2a}\right)e^{-a|p|}$$

Using Parseval identity, we have

$$\int_{-\infty}^{\infty}|f'(x)|^2dx = \int_{-\infty}^{\infty}|F\,f'(x)|^2dp$$

$$= \int_{-\infty}^{\infty}|-ip\,F\{f(x)\}|^2dp$$

Then, $$\int_{-\infty}^{\infty}\frac{x^2}{(a^2+x^2)^4}\,dx = \frac{\pi}{2}\int_{-\infty}^{\infty}\frac{p^2}{(2a)^2}\,e^{-2a|p|}\,dp$$

$$\Rightarrow \qquad \int_{-\infty}^{\infty}\frac{x^2}{(x^2+a^2)^4}\,dx = \frac{\pi}{8a^2}\int_{-\infty}^{\infty}p^2e^{-2ap}\,dp$$

$$= \frac{\pi}{8a^2}\cdot\frac{2}{(2a)^3} = \frac{2}{(2a)^5}.$$

EXERCISE

1. Show that $\int_{0}^{\infty}\frac{\sin(k_1t)\sin(k_2t)}{t^2}\,dt = \frac{\pi}{2}\min(k_1, k_2)$ where k_1 and k_2 are positive numbers.

2. Find the convolution of $f(x) = \cos x$ and $g(x) = e^{-a|x|}$, $a > 0$.

$$\left[\textbf{Ans. } \left(\sqrt{\frac{2}{\pi}}\right)\frac{a\cos x}{1+a^2}\right]$$

Example 9 Evaluate the Fourier cosine transform of e^{-x^2}.

Solution: Given that

$$f(x) = e^{-x^2} \qquad \text{...(i)}$$

Fourier cosine transform is defined as

$$F_c(p) = \sqrt{\frac{2}{\pi}}\int_{0}^{\infty}f(x)\cos px\,dx$$

or $$I = \sqrt{\frac{2}{\pi}} \int_0^\infty e^{-x^2} \cos px \, dx \qquad \text{...(ii)}$$

Differentiating equation (ii) with respect to p, we get

$$\frac{dI}{dp} = -\left(\sqrt{\frac{2}{\pi}}\right) \cdot \int_0^\infty x\, e^{-x^2} \sin px \, dx$$

$$= \frac{1}{\sqrt{2\pi}} \int_0^\infty \left(-2xe^{-x^2}\right) \sin px \, dx$$

$$= \frac{1}{\sqrt{2\pi}} \left[\left\{ (\sin px)\left(e^{-x^2} \right) \right\}_0^\infty - \int_0^\infty p \cos px \, (e^{-x^2}) \, dx \right]$$

$$= 0 - \frac{p}{2} I$$

$$\Rightarrow \qquad \frac{dI}{I} = -\frac{p}{2} dp$$

On integration

$$\log I = -\frac{p^2}{4} + \log c,$$

$$\Rightarrow \qquad I = c_1 \, e^{-p^2/4} \qquad \text{...(iii)}$$

at $$p = 0, I = \frac{1}{\sqrt{2}}$$

Using equation (iii), we get

$$\frac{1}{\sqrt{2}} = I = c_1$$

or $$c_1 = \frac{1}{\sqrt{2}}$$

Hence, $$F_c[e^{-x^2}] = \frac{1}{\sqrt{2}} e^{-p^2/4}.$$

Example 10 Find Fourier transform of $e^{-a^2x^2}$. Also, deduce cosine transform of $\cos^2 x/2$ and $\sin^2 x/2$.

Solution: We have the identity

$$x^2 - ipx = \left(x - \frac{1}{2} ip \right)^2 + \frac{1}{4} p^2 \qquad \text{...(i)}$$

and $$F(f(x)) = \frac{1}{\sqrt{2\pi}} \int_{-\infty}^{\infty} f(x)\, e^{ipx}\, dx$$

$$\Rightarrow \quad F\{e^{-x^2}\} = \frac{1}{\sqrt{2\pi}} \int_{-\infty}^{\infty} e^{-x^2} \cdot e^{ipx}\, dx$$

$$= \frac{1}{\sqrt{2\pi}} \int_{-\infty}^{\infty} e^{-(x^2 - ipx)} dx$$

$$= \frac{1}{\sqrt{2\pi}} \int_{-\infty}^{\infty} e^{\left[\left[x - \frac{1}{2}ip\right]^2 + \frac{1}{4}p^2\right]} dx \qquad \text{[Using equation (i)]}$$

$$= \frac{e^{-p^2/4}}{\sqrt{2\pi}} \int_{-\infty}^{\infty} e^{-t^2}\, dt \qquad \left[x - \frac{1}{2}ip = t\right]$$

$$= \frac{e^{-p^2/4}}{\sqrt{2\pi}} \sqrt{\pi}, \quad \left[\int_{-\infty}^{\infty} e^{-t^2}\, dt = \sqrt{\pi}\right] \qquad \text{...(ii)}$$

Applying the property of change of scale in Fourier transform, we have

$$F\left[e^{-a^2x^2}\right] = 2^{-1/2} \cdot a^{-1} e^{-p^2/4a^2}, a > 0$$

$$= \frac{1}{a\sqrt{2}} e^{-p^2/4a^2} \qquad \text{...(iii)}$$

Let us assume, $$a = \frac{1}{\sqrt{2}} e^{-i\pi/4}$$

$$= \frac{1-i}{2}$$

Thus, $$\frac{1}{a\sqrt{2}} = \frac{1+i}{2}$$

$$a^2 = -\frac{1}{2}i$$

Then, $$F\{e^{ix^2/2}\} = \frac{1}{\sqrt{2}}(1+i)e^{ip^2/2}$$

Now, $$F\{e^{ix^2/2}\} = \frac{1}{\sqrt{2\pi}} \int_{-\infty}^{\infty} e^{ipx} \cdot e^{ix^2/2}\, dx$$

$$= \frac{1}{\sqrt{2\pi}} \int_{-\infty}^{\infty} e^{i\left(\frac{x^2}{2}+px\right)} dx$$

$$= \frac{1}{\sqrt{2\pi}} \left[\int_{-\infty}^{0} e^{i\left(\frac{x^2}{2}+px\right)} dx + \int_{0}^{\infty} e^{i\left(\frac{x^2}{2}+px\right)} dx \right]$$

$$= \frac{1}{\sqrt{2\pi}} \left[\int_{0}^{\infty} e^{i\left(\frac{x^2}{2}-px\right)} dx + \int_{0}^{\infty} e^{i\left(\frac{x^2}{2}+px\right)} dx \right]$$

$$= \frac{1}{\sqrt{2\pi}} \left[\int_{0}^{\infty} e^{ix^2/2}(e^{ipx} + e^{-ipx})\, dx \right.$$

$$= \frac{1}{\sqrt{2\pi}} \left[\int_{0}^{\infty} 2e^{ix^2/2} \cos px\, dx \right]$$

$$= \sqrt{\frac{2}{\pi}} \int_{0}^{\infty} e^{ix^2/2} \cos px\, dx$$

$$= \sqrt{\frac{2}{\pi}} \int_{-\infty}^{\infty} (\cos^2 x/2 + i \sin^2 x/2) \cos px\, dx$$

$\Rightarrow$ We know that

$$F\{e^{ix^2/2}\} = F_c \{\cos^2 x/2\} + i\, F_c \{\sin^2 x/2\}$$

and $\frac{1}{\sqrt{2}}(1+i)e^{-ip^2/2} = \frac{1}{\sqrt{2}} (1 + i) [\cos^2 p/2 - i \sin^2 p/2]$

$$= \frac{1}{\sqrt{2}} [(\cos^2 p/2 + \sin^2 p/2) + i (\cos^2 p/2 - \sin^2 p/2)]$$

Separating real and imaginary parts, we get

$$F_c\{\cos^2 x/2\} = \frac{1}{\sqrt{2}} [\cos^2 p/2 + \sin^2 p/2]$$

$$F_c\{\sin^2 x/2\} = \frac{1}{\sqrt{2}} [\cos^2 p/2 + \sin^2 p/2].$$

Example 11 Find Fourier sine transform of

$$f(x) = x(a^2 - x^2)^{\nu-3/2}\, H(a - x),\ \nu = 1/2,\ a < 0$$

where, $H(a - x)$ is Heaviside's unit step function.

Solution: We know that

$$F_s\{f(x)\} = \sqrt{\frac{2}{\pi}} \int_0^\infty f(x) \sin px\, dx \qquad ...(i)$$

$$\Rightarrow \quad F_s\{x(a^2 - x^2)^{\nu-3/2}\, H(a - x)\}$$

$$= \sqrt{\frac{2}{\pi}} \int_0^\infty x(a^2 - x^2)^{\nu-3/2}\, H(a - x) \sin px\, dx$$

$$= \sqrt{\frac{2}{\pi}} \int_0^a x(a^2 - x^2)^{\nu-3/2} \sin px\, dx \qquad ...(ii)$$

and $$\sin px = \sum_{k=0}^{\infty} \frac{(-1)^k (px/2)^{2k+1}}{k!\,\Gamma(k+3/2)}$$

Then, equation (ii) becomes

$$F_s\{x(a^2 - x^2)^{\nu-3/2}\, H(a - x)\}$$

$$= 2^{1/2} \int_0^a x(a^2 - x^2)^{\nu-3/2} \left[\sum_{k=0}^{\infty} \frac{(-1)^k (px/2)^{2k+1}}{k!\,\Gamma(k+3/2)} \right] dx$$

$$= 2^{1/2} \sum_{k=0}^{\infty} \frac{(-1)^k (p/2)^{2k+1}}{k!\,\Gamma(k+3/2)} \int_0^a x^{2k+2} (a^2 - x^2)^{\nu-3/2}\, dx$$

On putting $x = a \sin t$

$$\Rightarrow \qquad dx = a \cos t\, dt$$

$$= 2^{1/2} \sum_{k=0}^{\infty} \frac{(-1)^k (p/2)^{2k+1}}{k!\,\Gamma(k+3/2)} a^{2\nu+2k} \int_0^{\pi/2} \sin^{2(k+1)} t \cos^{2\nu-2} t\, dt$$

$$= 2^{1/2} \sum_{k=0}^{\infty} \frac{(-1)^k (p/2)^{2k+1}}{k!\,\Gamma(k+3/2)} a^{2(\nu+k)} \cdot X \frac{\Gamma(\nu-1/2)\,\Gamma(k+3/2)}{r\Gamma(\nu+k+1)}$$

$$= \frac{1}{\sqrt{2}} \Gamma \nu - y_2 \sum_{k=0}^{\infty} \frac{(-1)^k (ap/2)^{\nu+2k}}{k!\,\Gamma(\nu+k+1)} a^\nu \cdot p^{1-\nu} \cdot 2^{\nu-1}$$

$$= 2^{\nu-3/2}\, \Gamma(\nu - y_2) a^\nu p^{1-\nu} \sum_{k=0}^{\infty} \frac{(-1)^k}{k!\,\Gamma(k+\nu+1)} \cdot (pa/2)^{\nu+2k}$$

$$= 2^{v-3/2}\Gamma(v-y_2)a^v p^{1-v}\sum_{k=0}^{\infty}\frac{(-1)^k}{k!\Gamma(k+v+1)}\cdot(pa/2)^{v+2k}$$

$$= av\, p^{1-v}\, 2^{v-3/2}\, J_v\,(ap)$$

8.2 FINITE FOURIER TRANSFORM

Finite Cosine Fourier transform: Let $f(x)$ be a function of variable x which is sectionally continuous in some finite interval $(0, m)$. The finite Fourier cosine transform of $f(x)$ in the given interval, denoted by $\phi_c(p)$, is defined as

$$\phi_c(p) = \int_0^m f(x)\cos\frac{\pi px}{m}dx,\ p \in x \qquad ...(i)$$

Further, let us assume a substitution such that end point of limit is changed to 0 and π, then,

$$\phi_c(p) = \int_0^{\pi} f(x)\cos px\, dx$$

In this case, $f(x)$ is said to be inverse finite Fourier cosine transform of $\phi_c(p)$, in another words

$$f(x) = \phi_c^{-1}\{\phi_c(p)\}$$

Then, $$f(x) = \frac{1}{m}\int_0^m f(x)dx + \frac{2}{m}\sum_{p=1}^{\infty}\phi_c(p)\cos\frac{\pi px}{m}$$

Let us assume

$$\int_0^m f(x)dx = \phi_c(0)$$

Then, $$f(x) = \frac{1}{m}\phi_c(0) + \frac{2}{m}\sum_{p=1}^{\infty}\phi_c(p)\cos\frac{\pi px}{m}$$

Further assume the substitution, such that upper limit of the integration be π, then we have

$$f(x) = \frac{1}{\pi}\int_0^{\pi} f(x)dx + \frac{2}{\pi}\sum_{p=1}^{\infty}\phi_c(p)\cos px$$

If $$\int_0^{\pi} f(x)dx = \phi_c(0)$$

Then, $$f(x) = \frac{1}{\pi}\phi_c(p) + \frac{2}{\pi}\sum_{p=1}^{\infty}\phi_c(p)\cos px\, dx$$

Finite Fourier Sine Transform: Let $f(x)$ be a function of variable x, which is sectionally continuous in some finite interval $(0, m)$. The finite Fourier sine transform of $f(x)$ in the given interval, denoted by $\phi_s(p)$, is defined as

$$\phi_s(p) = \int_0^m f(x)\sin\frac{\pi px}{m}\,dx\,,\; p \in z \qquad ...(i)$$

Further, let us assume a substitution such that end point of limit is changed to 0 and π, then

$$\phi_s(p) = \int_0^\pi f(x)\,\sin px\; dx$$

In this case, $f(x)$ is said to be inverse finite sine transform of $\phi_s(p)$, in another words

$$f(x) = \phi_s^{-1}[\phi_s(p)]$$

Then,
$$f(x) = \frac{2}{m}\sum_{p=1}^{\infty}\phi_s(p)\sin\frac{\pi px}{m}$$

and for the interval $(0, \pi)$, we have

$$f(x) = \frac{2}{\pi}\sum_{p=1}^{\infty}\phi_s(p)\sin px\,.$$

Convolution: Let $f_1(x)$ and $f_2(x)$ be two functions defined on the interval $-2\pi < x < 2\pi$, then

$$f_1(x) * f_2(x) = \int_{-\pi}^{\pi} f_1(x-y)\,f_2(y)\,dy$$

is said to be convolution of the given functions, $f_1(x)$ and $f_2(y)$.

Example 12 Evaluate $f(x)$ if its finite sine transform is given by

$$\phi_s(p) = \frac{1-\cos px}{\pi^2 p^2}, 0 < x < \pi.$$

Solution: We know that

$$f(x) = \frac{2}{\pi}\sum_{p=1}^{\infty}\phi_s(p)\sin px$$

$$= \frac{2}{\pi}\sum_{p=1}^{\infty}\frac{1-\cos px}{\pi^2 p^2}\sin px$$

$$= \frac{2}{\pi^3}\sum_{p=1}^{\infty}\frac{1}{p^3}(1-\cos px)\sin px.$$

Example 13 Evaluate $f(x)$ if its finite Fourier sine transform is

$$\phi_s(p) = \frac{(-1)^{p-1}2\pi}{p^3} \quad p = 1, 2, 3, \ldots \text{ and } 2 < x < \pi.$$

Solution: We know that

$$f(x) = \frac{2}{m}\sum_{p=1}^{\infty}\phi_s(p)\sin\frac{\pi px}{m}$$

$$= \frac{2}{\pi} \sum_{p=1}^{\infty} \frac{(-1)^{p-1} 2\pi}{p^3} \sin px$$

$$= 4 \sum_{p=1}^{\infty} \frac{(-1)^{p-1}}{p^3} \sin px .$$

Example 14 Prove that

$$\phi_c\{3x^2\} = \begin{cases} \dfrac{6\pi(-1)^p}{p^2} & , \quad p = 1, 2, ... \\ \pi^2 & , \quad p = 0 \end{cases}$$

Solution:

$$\phi_c\{p\} = \int_0^{\pi} f(x) \cos px \, dx$$

$$= \int_0^{\pi} 3x^2 \cos px \, dx$$

For $p = 0$, we have

$$= \int_0^{\pi} 3x^2 \, dx$$

$$= \left[x^3 \right]_p^{\pi} = \pi^3$$

Now, for $p = 1, 2, ...$

$$\phi_c(p) = \int_0^{\pi} 3x^2 \cos px \, dx$$

$$= 3 \int_0^{\pi} x^2 \cos px \, dx$$

$$= 3\left[\left\{ \frac{x^2 \sin px}{p} \right\}_0^{\pi} - 2\int_0^{\pi} x \frac{\sin px \, dx}{p} \right] \qquad \text{(On integration by parts)}$$

$$= -\frac{6}{p} \int_0^{\pi} x \sin px \, dx$$

$$= -\frac{6}{p} \left[-\frac{x}{p} \cos px + \frac{\sin px}{p^2} \right]_0^{\pi}$$

$$= \frac{6\pi}{p^2} \cos px = \frac{6\pi}{p^2} (-1)^n$$

Example 15 Evaluate the finite Fourier sine transform of

(i) e^{ax} (ii) cos ax

(iii) x^3 (iv) sin ax

Solution: We know that

$$\phi_s(p) = \int_0^\pi f(x)\sin px\, dx \qquad ...(i)$$

(i) If $f(x) = e^{ax}$, then using equation (i), we have

$$\phi_s(p) = \int_0^\pi e^{ax} \sin px\, dx$$

$$= \frac{1}{a^2+p^2}\left[e^{ax}(a\sin px - p\cos px)\right]_0^\pi$$

$$= \frac{1}{(a^2+p^2)}[e^{a\pi}(a \sin p\pi - p \sin p\pi - e^o(a \sin o - p \cos o)]$$

$$= \frac{1}{a^2+p^2}[\{1 - e^{a\pi}\cos p\pi\}]$$

$$= \frac{p}{a^2+p^2}[1-(-1)^p e^{a\pi}]$$

(ii) If $f(x) = x^3$, using equation (i), we have

$$\phi_s(p) = \int_0^\pi x^2 \sin px\, dx$$

$$= \left[\frac{-x^3\cos px}{p}\right]_0^\pi + \frac{3}{p}\int_0^\pi x^2\cos px\, dx$$

$$= -\frac{\pi^3\cos\pi p}{p} + \frac{3}{p}\left[\left\{\frac{x^2\sin px}{p}\right\}_0^\pi - \frac{2}{p}\int_0^\pi x\sin px\, dx\right]$$

$$= -\frac{\pi^3}{p}\cos\pi p + \frac{6\pi}{p^3}\cos\pi p$$

$$= \frac{\pi}{p}(-1)^p\left[-\pi^2 + \frac{6}{p^2}\right]$$

(iii) If $f(x) = \cos ax$. Using equation (i), we have

$$\phi_s(p) = \int_0^\pi \cos ax \sin px\, dx$$

$$= \frac{1}{2}\int_0^{\pi} 2[\sin(a+p)x + \sin(p-a)x]dx$$

$$= \frac{1}{2}\left[\frac{-\cos(a+p)x}{a+p} - \frac{\cos(p-a)x}{p-a}\right]_0^{\pi}$$

$$= \frac{1}{2}\left[\frac{1}{a+p} - \frac{\cos(a+b)\pi}{a+p} + \frac{1}{b-a} - \frac{\cos(p-a)\pi}{p-a}\right]$$

$$= \frac{1}{2(p^2-a^2)}[p-a+a+p-(p-a)\cos(a+p)\pi-(p+a)\cos(p-a)\pi]$$

$$= -\frac{1}{p^2-a^2}[1-(-1)^p \cos a\pi],\ p > 0$$

(iv) If $f(x) = \sin ax$, then using equation (i), we have

$$= \int_0^{\pi} \sin ax \sin px\, dx$$

$$= \frac{1}{2}\int_0^{\pi} [\cos(a-p)x - \cos(a+p)x]dx$$

$$= \frac{1}{2}\left[\frac{\sin(a-p)x}{a-p} - \frac{\sin(a+p)x}{a+p}\right]$$

$$= \frac{1}{2}[0]$$

$$= 0, \text{ if } a \neq p\ ;\ a, b \in Z$$

Let $a = p$, then

$$\phi_s(p) = \int_0^{\pi} \sin px \sin px\, dx$$

$$= 2\int_0^{\pi} 2\sin^2 px\, dx$$

$$= 4\int_0^{\pi} \sin^2 px\, dx = \pi/2$$

$$\phi_s(p) = 0 \quad , \quad a \neq p$$

$$= \pi/2 \quad , \quad a = p$$

Example 16 Find the finite Fourier sine and cosine transform of the function

$$f(x) = 2x,\ 0 < x < 4.$$

Solution: We know that

$$\phi_s(p) = \int_0^m f(x)\sin\frac{p\pi x}{m}\,dx$$

$$= \int_0^4 2x\cdot\sin\frac{p\pi x}{m}\,dx, \quad m=4$$

$$= \left[\frac{-2x\cos\frac{p\pi x}{4}}{\frac{p\pi}{4}}\right]_0^4 + 2\int_0^4 \frac{\cos(p\pi x)/4}{p\pi/4}\,dx$$

$$= -\frac{32}{p\pi}\cos p\pi + \frac{8}{p\pi}\left(\frac{-2x\cos\frac{p\pi x}{4}}{\frac{p\pi}{4}}\right)_0^4$$

$$= -\frac{32}{p\pi}\cos p\pi$$

and
$$\phi_c(p) = \int_0^4 f(x)\cos\frac{p\pi x}{m}\,dx$$

$$= 2\int_0^4 x\cos\frac{p\pi x}{m}\,dx$$

$$= 2\left[\frac{x\sin\frac{p\pi x}{4}}{p\pi/4}\right]_0^4 - 2\int_0^4 \frac{\sin(p\pi x)/4}{p\pi/4}\,dx$$

$$= \frac{32}{p^2\pi^2}(\cos p\pi - 1),\ p = 0.$$

Example 17 Find the finite sine and cosine transform of $f(x) = 1$.

Solution: We know that

$$\phi_s(p) = \int_0^\pi f(x)\sin px\,dx$$

$$= \int_0^\pi \sin px\,dx$$

$$= \left(\frac{-\cos px}{p}\right)_0^\pi$$

$$= \frac{1-(-1)^p}{p}$$

$$\phi_c(p) = \int_0^{\pi} f(x) \cos px \, dx$$

$$= \int_0^{\pi} \cos px \, dx$$

$$= \left(\frac{\sin px}{p}\right)_0^{\pi}$$

$$\begin{cases} = 0 & , \text{ if } = p = 1, 2, 3,.. \\ = \pi & , \text{ if } = p = 0 \end{cases}$$

As $\qquad \phi_c(p) = \int_0^{\pi} dx = 0.$

Example 18 Evaluate the finite Fourier cosine transform of

(i) $\qquad f(x) = x,$

(ii) $\qquad f(x) = \left(1 - \frac{x}{\pi}\right)^2$, and

(iii) $\qquad f(x) = \frac{\pi}{3} - x + \frac{x^2}{2\pi}$

Solution We know that

$$\phi_c(p) = \int_0^{m} f(x) \cos \frac{p\pi x}{m} dx$$

or $\qquad \phi_c(p) = \int_0^{\pi} f(x) \cos px \, dx$

(i) $f(x) = x$, then

$$\phi_c(p) = \int_0^{\pi} x \cos px \, dx$$

$$= \left[\frac{x \sin px}{p}\right]_0^{\pi} - \frac{1}{p}\int_0^{\pi} \sin px \, dx$$

$$= \left[\frac{\cos px}{p^2}\right]_0^{\pi}$$

$$= \frac{(-1)^p - 1}{p^2}, \; p = 1, 2, 3, \ldots$$

(ii) If $$f(x) = \left(1 - \frac{x}{\pi}\right)^2$$

Then, $$\phi_c(p) = \int_0^{\pi} \left(1 - \frac{x}{\pi}\right)^2 \cos px \, dx$$

$$= \left[\left(1 - \frac{x}{\pi}\right)^2 \cdot \frac{\sin px}{p}\right]_0^{\pi} + \frac{2}{p\pi} \int_0^{\pi} \left(1 - \frac{x}{\pi}\right) \sin px \, dx$$

$$= \frac{2}{\pi p}\left[\left(1 - \frac{x}{\pi}\right)^2 \frac{\cos px}{p}\right]_0^{\pi} - \frac{2}{p^2 \pi^2} \int_0^{\pi} \cos px \, dx$$

$$= \frac{2}{\pi p^2}, \; p > 0$$

If $$p = 0$$

$$\phi_c(p) = \int_0^{\pi} \left(1 - \frac{x}{\pi}\right)^2 dx = \frac{\pi}{3}$$

(iii) If $$f(x) = \frac{\pi}{3} - x + \frac{x^2}{2\pi}$$

Then, $$\phi_c(p) = \int_0^{\pi} \left(\frac{\pi}{3} - x - \frac{x^2}{2\pi}\right) \cos px \, dx$$

$$= \left[\frac{1}{p}\left(\frac{\pi}{3} - x - \frac{x^2}{2\pi}\right) \sin px\right]_0^{\pi} - \frac{1}{p} \int_0^{\pi} \left(\frac{x}{\pi} - 1\right) \sin px \, dx$$

$$= -\frac{1}{p}\left[\left(\frac{x}{\pi} - 1\right) \frac{\cos px}{p}\right]_0^{\pi} - \frac{1}{p^2} \int_0^{\pi} \cos px \, dx$$

$$= \frac{1}{p^2} - \frac{1}{\pi p^2} (\sin px)_0^{\pi}$$

$$= \frac{1}{p^2}, \; p > 0$$

$$\phi_c(p) = \int_0^{\pi} \left(\frac{\pi}{3} - x + \frac{x^2}{2\pi}\right) dx$$

$$= 0$$

Parseval's Identity for Fourier Transform (Rayleigh's Theorem):

(i) $$\frac{1}{2\pi}\int_{-\infty}^{\infty} f_1(p)\bar{f}_2(p)dp = \int_{-\infty}^{\infty} f_1(x)\bar{f}_2(x)dx$$

(ii) $$\int_{-\infty}^{\infty} |f(p)|^2 dp = \int_{-\infty}^{\infty} |f(x)|^2 dx$$

Reimann – Lebesgue Theorem: Let us assume a function $f(x)$ which is uniformly converges at every point of the interval $-m < x < m$, and if the Fourier series corresponding to $f(x)$ converges to $f(x)$, then

$$\lim_{n\to\infty} a_n = 0, \quad \lim_{n\to\infty} b_n = 0$$

where, a_n and b_n are Fourier coefficients.

Finite Fourier Cosine Transform of Derivatives: (1) If $f(x)$ and $f'(x)$ are continuous and if $f''(x)$ is sectionally continuous, the finite Fourier cosine transformation resolve $f''(x)$ into an algebraic form in $\phi_c(p)$ and boundary values $f'(0)$ and $f'(\pi)$,

$$\phi_c[f''(x)] = -p^2\,\phi_c(p) - f'(0) + (-1)^p f'(\pi)$$

Proof: (1) We know that

$$\begin{aligned}
\phi_c\{f''(x)\} &= \int_0^{\pi} f''(x)\cos px\,dx \\
&= \int_0^{\pi} \cos px\, f''(x)\,dx \\
&= \left[\cos px\, f'(x)\right]_0^{\pi} + p\int_0^{\pi} f'(x)\sin px\,dx \\
&= f'(\pi)\cos p\pi - f'(0) + p\left[f(x)\sin px\right]_0^{\pi} - p^2\int_0^{\pi} f(x)\cos px\,dx \\
&= (-1)^p f'(\pi) - f'(0) - p^2\,\phi_c(p)
\end{aligned}$$

$$\Rightarrow \quad \phi_c\{f''(x)\} = -p^2\,\phi_c(p) - f'(0) + (-1)^p f'(\pi)$$

Similarly, we can find

$$\phi_c\{f^{\text{iv}}(x)\} = p^4\,\phi_c(p) + p^2[f'(0) - (-1)^p f'(\pi)] - f'''(0) + (-1)^p f'''(\pi)$$

(2) If $\phi_c(p)$ is the cosine transform of sectionally continuous function $f(x)$, $0 \le x \le \pi$, then

$$\phi_c^{-1}\left\{\frac{\phi_c(p)}{p^2}\right\} = \int_0^{x}\int_t^{\pi} f(r)\,dr\,dt + \frac{\phi_c(0)}{2\pi}(x-\pi)^2 + c_1$$

where, c_1 is an arbitrary constant.

Finite Fourier Sine Transform of Derivatives. 1. The Finite Fourier sine transformation resolves the differential form $f''(x)$ into a linear algebraic form in the transform and the boundary values $f(0)$ and $f(\pi)$ in the form

$$\phi_s[f''(x)] = -p^2\,\phi_s(p) - p[f(0) - (-1)^p f(\pi)]$$

where, $f(x)$ and $f'(x)$ are continuous and $f''(x)$ is sectionally continuous in the interval $[0, \pi]$.

Proof: 1. We know that

$$\phi_s\{f''(x)\} = \int_0^{\pi} f''(x)\sin px\,dx$$

$$= \left[\left[(\sin px)f'(x)\right]_0^{\pi} + p\int_0^{\pi} f(x)\cos px\,dx\right]$$

$$= 0 - \left(p\,f(x)\cos px\right)_0^{\pi} - p^2\int_0^{p} f(x)\sin px\,dx$$

$$-p\,f(\pi)\cos p\pi + p\,f(0) - p^2\,\phi_s(p)$$

$$\Rightarrow \quad \phi_s\{f''(x)\} = -p^2\,\phi_s(p) + p[f(0) - (-1)^p f(\pi)]$$

2. If $\phi_s(p)$ is the sine transform of a sectionally continuous function $f(x)$, $x \in \{0, \pi\}$ then

$$\phi_s^{-1}\left\{\frac{\phi_s(p)}{p^2}\right\} = \frac{2}{\pi}\int_0^{\pi}(\pi - r)f(r)dr - \int_0^{x}(x - r)f(r)dr$$

Combined Properties of Finite Fourier Sine and Cosine Transform: Let $f(x)$ be a continuous and $f'(x)$ is sectionally continuous function, then

(a) $\phi_s\{f'(x)\} = -p\,\phi_c\{f(x)\}$, $p = 1, 2, \ldots$

(b) $\phi_c\{f'(x)\} = p\,\phi_s\{f(x) - f(0) + (-1)^p f(\pi)\}$, $p = 0, 1, 2, \ldots$

Solution:

(a) We know that

$$\phi_s\{f'(x)\} = \int_0^{\pi} f'(x)\sin px\,dx$$

$$= \left[\sin px\,f(x)\right]_0^{\pi} - p\int_0^{\pi} f(x)\cos px\,dx$$

$$= -p\,\phi_c\{f(x)\},\ p = 1, 2, \ldots$$

(b) We know that

$$\phi_c\{f(x)\} = \int_0^{\pi} f'(x)\cos px\,dx$$

$$= \left[\cos px\, f(x)\right]_0^{\pi} - p\int_0^{\pi} \sin px\, f(x)dx$$

$$= \cos p\pi\, f(\pi) - f(0) + p\, \phi_s(p)$$

$$= (-1)^p f(\pi) - f(0) + p\, \phi_s(p)$$

Example 19 Find the finite Fourier transform of the function

$$f(x) = \begin{matrix} -x & , & x < c \\ x - c & , & x > c \end{matrix}$$

where $0 \le c \le \pi$.

Solution: We know that

Finite Fourier sine transform of function $f(x)$ is

$$\phi_s(p) = \int_0^{\pi} f(x) \sin px\, dx$$

$$= \int_0^{c} -x \sin px\, dx + \int_c^{\pi} (\pi - x) \sin px\, dx$$

$$= \left[\frac{+x\cos px}{p}\right]_0^c - \frac{1}{p}\int_0^c \cos px\,dx - \left[\frac{(\pi - x)\cos px}{p}\right]_c^{\pi} - \frac{1}{p}\int_c^{\pi} \cos px\,dx$$

$$= \frac{c\cos pc}{p} - \frac{1}{p^2}\left[\sin px\right]_0^c + \frac{(\pi - x)\cos pc}{p} - \frac{1}{p^2}\left[\sin px\right]_c^{\pi}$$

$$= \frac{c\cos pc}{p} - \frac{1}{p^2}\sin pc + \frac{(\pi - c)\cos pc}{p} + \frac{1}{p^2}\sin pc$$

$$= \frac{\pi}{p}\cos pc\,.$$

8.2.1 Initial and Boundary Value Problems

Fourier Transforms of Partial Derivatives

$$F_c\left\{\frac{\partial^2 U}{\partial x^2}\right\} = \sqrt{\frac{2}{\pi}}\int_0^{\infty} \frac{\partial^2 U}{\partial x^2} \cos px\, dx$$

$$= \sqrt{\frac{2}{\pi}}\left[\left(\cos px\frac{\partial U}{\partial x}\right)_0^{\infty} + p\int_0^{\infty} \sin px\frac{\partial U}{\partial x}dx\right]$$

$$= \sqrt{\frac{2}{\pi}}\left[-\left(\frac{\partial U}{\partial x}\right)_{x=0} + p\int_0^{\infty} \sin px\frac{\partial U}{\partial x}dx\right]$$

$$\left[\text{Therefore, } \frac{\partial U}{\partial x} \to 0 \text{ as } x \to \infty\right]$$

$$= \sqrt{\frac{2}{\pi}}\left[-\left(\frac{\partial U}{\partial x}\right)_{x=0} + p\left[(U \sin px)_0^\infty - p\int_0^\infty U \cos px\, dx\right]\right]$$

$$= \sqrt{\frac{2}{\pi}}\left[-\left(\frac{\partial U}{\partial x}\right)_{x=0} - p^2 \int_0^\infty U(x,t) \cos px\, dx\right]$$

Suppose $U(x, t) \to 0$ as $\sin x \to \infty$

$$\Rightarrow \qquad F_c\left\{\frac{\partial^2 U}{\partial x^2}\right\} = -\sqrt{\frac{2}{\pi}}\left[\left(\frac{\partial U}{\partial x}\right)_{x=0} - p^2 u_c(p,t)\right]$$

$$\left[u_c(p,t) = \int_0^\infty U(x,t) \cos px\, dx\right]$$

and

$$F_s\left\{\frac{\partial^2 U}{\partial x^2}\right\} = \sqrt{\frac{2}{\pi}} \int_0^\infty \frac{\partial^2 U}{\partial x^2} \sin px\, dx$$

$$= \sqrt{\frac{2}{\pi}} \int_0^\infty \sin px \cdot \frac{\partial^2 U}{\partial x^2} dx$$

$$= \sqrt{\frac{2}{\pi}}\left[\sin px \cdot \frac{\partial U}{\partial x} - p \int_0^\infty \cos px\, x \frac{\partial U}{\partial x} dx\right]$$

$$= -p\sqrt{\frac{2}{\pi}} \int_0^\infty \frac{\partial U}{\partial x} \cos px\, dx$$

$$\left[\because \frac{\partial U}{\partial x} \to 0, x \to \infty\right]$$

$$= -p\sqrt{\frac{2}{\pi}}\left[[U \cos px]_0^\infty + p\int_0^\infty U \sin px\, dx\right]$$

$$\Rightarrow \qquad F_s\left(\frac{\partial^2 U}{\partial x^2}\right) = \sqrt{\frac{2}{\pi}}\, [p\; U(0, t) - p^2 u_s(p, t)] \quad \left[\text{Taking, } u_s(p,t) = \int_0^\infty U \sin px\, dx\right]$$

To eliminate the term $\frac{\partial^2 U}{\partial x^2}$ from a partial differential equation, we have to do.

(i) $U(0, t)$ or $U(x, t)$ at $x = 0$ in sine transform, and

(ii) $U_x(0, t)$ or $\dfrac{\partial U}{\partial x}$ at $x = 0$ is cosine transform.

We are giving some examples to apply these results.

Example 20 The temperature U in the semi-infinite rod $x \in [0, \infty)$ is defined as

$$\frac{\partial U}{\partial x} = k\,\frac{\partial^2 U}{\partial x^2}$$

Such that

(i) $\dfrac{\partial U}{\partial x} = -\,\mu$ (a constant) when $x = 0,\ t > 0$

(ii) $U = 0$ at $t = 0,\ x \geq 0$

Then, prove that by applying cosine transform,

$$U(x, t) = \frac{2\mu}{\pi}\int_0^\infty \frac{1}{p^2}\cos px\left(1 - e^{-kp^2 t}\right)dp$$

Solution: Given that

$$\frac{\partial U}{\partial x} = k\,\frac{\partial^2 U}{\partial x^2} \qquad \text{...(i)}$$

Applying Fourier cosine transform both sides to the equation (i), we get

$$\sqrt{\frac{2}{\pi}}\int_0^\infty \cos px\frac{\partial U}{\partial t}dx = k\cdot\sqrt{\frac{2}{\pi}}\int_0^\infty \cos px\frac{\partial^2 U}{\partial x^2}dx$$

$$\sqrt{\frac{2}{\pi}}\frac{d}{dt}\int_0^\infty U\cos px\,dx = k\sqrt{\frac{2}{\pi}}\left[\cos px\frac{\partial U}{\partial x}\right]_0^\infty + kp\sqrt{\frac{2}{\pi}}\int_0^\infty \frac{\partial U}{\partial x}\sin px\,dx$$

If $\dfrac{\partial U}{\partial x} \to 0$ as $x \to \infty$

$$\Rightarrow \quad \frac{d}{dt}[F_c\{U\}] = -k\sqrt{\frac{2}{\pi}}\cdot\left(\frac{\partial U}{\partial x}\right)_{x=0} + kp\sqrt{\frac{2}{\pi}}\,[U\sin px]_0^\infty - kp^2\sqrt{\frac{2}{\pi}}\int_0^\infty U\cos px\,dx$$

Let $F_c(U) = V$ and if $U \to 0$ as $x \to \infty$

Then,
$$\frac{dV}{dt} = \sqrt{\frac{2}{\pi}}\cdot k\mu - kp^2 V$$

$$\Rightarrow \quad \frac{dV}{dt} + kp^2 V = k\mu\sqrt{\frac{2}{\pi}} \qquad \text{...(ii)}$$

Equation (ii) is a first order linear differential equation, then solution

$$\text{I·F·} = e^{\int skp^2 dt}$$
$$= e^{kp^2 t}$$

Thus, the solution

$$V \cdot e^{kp^2 t} = k\mu\sqrt{\frac{2}{\pi}}\int e^{kp^2 t}dt$$

$$= \frac{\mu}{p^2}\sqrt{\frac{2}{\pi}}\; e^{kp^2 t} + c_1$$

Since, at $t = 0$, $V = 0$

Then,
$$0 = \sqrt{\frac{2}{\pi}} \cdot \frac{\mu}{p^2} + c_1$$

$$\Rightarrow \qquad c_1 = -\frac{\mu}{p^2}\sqrt{\frac{2}{\pi}}$$

Then,
$$V \cdot e^{kp^2 t} = \frac{\mu}{p^2}\left(e^{kp^2 t} - 1\right)\sqrt{\frac{2}{\pi}}$$

$$\Rightarrow \qquad V = \frac{\mu}{p^2} \cdot \sqrt{\frac{2}{\pi}}\left(1 - e^{kp^2 t}\right) \qquad \text{...(iii)}$$

Taking inverse Fourier cosine transform of the equation (iii), we get

$$U_c = \sqrt{\frac{2}{\pi}}\int_0^\infty V \cos px \, dp$$

$$\Rightarrow \qquad U_c = \frac{2\mu}{\pi}\int_0^\infty \frac{\cos px}{p^2}\left(1 - e^{-kp^2 t}\right)dp$$

8.3 DISCRETE WAVELET TRANSFORM

Discrete wavelet transform can be defined as the continuous wavelets with the discrete scale and translation factors. The wavelet transform is then evaluated at discrete scale and translation. Consider the function $f(t)$ with the Fourier transform given by

$$F(w) = \begin{cases} 1, & 2 \le |w| \le 3 \\ 0, & \text{Otherwise} \end{cases} \qquad \text{...(i)}$$

Let, we find the continuous wavelet transform (CWT) of $f(t)$ using wavelet $\varphi(t)$ with the Fourier transform

$$\psi(w) = \begin{cases} 1, & 2 \le |w| \le 4 \\ 0, & \text{Otherwise} \end{cases} \qquad \text{...(ii)}$$

Suppose $W(a, b)$ is obtained CWT and $W_a(w)$ its Fourier transform in the variable b, then

$$W_a(w) = \int_{b=-\infty}^{\infty} W(a, b)\, e^{-jwb}\, db \qquad \text{...(iii)}$$

But, we have

$$F[\psi_{(a,0)}(t)] = \sqrt{|a|}\ \psi(a, w) \qquad \text{...(iv)}$$

Then, $$W_o(w) = \sqrt{|a|}\ F(w)\ \varphi(a, w) \qquad \text{...(v)}$$

Using equation (v), region of $W(a, b)$ can be defined. Then,

$$W_a(w) = \sqrt{|a|}\, F(w)\ \frac{1}{2} \le |a| \le \frac{4}{3} \qquad \text{...(vi)}$$

Now, from equation (vi) region defined by $F(w)$ can be obtained.

Here, non-redundant wavelet formulation can be defined as

$$f(t) = \sum_{k=-\infty}^{\infty} \sum_{t=-\infty}^{\infty} d(k, l) 2^{-k/2}\, \psi(2^{-k}t - l)$$

which, involves a continuum of dilation and translation.

dilation $a = 2^k$, k is an integer.

Translation (corresponding to 2^k) = $2^k l$. l is an integer.

In this formation the values $d(k, l)$ are corresponding to the values of wavelet transform

$$W(a, b) = W(f(t))$$

where, $a = 2^k$ and $b = 2^k l$. This is related to coordinates (a, b) on a graph shown in Fig. 8.1

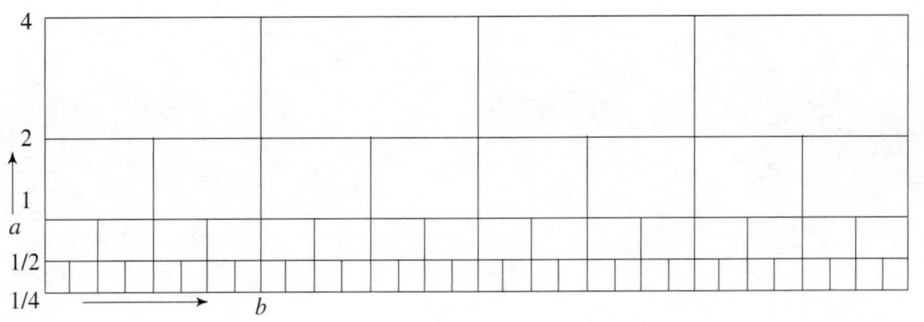

Fig. 8.1

This method is known as dyadic sampling. Here, consecutive values on discrete scales as well as the related sampling intervals differ by a factor of 2. $d(k, l)$, a two-dimensional sequences is commonly referred to discrete wavelet transform (DWT) of $f(t)$.

Wavelet

A wavelet can be defined as wave-like oscillation that is localised in time. Thus, it has two basic properties, i.e., scale and location. Scale or dilation defines how stretched or squished a wavelet is. This property is corresponding to frequency as defined for waves while, the location corresponding to the position of wavelet time.

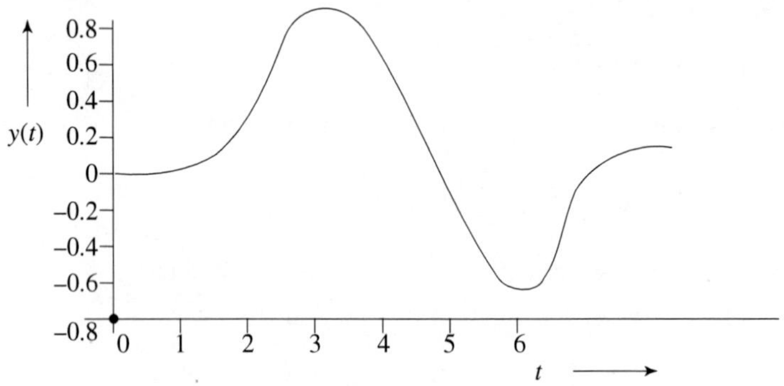

Fig. 8.2 First derivation of Gaussian function

First derivation of the Gaussian function

$$-(x-b)\,e^{\frac{-(x-b)^2/(2a^2)}{\sqrt{2\pi a^3}}}$$

The parameter a in the above formulation sets the scale of the wavelet. On increasing the value the wavelet look stretched while for smaller value, it is squished. A wavelet capture low frequency information in stretched form which is the result of increasing of a.

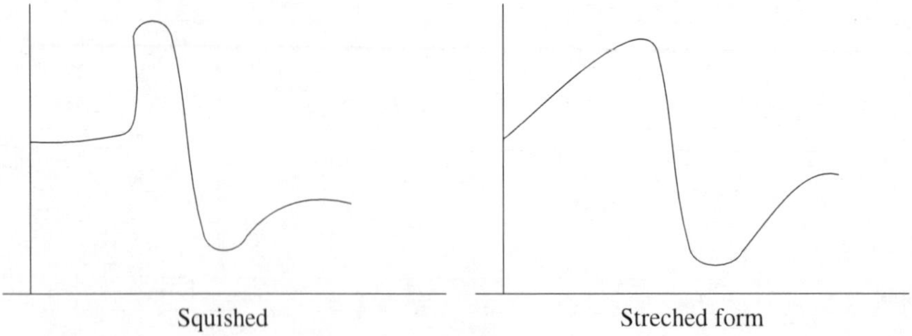

Fig. 8.3 First derivation of Gaussian function

8.4 CONTINUOUS WAVELET TRANSFORMS

Grossman and Morlet (1984) introduced wavelet transform which is an important tool in the field of science and technology. It is very important in local analysis of non-stationary and fast transient wide band signals. The wavelet transform is mapping between time signal and time scale joint representation which is used in short time Fourier transform, the Wigner distribution and in the ambiguity function. It also provides multiresolution analysis with dilated windows. It is also noted here that, the lower frequency analysis is done by using wider windows while higher frequency analysis is done by using narrower windows. Thus, in other words, the wavelet transform is a stant Q analysis.

Wavelets are basic function of wavelet transform which are generated from a basic wavelet function by translations and dilation. They satisfy an admissible condition so that the original signal may be reconstructed by using the inverse transform methods. Here, wavelets satisfy also the regularity condition such that wavelet coefficients decreases fast with the decreasing of the scale.

The wavelet transform is very important formulative resolution local spectrum analysis of non-stationary signals like seismic, radar, sonar, electrocardiographic, image processing and pattern recognition, etc. In this chapter, a brief introduction of wavelet transform introduced.

8.4.1 Continuous-Time Wavelets

For a real or complex-value continuous time function $f(t)$, if

$$\int_{-\infty}^{\infty} f(t)\, dt = 0 \qquad \text{...(i)}$$

and

$$\int_{-\infty}^{\infty} |f(t)|^2\, dt < 0 \qquad \text{...(ii)}$$

Then, the function $f(t)$ is known as mother wavelet or wavelet if $f(t)$ satisfies admissibility condition along with both the above properties.

Property (i) is suggestive of a function that is wavy or oscillatory in appearance while A function does not necessarily have to oscillate to satisfy this property. Property (ii) implies that most of the energy in $f(t)$ is confined to a finite duration. Thus, for a squave integrable function $\phi(t)$, the continuous-time wavelet transform of $\phi(t)$ with respect to a wavelet $\phi(t)$ is defined as

$$W(a, b) = \int_{-\infty}^{\infty} \phi(t) \frac{1}{\sqrt{a}} f^*(t - b/a)dt \qquad \text{...(iii)}$$

where, a and b are real numbers and $*$ denotes complex conjugate. Therefore, the wavelet transform is a function of two variables over $L^2(R)$, where $L^2(R)$ is a set of energy signals or can be termed set of square integrable functions.

Let us define

$$f_{a,b}(t) \equiv \frac{1}{\sqrt{a}} f\left(\frac{t-b}{a}\right) \qquad \text{...(iv)}$$

Using equations (iii) and (iv), we have

$$W(a, b) = \int_{-\infty}^{\infty} \phi(t) f^*_{a,b}(t)dt \quad \text{...(v)}$$

Thus, $f_{1,0}(t) = f(t)$

Here, $1/\sqrt{|a|}$ is known as normalising factor which keeps energy same for a and b. Thus,

$$\int_{-\infty}^{\infty} |f_{a,b}(t)|^2 dt = \int_{-\infty}^{\infty} |f(t)|^2 dt \quad \text{...(vi)}$$

Now, for a given value of a, the function $f_{a,b}(t)$ is a shift of $f_{a,0}(b)$ by b on the time axis. The variable (b) known as translation or time shift.

Therefore, $$f_{a,0}(t) = \frac{1}{\sqrt{|a|}} f\left(\frac{t}{a}\right) \quad \text{...(vii)}$$

Note: (i) Variable a is known as dilation or scale variable.

(ii) Negative value of variable a results in a time reversal in combination with dilation.

8.4.2 The CWT as a Correlation

We know that the set of square integrable functions forms a linear vector space under the operations addition and scalar multiplication, and also, vector space comes with well-defined inner product denoted by $x(t) \cdot y(t)$.

For any two given energy signals $x(t)$ and $y(t)$, we have

$$<x(t) \cdot y(t)> = \int_{-\infty}^{\infty} x(t)\, y^*(b)dt \quad \text{...(viii)}$$

and the total energy in energy signal, $x(t)$, can be given as $\int_{-\infty}^{\infty} |x(t)|^2\, dt$ ...(ix)

We know that (using Hoftman and Kunze (1971), the square root of energy is known as norm of energy signal $x(t)$ which can be denoted by $\|x(t)\|$. By equation (viii) it can be noted that CWT is a collection of inner products of a signal $\phi(t)$, thus, dilated and translated wavelet $f_{a,b}(t)$ is

$$W(a, b) = <\phi(t), f_{a,b}(t)> \quad \text{...(x)}$$

Cross-correlation interpretations, $R_{x,y}(u)$, of two functions $x(t)$ and $y(t)$ can be defined as

$$\begin{aligned} R_{x,y}(u) &= \int x(t)\, y^*(t-u) \\ &= <x(t), y^*(t-u)> \end{aligned} \quad \text{...(xi)}$$

Here, u is known as shift parameter or lag.

Now, using equations (iv), (viii) and (x), we have

$$\begin{aligned} W(a, b) &= <\phi(t), \phi_{a,0}(t-b)> \\ &= R_{f}, f_{a,0}(b) \end{aligned} \quad \text{...(xii)}$$

Thus, CWT is the cross-correlation at lag b between $\phi(t)$ and the wavelet dilated to scale factor, a. By using Cauchy–Schwarz inequality, we have

$$|W(a, b)|^2 \leq \|\phi(t)\|^2 \, \|f_{a,b}(t)\|^2 \qquad \text{...(xiii)}$$

which implies, that $W(a, b)$ always exists since the function and the wavelet have finite norm.

Inequatity Eq. (xiii) turns into equality if and only if

$$f_{a,b}(t) = a\phi(t)$$

8.5 THE CWT AS AN OPERATOR

The CWT takes a member of the set of square integrable functions of one real variable in $L^2(R)$ and transforms it to a member of the set of functions of two real variables. A mapping operator from $L^2(R)$ to the latter set can be defined as

$$M_f[\phi(t)] = W(a, b) \qquad \text{...(1)}$$

Here, $M_f[\ ]$ stands for "CWT with respect to $f(t)$ of".

Properties

1. *Linear Property*

   $M_f[c_1\phi(t) + c_2 g(t)] = c_1 M_f[\phi(t)] + c_2 M_f[g(t)]$

   For c_1, c_2, $\phi(t)$ and $g(t) \in L^2(R)$
2. *Scaling Property*

   $M_f[\sqrt{c_1}\ \phi(t/c_1)] = M\left(\dfrac{a}{c_1}, \dfrac{b}{c_1}\right)$

   For $c_1 > 0$
3. *Translation Property*

   $M_f[\phi(t - u)] = M(a, b - u)$

Wavelet Scaling Property

Let us assume

$$\hat{f}(t) = f(t - t/c_1)/\sqrt{|c_1|}, \text{ then}$$

$$M_{\hat{f}}[\phi(t)] = W(aa, b)$$

8.5.1 Wavelet Shifting Property

Let us assume

$$\hat{f}(t) = f(t - u), \text{ then}$$

$$M_{\hat{f}}[\phi(t)] = W(a, b + au)$$

Note: For any two wavelets, linear combination of wavelets is also a wavelet. It can be generalised for a finite number of wavelets.

Inverse CWT

If the mother wavelet satisfies the sufficient conditions

$$K \equiv \int_{-\infty}^{\infty} \frac{|f(w)|^2}{|w|} dw \qquad ...(i)$$

and $0 < K < \infty$, then

$$\phi(t) = \frac{1}{K} \int_{a=-\infty}^{\infty} \int_{b=-\infty}^{\infty} \frac{1}{|a|} W(a, b) f_{a,b}(t)\, da\, db \qquad ...(ii)$$

Then the equation (i) is called admissibility condition on $f(t)$.

EXERCISE

1. Prove that the mexican hat wavelet satisfies the admissibility condition.
2. Prove that if

$$W(a, b) = M_f[\phi(t)]$$

for a wavelet that safisfies the admissibility conditions then

$$\|\phi(t)\|^2 = \frac{1}{K} \int_{-\infty}^{\infty} \int_{-\infty}^{\infty} \frac{|W(a,b)|^2}{|a|^2} da\, db$$

Chapter 9

The Mellin Transform

9.1 THE MELLIN TRANSFORM

Let $f(t)$ be a function defined for $0 < t < \infty$, then the Mellin transform of $f(t)$ is denoted by $M\{f(t); p\}$ or $F(p)$ is defined by

$$M\{f(t); p\} = F(p) = \int_0^\infty t^{p-1} f(t)\, dt$$

Here t^{p-1} is known to be the kernel of the Mellin transform and M stands for Mellin transform operator.

9.1.1 Some Elementary Properties of the Mellin Transform

1. Linearity Property: If $f(t)$ and $g(t)$ be two functions then

$$M\{c_1 f(t) \pm c_2 g(t); p\} = c_1 M\{f(t); p\} \pm c_2 M\{g(t); p\}$$

where, c_1 and c_2 are constants.

Solution: We have

$$M\{c_1 f(t) \pm c_2\, g(t); p\} = \int_0^\infty t^{p-1} \{c_1 f(t) \pm c_2 g(t)\} dt$$

$$= c_1 \int_0^\infty t^{p-1} f(t)\, dt \pm c_2 \int_0^\infty t^{p-1} g(t)\, dt$$

$$= c_1\, M\{f(t); p\} \pm c_2\, M\{g(t); p\}$$

2. Change of Scale Property

If $\quad M\{f(t); p\} = F(p)$, then

$$M\{f(at); p\} = a^{-p} M\{f(t); p\} = a^{-p} F(p),\ a > 0$$

Solution: By definition, we have

$$M\{f(at);p\} = \int_0^\infty t^{p-1} f(at)\,dt$$

Put $at = u$ so that $adt = du$

$$= \int_0^\infty \left(\frac{u}{a}\right)^{p-1} f(u)\,\frac{du}{a} = a^{-p}\int_0^\infty u^{p-1} f(u)du$$

$$= a^{-p} M\{f(u);p\} = a^{-p} F(p);\ a > 0$$

3. If $M\{f(t);p\} = F(p)$, then

$$M\{f(a);p\} = \frac{1}{a}\,F\left(\frac{p}{a}\right),\ a > 0$$

Proof: By definition, we have

$$M\{f(a)\,dt\} = \int_0^\infty t^{p-1} f(t^a)\,dt$$

Put $t^a = u$ or $t = u^{1/a}$, $dt = \frac{1}{a}\int_0^\infty u^{\frac{1}{a}-1}\,du$

$$= \int_0^\infty u^{\frac{p-1}{a}} f(u)\,\frac{1}{a}\,u^{\frac{1}{a}-1}\,du = \frac{1}{a}\int_0^\infty u^{\frac{p}{a}-1} f(u)du$$

$$= \frac{1}{a}F\left(\frac{p}{a}\right)$$

4. If $M\{f(t);p\} = F(p)$, then

$$M\{t^a f(t);p\} = F(p+a).$$

Proof: By definition, we have

$$M\{t^a f(t);p\} = \int_0^\infty t^{p-1}\,t^a f(t)\,dt$$

$$= \int_0^\infty t^{p+a-1} f(t)\,dt = F(p+a)$$

5. If $M\{f(t);p\} = F(p)$, then

$$M\{t^c f(t^a);p\} = \frac{1}{a}F\left(\frac{p+c}{a}\right).$$

Proof: By definition, we have

$$M\{t^c f(t^a);p\} = \int_0^\infty t^{p-1}\,t^c f(t^a)\,dt$$

Let $t^a = u$ or $t = u^{1/a}$, $dt = \frac{1}{a} u^{\frac{1}{a}-1} du$

$$= \frac{1}{a} \int_0^\infty u^{\frac{p+c-1}{a}} f(u)\, u^{\frac{1}{a}-1} du$$

$$= \frac{1}{a} \int_0^\infty u^{\frac{p}{a}+\frac{c}{a}-1} f(u) du$$

$$= \frac{1}{a} \int_0^\infty u^{\frac{p+c}{a}-1} f(u) du = \frac{1}{a} F\left(\frac{p+c}{a}\right).$$

Note: By combining property (3) and (4), we can directly have the above result.

6. If $\quad M\{f(t); p\} = F(p)$, then

$$M\left\{\frac{1}{t^m} f\left(\frac{1}{t^n}\right); p\right\} = \frac{1}{n} F\left(\frac{m-p}{n}\right).$$

Proof: By definition, we have

$$M\left\{\frac{1}{t^m} f\left(\frac{1}{t^n}\right); p\right\} = \int_0^\infty t^{p-1} \frac{1}{t^m} f\left(\frac{1}{t^n}\right) dt$$

$$= \int_0^\infty t^{p-m-1} f\left(\frac{1}{t^n}\right) dt$$

[Let $\frac{1}{t^n} = u$ or $t = \frac{1}{u^{1/n}}$ so that $dt = -\frac{1}{n} u^{\frac{1}{n}-1} du$]

$$= \frac{1}{n} \int_0^\infty u^{\frac{m-p}{n}-1} f(u) du$$

$$= \frac{1}{n} F\left(\frac{m-p}{n}\right).$$

Special Case: When $m = n = 1$, we have

$$M\left\{\frac{1}{t} f\left(\frac{1}{t}\right); p\right\} = F(1-p)$$

7. If $\quad M\{f(t); p\} = F(p)$, then

$$M\{t^m f(at^n); p\} = \frac{1}{n} a^{-\left(\frac{p+m}{n}\right)} F\left(\frac{p+m}{n}\right), \; a > 0$$

Proof: By definition, we have

$$M\{t^m f(at^n); p\} = \int_0^\infty t^{p-1} \{t^m f(at^n)dt$$

$$= \int_0^\infty t^{p+m-1} f(at^n)dt$$

[Let $at^n = u$, $t = \left(\frac{u}{a}\right)^{1/n}$ and $dt = \frac{1}{n a^{1/n}} u^{\frac{1}{n}-1} du$]

$$= \int_0^\infty \left(\frac{u}{a}\right)^{\frac{p+m-1}{n}} f(u) \frac{1}{n a^{1/n}} u^{\frac{1}{n}-1} du$$

$$= \frac{1}{n} a^{-\left(\frac{p+m}{n}\right)} \int_0^\infty u^{\frac{p+m}{n}-1} f(u)du$$

$$= \frac{1}{n} a^{-\left(\frac{p+m}{n}\right)} F\left(\frac{p+m}{n}\right)$$

8. If $M\{f(t); p\} = F(p)$, then

$$M\{\log t\, f(t); p\} = \frac{d}{dp} F(p)$$

Proof: By definition, we have

$$M\{\log t \cdot f(t); p\} = \int_0^\infty t^{p-1} \log t\, f(t)dt \qquad ...(1)$$

and $$F(p) = M\{f(t); p\} = \int_0^\infty t^{p-1} f(t)dt \qquad ...(2)$$

Differentiating equation (2) w.r.t. p with the help of Leibnitz rule of differentiation under the sign of integration

$$\frac{d}{dp} F(p) = \int_0^\infty t^{p-1} \log t\, f(t)dt \qquad ...(3)$$

$$\left(\because \frac{d}{dp} t^{p-1} = t^{p-1} \log t\right)$$

Using equation (3) in equation (1), we have

$$M\{\log t\, f(t); p\} = \frac{d}{dp} F(p)$$

Example 1 Find

(i) $M\{\sin t; p\}$

(ii) $M\{\cos t; p\}$

Solution: Consider the integral $\int_0^\infty t^{p-1} e^{-it}\, dt$

$$\left[\text{Let } it = u,\ t = \frac{u}{i} \text{ so that } dt = \frac{du}{i}\right]$$

$$= \int_0^\infty \left(\frac{u}{i}\right)^{p-1} e^{-u} \frac{du}{i} = \frac{1}{i^p} \int_0^\infty u^{p-1} e^{-u}\, du$$

$$= \frac{\Gamma(p)}{i^p} \quad \text{(Using the definition of Gamma function)}$$

where, $\text{Re}(p) > 0$

or $\int_0^\infty t^{p-1} (\cos t - i \sin t) dt = \Gamma(p)\, i^{-p}$

$$= \Gamma(p) \left(e^{\frac{i\pi}{2}}\right)^{-p}$$

$$= \Gamma(p) \left[\cos\frac{\pi}{2} - i \sin\frac{p\pi}{2}\right]$$

Now equating real and imaginary parts,

$$\int_0^\infty t^{p-1} \cos t\, dt = M\{\cos t; p\} = \Gamma(p) \cos \frac{p\pi}{2};\ 0 < \text{Re } p < 1$$

and $$\int_0^\infty t^{p-1} \sin t\, dt = M\{\sin t; p\} = \Gamma(p) \sin \frac{p\pi}{2};\ 0 < \text{Re } p < 1$$

Example 2 Prove that $M\{e^{-at}; p\} = \dfrac{\Gamma(p)}{a^p}$, Re $(p) > 0$

Solution: We have

$$M\{e^{-at}; p\} = \int_0^\infty t^{p-1} e^{-at}\, dt$$

$$= \frac{\Gamma(p)}{a^p}, \text{ (By definition of Gamma function)}$$

where, Re $(p) > 0$

Special Case: Taking $a = 1$, we have

$$M\{e^{-t}; p\} = \Gamma(p), \text{ Re } (p) > 0$$

Example 3 Prove that

$$M\{t^m f(at^{-n}); p\} = \frac{1}{n} a^{\left(\frac{p+m}{n}\right)} F\left(-\frac{p+m}{n}\right).$$

Solution: By definition, we have

$$M\{t^m f(at^{-n}); p\} = \int_0^\infty t^{m+p-1} f(at^{-n})dt$$

$$\left[\text{Let } at^{-n} = u, \ t = \left(\frac{a}{u}\right)^{1/n} = a^{1/n} u^{-1/n}, \ dt = -\frac{1}{n} a^{1/n} u^{-\frac{1}{n}-1}\right]$$

$$= \int_0^\infty a^{\frac{m+p-1}{n}} u^{-\left(\frac{m+p-1}{n}\right)} f(u) \frac{1}{n} a^{\frac{1}{n}} u^{-\frac{1}{n}-1} du$$

$$= \frac{a^{\frac{m+p}{n}}}{n} \int_0^\infty u^{-\left(\frac{m+p}{n}\right)-1} f(u)du = \frac{a^{\frac{m+p}{n}}}{n} F\left(-\frac{m+p}{n}\right).$$

Example 4 Show that if n is a positive integer, then

$$M\{t^n P_n(t) H(1-t); p\} = (p+n)^{-1} {}_2F_1\left(-n, n+1; n+p+1; \frac{1}{2}\right)$$

Solution: We know that

$$P_n(t) = {}_2F_1\left(-n, n+1; 1; \frac{1}{2} - \frac{t}{2}\right)$$

$$M\{t^n P_n(t)H(1-t); p\} = \int_0^\infty t^{p-1} t^n {}_2F_1\left(-n, n+1; 1; \frac{1}{2} - \frac{t}{2}\right)H(1-t)dt$$

$$= \int_0^1 t^{p+n-1} {}_2F_1\left(-n, n+1; 1; \frac{1}{2} - \frac{t}{2}\right)dt$$

(By the definition of Heaviside unit function)

$$= \int_0^1 t^{p+n-1} \sum_{r=0}^\infty \frac{(-n)_r (n+1)_r}{(1)_r r!} \left(\frac{1-t}{2}\right)^r dt$$

$$= \sum_{r=0}^{\infty} \frac{(-n)_r\,(n+1)_r}{\Gamma(1+r)r!} \frac{1}{2^r} \int_0^1 t^{p+n-1}(1-t)^{r+1-1}\,dt$$

$$= \sum_{r=0}^{\infty} \frac{(-n)_r\,(n+1)_r}{r!\,\Gamma(1+r)2^r} \frac{\Gamma(p+n)\,\Gamma(r+1)}{\Gamma(p+n+r+1)} \quad \text{(By the definition of Gamma function)}$$

$$= \sum_{r=0}^{\infty} \frac{(-n)_r\,(n+1)\,\Gamma(p+n)}{r!\,2^r \cdot \Gamma(p+n+1)(p+n+1)_r}$$

$$= (p+n)^{-1} \sum_{r=0}^{\infty} \frac{(-n)_r\,(n+1)_r}{\Gamma(1+r)\cdot 2^r \cdot (p+n+1)_r}$$

$$= (p+n)^{-1}\, {}_2F_1\left(-n,\ n+1;\ n+p+1;\ \frac{1}{2}\right)$$

Example 5 Find the Mellin transforms of (i) $E_i(t)$ and (ii) $t^m\,H(t-a)$

Solution: $M\{E_i(t); p\} = \int_0^{\infty} t^{p-1}\,E_i(t)\,dt$

$$= \int_0^{\infty} t^{p-1} \left\{\int_t^{\infty} \frac{e^{-u}}{u}\,du\right\} dt$$

($\because$ Exponential integral function)

$$= E_i\,(t) = \int_t^{\infty} \frac{e^{-u}}{u}\,du$$

Now let $u = vt$ so that $du = tdv$

$$= \int_0^{\infty} t^{p-1} \left\{\int_1^{\infty} \frac{e^{-vt}}{vt}\,t\,dv\right\} dt$$

$$= \int_1^{\infty} \frac{1}{v} \left\{\int_0^{\infty} e^{-vt}\,t^{p-1}\,dt\right\} dv$$

On changing the order of integration, we have
Let $vt = x$ so that $v\,dt = dx$.

$$= \int_1^{\infty} \frac{1}{v} \left\{\int_0^{\infty} e^{-x} \left(\frac{x}{v}\right)^{p-1} \frac{dx}{v}\right\} dv$$

$$= \int_1^\infty \frac{1}{v^{p+1}} \left\{ \int_0^\infty e^{-x} x^{p-1} dx \right\} dv$$

$$= \int_1^\infty \frac{1}{v^{p+1}} \Gamma(p)\, dv \qquad \text{(By definition of Gamma function)}$$

$$= \Gamma(p) \int_1^\infty v^{-p-1} dv$$

$$= \Gamma(p) \left[\frac{v^{-p}}{-p} \right]_1^\infty$$

$$= -\frac{\Gamma(p)}{p} \left[\frac{1}{v^p} \right]_1^\infty = \frac{\Gamma(p)}{p}$$

(ii) $M\{t^m H(t-a); p\} = \int_0^\infty t^{p-1} \{t^m H(t-a)\} dt$

$$= \int_a^\infty t^{p+m-1} dt \qquad \text{(By definition of Heaviside unit function)}$$

$$= \left[\frac{t^{p+m}}{p+m} \right]_a^\infty$$

$$= -\frac{a^{p+m}}{p+m}, \ \text{Re}\,(p) < -\text{Re}\,(m).$$

Example 6 Prove that

$$M\{(1+t)^{-a}; p\} = \frac{\Gamma(p)\Gamma(a-p)}{\Gamma(a)}, \ 0 < \text{Re}\,(p) < \text{Re}\,(a)$$

Solution: $M\{(1+t)^{-a}; p\} = \int_0^\infty t^{p-1}(1+t)^{-a} dt$

$$= \int_0^\infty \frac{t^{p-1}}{(1+t)^a} dt$$

$$\left[\text{Let } \frac{t}{1+t} = u \text{ so that } \frac{(1+t-t)}{(1+t)^2} dt = du \text{ and } t = \frac{u}{1-u} \right]$$

$$= \int_0^1 \left(\frac{u}{1-u}\right)^{p-1} \frac{1}{\left(1+\frac{u}{1-u}\right)} \frac{du}{(1-u)^2}$$

$$= \int_0^1 u^{p-1}(1-u)^{a-p-1}\, du$$

$$= \frac{\Gamma(p)\,\Gamma(a-p)}{\Gamma(a)}, \qquad \text{(By definition of Beta function)}$$

$$0 < \text{Re } p < \text{Re } (a)$$

Particular Case: If $a = 1$, and making use of the relation $\Gamma(p)\,\Gamma(1-p) = \dfrac{\pi}{\sin \pi p}$, $0 < \text{Re } (p) < 1$

then $M\{(1+t)^{-1}; p\} = \dfrac{\Gamma(p)\Gamma(1-p)}{\Gamma(1)} = \dfrac{\pi}{\sin \pi p}$

$$= \pi \text{ cosec } (\pi p),\ 0 < \text{Re } (p) < 1.$$

Example 7 Show that $M\{(1+t^a)^{-b}; p\} = \dfrac{\Gamma(p/a)\Gamma(b-p/a)}{a\Gamma(b)}$, provided that

$$0 < \text{Re } (b) < \text{Re } (ab).$$

Solution: We know that

$$M\{(1+t)^{-b}; p\} = \frac{\Gamma(p)\,\Gamma(b-p)}{\Gamma(b)},\ 0 < \text{Re } (p) < \text{Re } (b) \qquad \text{...(i)}$$

Also we have

$$M\{f(t^a); p\} = \frac{1}{a} F\left(\frac{p}{a}\right),\ a > 0\ ; \qquad \text{...(ii)}$$

where, $F(p) = M\{f(t); p\}$

Using equations (i) and (ii)

$$M\{(1+t^a)^{-b}; p\} = \frac{1}{a} \frac{\Gamma\left(\frac{p}{a}\right)\Gamma\left(b-\frac{p}{a}\right)}{\Gamma(b)}, \text{ provided that } 0 < \text{Re } (p) < \text{Re } (ab) \qquad \text{...(iii)}$$

Special Cases:

(i) Let $b = 1$ in equation (iii)

$$M\{(1+t^a)^{-1}; p\} = \frac{1}{a}\ \Gamma(p/a)\ \Gamma(1-p/a)$$

$$= \frac{\pi}{a} \text{ cosec}\left(\frac{\pi p}{a}\right),\ 0 < \text{Re } (p) < \text{Re } (a) \qquad \text{...(iv)}$$

(ii) Let $a = 2$, $b = 1$ in (iii)

$$M\{(1+t^2)^{-1}; p\} = \frac{\pi}{2} \operatorname{cosec}\left(\frac{\pi p}{2}\right), \ 0 < \operatorname{Re} p < 2$$

Example 8 Prove that

$$M\left\{\left(t^{1-\alpha}\frac{d}{dt}\right)^m t^\beta g(t); p\right\} = \frac{(-1)^m a^m \Gamma(p/\alpha)}{\Gamma\{(p/\alpha)-m\}} M\{g(x); p - m\alpha + \beta\}$$

Solution: We prove the result by using principle of mathematical induction. By definition of Mellin transform, we have for $m = 1$

$$M\left\{\left(t^{1-\alpha}\frac{d}{dt}\right) t^\beta g(t); p\right\} = \int_0^\infty t^{p-1}\left(t^{1-\alpha}\frac{d}{dt}\right) t^\beta g(t)\, dt$$

$$= \int_0^\infty t^{p-\alpha} \frac{d}{dt}\{t^\beta g(t)\} dt$$

Integrating by parts

$$= [t^{p-\alpha} t^\beta g(t)]_0^\infty - (p-\alpha)\int_0^\infty t^{p-\alpha-1}\{t^\beta g(t)\}dt$$

$$= -(p-\alpha) M\{g(t); p - \alpha + \beta\} \qquad ...(1)$$

Provided that $\lim t^{p-\alpha} t^\beta g(t) = 0$

as $t \to 0$ and $t \to \infty$

Let the result be true for $m = k$

(a fixed positive integer), then

$$M\left\{\left(t^{1-\alpha}\frac{d}{dt}\right)^k t^\beta g(t); p\right\} = (-1)^k \frac{\alpha^k \Gamma\left(\frac{p}{\alpha}\right)}{\Gamma\left(\frac{p}{\alpha}-k\right)} M\{g(t); p - k\alpha + \beta\} \qquad ...(2)$$

Taking $\left(t^{1-\alpha}\frac{d}{dt}\right)^k g(t) = f(t)$, then $M\left\{\left(t^{1-\alpha}\frac{d}{dt}\right)^{k+1} t^\beta g(t); p\right\}$

$$= -(p-\alpha) M\left\{\left(t^{1-\alpha}\frac{d}{dt}\right)^k t^\beta g(t); p-\alpha\right\} \text{ (by equation (1))}$$

$$= -(p-\alpha)(-1)^k \alpha^k \frac{\Gamma\left(\frac{p-\alpha}{\alpha}\right)}{\Gamma\left(\frac{p-\alpha}{\alpha}-k\right)} M\{g(t); p-\alpha-k\alpha+\beta\}$$

(Using equation (2))

$$= -\alpha\left(\frac{p}{\alpha}-1\right)\alpha^k (-1)^k \frac{\Gamma\left(\frac{p}{\alpha}-1\right)}{\Gamma\left(\frac{p}{\alpha}-1-k\right)} M\{g(t); p-\alpha-k\alpha+\beta\}$$

$$= (-1)^{k+1}\alpha^{k+1} \frac{\Gamma\left(\frac{p}{\alpha}\right)}{\Gamma\left[\frac{p}{\alpha}-(k+1)\right]} M\{g(t); p-(k+1)\alpha+\beta\}$$

Hence, the result is true for $m = k + 1$. Hence, by principle of mathematical induction the result is true for all $m \in N$.

Example 9 Show that

$$M\{e^{-at} J_n(bt); p\} = \frac{b^n 2^{p-1}}{\sqrt{\pi}\,\Gamma(n+1)}(a^2+b^2)^{-(p+n)/2} \times \Gamma\left(\frac{n+p}{2}\right)\Gamma\left(\frac{p+n+1}{2}\right)$$

$$\times\ {}_2F_1\left(\frac{p+n}{2}, \frac{p+n+1}{2}; n+1; \frac{b^2}{a^2+b^2}\right)$$

$$\left[\text{Re}(a) > 0, n > \frac{1}{2}\right]$$

Hence, deduce that

(i) $M\{x^{-n} J_n(t); p\} = \dfrac{2^{p-n-1}\,\Gamma(p/2)}{\Gamma(n-p/2+1)}, \ 0 < \text{Re}(p) < 1, n > -\dfrac{1}{2}$

(ii) $M\{J_n(bt); p\} = \dfrac{b^{-p} 2^{p-1}\,\Gamma\left(\frac{p+n}{2}\right)}{\Gamma\left(\frac{n-p+2}{2}\right)}, \ -n < p < n+2$

Solution: $M\{x^{-at}\, J_n(bt);\, p\} = \int_0^\infty t^{p-1}\, e^{-at}\, J_n\,(bt)dt$

$$= \int_0^\infty t^{p-1}\, e^{-at} \sum_{r=0}^{\infty} \frac{(-1)^r}{r!\,\Gamma(n+r+1)} \left(\frac{bt}{2}\right)^{n+2r} dt$$

$$= \sum_{r=0}^{\infty} \frac{(-1)^r}{r!\,\Gamma(n+r+1)} \left(\frac{b}{2}\right)^{n+2r} \int_0^\infty e^{-at}\, t^{p+n+2n-1}\, dt$$

Changing the order of integration and summation

$$= \sum_{r=0}^{\infty} \frac{(-1)^r}{r!\,\Gamma(n+r+1)} \left(\frac{b}{2}\right)^{n+2r} \frac{\Gamma(p+n+2r)}{a^{p+n+2r}}$$

$$= \sum_{r=0}^{\infty} \frac{(-1)^r}{r!\,\Gamma(n+r+1)} \left(\frac{b}{2}\right)^{n+2r} \frac{2^{p+n+2r-1}\,\Gamma\left(\frac{p+n}{2}+r\right)\Gamma\left(\frac{p+n+1}{2}+r\right)}{\sqrt{\pi}\, a^{p+n+2r}}$$

(Applying duplication formula for gamma function)

$$= \frac{b^n\, 2^{p-1}}{\sqrt{\pi}\, a^{p+v}} \sum_{r=0}^{\infty} \frac{(-1)^r}{r!(n+1)_r\,\Gamma(n+1)} \left(\frac{b^2}{a^2}\right)^r \left(\frac{p+n}{2}\right)_r$$

$$\times\, \Gamma\left(\frac{p+n}{2}\right)\left(\frac{p+n+1}{2}\right)_r \Gamma\left(\frac{p+n+1}{2}\right)$$

$$= \frac{b^n\, 2^{p-1}\,\Gamma\left(\frac{p+n}{2}\right)\Gamma\left(\frac{p+n+1}{2}\right)}{\sqrt{\pi}\, a^{p+v}\,\Gamma(n+1)}\; {}_2F_1\left(\frac{p+n}{2}, \frac{p+n+1}{2}; n+1, \frac{b^2}{a^2}\right) \qquad \text{...(i)}$$

Applying the transformation formula

$${}_2F_1(\alpha, \beta;\, y,\, z) = (1-z)^{-\alpha}\; {}_2F_1\left(a, \gamma-\beta; \gamma, \frac{z}{z-1}\right)$$

in equation (i), we get

$$M\{e^{-at}\, J_n(t);\, p\} = \frac{b^n\, 2^{p-1}\,\Gamma\left(\frac{p+n}{2}\right)\Gamma\left(\frac{p+n+1}{2}\right)}{\sqrt{\pi}\,\Gamma(n+1)}$$

$$\times \left(a^2+b^2\right)^{-\left(\frac{p+n}{2}\right)}\; {}_2F_1\left(\frac{p+n}{2}, \frac{n-p+1}{2}; n+1; \frac{b^2}{b^2+a^2}\right) \qquad \text{...(ii)}$$

Deductions

(i) From property of

$$M[t^{\alpha} f(t); p] = F(p + \alpha), \text{ where, } F(p) = M[f(t); p]$$

Applying this in equation (ii) property

$$M(x^{-\nu} e^{-at} J_n(bt); p) = \frac{b^n\, 2^{p-\nu-1}\, \Gamma\left(\frac{p-\nu+n}{2}\right)\Gamma\left(\frac{p-\nu+n+1}{2}\right)}{\sqrt{\pi}\,\Gamma(n+1)}$$

$$\times \left(a^2+b^2\right)^{-\left(\frac{p+\nu+n}{2}\right)} {}_2F_1\left(\frac{p-\nu+n}{2}, \frac{n-p+\nu+1}{2}; n+1; \frac{b^2}{b^2+a^2}\right)$$

Substituting $a = 0$ in above result

$$M[x^{-\nu} J_n(bt); p] = \frac{b^n\, 2^{p-\nu-1}\, \Gamma\left(\frac{p-\nu+n}{2}\right)\Gamma\left(\frac{p-\nu+n+1}{2}\right)}{\sqrt{\pi}\,\Gamma(n+1)}$$

$$\times\, b^{-(p+\nu+n)}\, {}_2F_1\left(\frac{p-\nu+n}{2}, \frac{n-p+\nu+1}{2}; n+1; 1\right)$$

Applying Gauss' theorem

$${}_2F_1(\alpha, \beta; \gamma; 1) = \frac{\Gamma(y)\,\Gamma(\gamma-\alpha-\beta)}{\Gamma(\gamma-\alpha)\,\Gamma(\gamma-\beta)}$$

in the above result, we get

$$M[x^{-\nu} J_n(bt); p] = \frac{b^{-(p+\nu)}\, 2^{p-\nu-1}\, \Gamma\left(\frac{p-\nu+n}{2}\right)}{\Gamma\left(1-\frac{p}{2}+\frac{\nu}{2}+\frac{n}{2}\right)} \qquad \text{...(i)}$$

For $b = 1$, in the above result

$$M[x^{-\nu} J_n(t); p] = \frac{2^{p-\nu-1}\, \Gamma\left(\frac{p-\nu+n}{2}\right)}{\Gamma\left(1-\frac{p}{2}+\frac{\nu}{2}+\frac{n}{2}\right)}$$

and for $\nu = n$ in the above result.

$$M[x^{-n} J_n(t); p] = \frac{2^{p-n-1}\, \Gamma\left(\frac{p}{2}\right)}{\Gamma\left(1-\frac{p}{2}+n\right)}, \quad 0 < \text{Re } p < 1,\ n > -\frac{1}{2}$$

(ii) Let $v = 0$ in equation (i), we get

$$M[J_n(bt); p] = \frac{b^{-p}\, 2^{p-1}\, \Gamma\left(\frac{p+n}{2}\right)}{\Gamma\left(\frac{2-p+n}{2}\right)}, \quad -n < p < n + 2$$

Mellin Transform of Derivatives

If $\qquad M\{f(t); p\} = F(p)$ then

$$M\{f'(t); p\} = -(p-1)\, F(p-1) \qquad \text{...(A)}$$

Proof: By definition, we have

$$M\{f'(t); p\} = \int_0^\infty t^{p-1} f'(t)\, dt$$

$$= \left[t^{p-1} f(t)\right]_0^\infty - \int_0^\infty (p-1)\, t^{p-2} f(t)\, dt$$

If there exist α_1, α_2 such that $\lim_{t\to 0} t^{p-1} f(t) = 0$ and $\lim_{t\to\infty} t^{p-1} f(t) = 0$ when $\alpha_1 < \text{Re}\,(p) < \alpha_2$ and if $F(p-1)$ exists in the band, then

$$M\{f'(t); p\} = -(p-1) \int_0^\infty t^{(p-1)-1} f(t)\, dt$$

$$= -(p-1)\, F(p-1)$$

where, $\alpha_1 < \text{Re}\,(p) < \alpha_2$

Theorem: $M\{f^{(n)}(t); p\} = \dfrac{(-1)^n\, \Gamma(p)}{\Gamma(p-n)}\, M\{f(t); p-n\}$

Proof: By last result, we have

$$M\{f'(t); p\} = -(p-1)\, F(p-1) = -(p-1)\, M\{f(t); p-1\} \qquad \text{...(i)}$$

Applying this formula again

$$M\{f''(t); p\} = -(p-1)\, M\{f'(t); p-1\}$$

$$= -(p-1)[(p-2)\, M\{f(t); p-2\}] \qquad \text{[Using equation (i)]}$$

$$= (-1)^2 (p-1)[(p-2)\, M\{f(t); p-2\}]$$

$$= \frac{(-1)^2\, \Gamma(p)}{\Gamma(p-2)} F(p-2) \qquad \text{...(ii)}$$

Obviously, we can prove by induction over n that for all values of p for which $F(p-n)$ exists

$$M\{f^{(n)}(t); p\} = (-1)^n (p-1)(p-2) \ldots (p-n) F(p-n)$$

$$= \frac{(-1)^n \Gamma(p)}{\Gamma(p-n)} M\{f(t); p-n\}$$

provided that $\lim_{t \to 0} t^{p-r-1} f^{(r)}(t) = 0.$

$r = 0, 1, 2, \ldots n-1$ and $F(p-1)$ exists

Remark: $M[t^n f(t); p] = F(p+n) = M\{f(t); p+n\}$
(using the property) of Mellin transform

Theorem: $M[t^n f^n(t); p] = (-1)^n \dfrac{\Gamma(p+n)}{\Gamma(p)} M[f(t); p]$ ...(i)

Proof: By definition

$$M[t^n f^n(t); p] = \int_0^\infty t^{p+n-1} f^{(n)}(t) dt$$

$$= M[f^{(n)}(t); p+n]$$

$$= (-1)^n \frac{\Gamma(p+n)}{\Gamma(p)} F(p) \quad \text{[Using theorem (I)]}$$

Theorem: If n is a positive integer then show that

$$M\left[\left(t\frac{d}{dt}\right)^n f(t); p\right] = (-1)^n p^n F(p) \quad \text{...(ii)}$$

where, $M[f(t); p] = F(p)$

Proof: We shall prove the result by applying principle of mathematical induction. For $n = 1$

$$M\left[t\frac{d}{dt} f(t); p\right] = M[t f'(t); p]$$

$$= M[f'(t); p+1] \quad \text{[Using prop. (4)]}$$

$$= -[(p+1)-1] M[f(t); p+1-1] \quad \text{[Using result (A)]}$$

or $$M[t f'(t); p] = -p M[f(t); p] = -p F(p)$$

or $$M\left[\left(t\frac{d}{dt}\right) f(t); p\right] = -p F(p)$$

Let the result (ii) be true for $n = m$ (a fixed integer) then

$$M\left[\left(t\frac{d}{dt}\right)^m f(t); p\right] = (-1)^m p^m F(p)$$

Also let $$\left(t\frac{d}{dt}\right)^m f(t) = g(t)$$

Now, $$M\left[\left(t\frac{d}{dt}\right) g(t); p\right] = -p\, M[g(t); p]$$

or $$M\left[\left(t\frac{d}{dt}\right)\left(t\frac{d}{dt}\right)^m f(t); p\right] = -p\, M\left[\left(t\frac{d}{dt}\right)^m f(t); p\right]$$

$$= (-1)^{m+1} p^{m+1} F(p)$$

Thus, the result is true for $n = m + 1$. Hence, by principle of mathematical induction the result (ii) is true for all $n \in N$.

Example 10 Prove that if n is a positive integer $\alpha \neq 0$

(i) $$M\left[\left(t^{1-\alpha}\frac{d}{dt}\right)^n f(t); p\right] = (-1)^n a^n \frac{\Gamma(p/\alpha)}{\Gamma\left(\frac{p}{\alpha} - n\right)} F/(p - n\alpha) \quad \text{...(B)}$$

(ii) $$M\left[\left(\frac{d}{dt}t\right)^n f(t); p\right] = (1 - p)^n F(p) \quad \text{...(C)}$$

where, $M[f(t); p] = F(p)$

Solution: We prove the results by using principle of mathematical induction

(i) Let $n = 1$, then

$$M\left[\left(t^{1-\alpha}\frac{d}{dt}\right) f(t); p\right] = \int_0^\infty t^{p-1}\left(t^{1-\alpha}\frac{d}{dt}\right) f(t)\, dt$$

$$= \int_0^\infty t^{p-\alpha}\frac{d}{dt} f(t)\, dt$$

Integrating by parts, we get

$$= \left[t^{p-\alpha} f(t)\right]_0^\infty - \int_0^\infty (p - \alpha)\, t^{p-\alpha-1} f(t)\, dt$$

$$= -(p - \alpha)\int_0^\infty t^{p-\alpha-1} f(t)\, dt$$

or $$M\left[\left(t^{1-\alpha}\frac{d}{dt}\right)f(t);p\right] = -(p-\alpha)\,M[f(t);p-\alpha] \qquad \text{...(D)}$$

Provided that $t^{p-\alpha}f(t) \to 0$ and $t \to 0$

Let the result (B) is true for $n = m$ then

$$M\left[\left(t^{1-\alpha}\frac{d}{dt}\right)^m f(t);p\right] = (-1)^m\,\alpha^m\,\frac{\Gamma\left(\frac{p}{\alpha}\right)}{\Gamma\left(\frac{p}{\alpha}-m\right)}\,F(p-m\alpha) \qquad \text{...(E)}$$

Now, $$M\left[\left(t^{1-\alpha}\frac{d}{dt}\right)^{m+1} f(t);p\right] = M\left[\left(t^{1-\alpha}\frac{d}{dt}\right)\left\{\left(t^{1-\alpha}\frac{d}{dt}\right)^m f(t)\right\};p\right] \qquad \text{...(F)}$$

Let $$\left(t^{1-\alpha}\frac{d}{dt}\right)^m f(t) = g(t) \qquad \text{...(G)}$$

Applying result (G) in (F)

$$M\left[\left(t^{1-\alpha}\frac{d}{dt}\right)^{m+1} f(t);p\right] = M\left[t^{1-\alpha}\frac{d}{dt}\,g(t);p\right]$$

$$= -(p-\alpha)\,M[g(t);p-\alpha] \qquad \text{[Applying result (D)]}$$

$$= -(p-\alpha)\,\frac{(-1)^m\,\alpha^m\,\Gamma\left(\frac{p-\alpha}{\alpha}\right)}{\Gamma\left(\frac{p-\alpha}{\alpha}-m\right)}\,F(p-\alpha-m\alpha)$$

[Using E]

$$= (-1)^{m+1}\,\frac{\left(\frac{p-\alpha}{\alpha}\right)\cdot\alpha^{m+1}\,\Gamma\left(\frac{p-\alpha}{\alpha}\right)}{\Gamma\left(\frac{p-\alpha}{\alpha}-m\right)}\,F[p-\alpha(m+1)]$$

$$= (-1)^{m+1}\,\frac{\alpha^{m+1}\,\Gamma\left(\frac{p}{\alpha}\right)}{\Gamma\left(\frac{p}{\alpha}-m-1\right)}\,F[p-\alpha(m+1)\,\alpha]$$

Thus, the result is true for $n = m + 1$. Hence, by principle of mathematical induction statement (B) is true for all $n \in N$.

(ii) Let $n = 1$, then

$$M\left[\left(t\frac{d}{dt}\right)f(t); p\right] = \int_0^\infty t^{p-1}\left(t\frac{d}{dt}\right)f(t)\,dt$$

$$= \int_0^\infty t^{p-1}\frac{d}{dt}[t\,f(t)]\,dt$$

$$= \left[t^{p-1}\,t\,f(t)\right]_0^\infty - (p-1)\int_0^\infty t^{p-2}\,t\,f(t)\,dt$$

$$= 0 - (p-1)\int_0^\infty t^{p-1}f(t)\,dt$$

or $$M\left[\left(t\frac{d}{dt}\right)f(t); p\right] = (1-p)\,M\{f(t); p\}$$

$$= (1-p)\,F(p) \qquad \text{...(H)}$$

Provided that $t^p f(t) \to 0$ as $t \to 0$ and $t \to \infty$. Let the given result is transfer a positive integer $n = m$, then

$$M\left[\left(t\frac{d}{dt}\right)^m f(t); p\right] = (1-p)^m F(p) \qquad \text{...(I)}$$

Now, $$M\left[\left(t\frac{d}{dt}\right)^{m+1} f(t);p\right] = M\left[\left(t\frac{d}{dt}\right)\left[\left(t\frac{d}{dt}\right)^m f(t)\right];p\right] \qquad \text{...(J)}$$

Let $$\left(t\frac{d}{dt}\right)^m f(t) = g(t) \qquad \text{...(K)}$$

Applying result (K) in (J)

$$M\left[\left(t\frac{d}{dt}\right)^{m+1} f(t);p\right] = M\left[\left(t\frac{d}{dt}\right)g(t); p\right]$$

$$= (1-p)\,M\{g(t); p\} \qquad \text{[Applying result (H)]}$$

$$= (1-p)\,M\left[\left(t\frac{d}{dt}\right)^m f(t); p\right]$$

$$= (1-p)\,(1-p)^m\,M\{f(t); p\} \qquad \text{[Using result (I)]}$$

$$= (1-p)^{m+1}\,M\{f(t); p\} = (1-p)^{m+1}\,F(p)$$

Thus, the result is true for $n = m + 1$. Hence, by principle of mathematical induction the result (C) is true for any positive integer, n.

Example 11 Prove that

$$M\left[t^2\frac{d^2f}{dt^2}+t\frac{df}{dt}; p\right] = p^2M\{f(t); p\}$$

$$= p^2F(p)$$

Solution: We have $M\left[t^2\frac{d^2f}{dt^2}+t\frac{df}{dt}; p\right] = M\left[\left(t^2\frac{d^2}{dt^2}+t\frac{d}{dt}\right)f(t); p\right]$

$$= M\left[t\frac{d}{dt}\left(t\frac{d}{dt}\right)f(t); p\right]$$

$$= M\left[\left(t\frac{d}{dt}\right)^2 f(t); p\right]$$

$$= p^2F(p) \qquad \text{(Using theorem } Z)$$

Example 12 The Mellin transform of integral. If $M[f(t); p] = F(p)$, then prove that

(i) $M\left[\int_0^t f(u)\,du; p\right] = \frac{-1}{p}F(p+1)$

(ii) $M\left[\int_0^t ds\int_0^s f(u)du; p\right] = \frac{1}{p(p+1)}F(p+2)$

(iii) $M[I_n f(t); p] = (-1)^n \frac{\Gamma(p)}{\Gamma(p+n)} F(p+n)$

where, $I_n f(t)$ denotes the nth repeated integral and can be written as

$$I_n f(t) = \int_0^t I_{n-1} f(u)\,du$$

Proof: Let $$g(t) = \int_0^t f(u)\,du \qquad ...(i)$$

Differentiating both sides of equation (i) w.r.t. t using Leibnitz rule

$$g'(t) = f(t) \qquad ...(ii)$$

Taking Mellin transform of equation (ii)

$$M[g'(t); p] = M[f(t); p] \qquad ...(iii)$$

Since we know that

$$M[g'(t); p] = -(p-1)\, M[g(t); p-1] \qquad \text{...(iv)}$$

Using equation (iv) in equation (iii)

$$-(p-1)\, M[g(t); p-1] = M[f(t); p]$$

or
$$M[g(t); p-1] = -\frac{1}{(p-1)}\, M[f(t); p]$$

or
$$M\left[\int_0^t f(u)du; p-1\right] = -\frac{1}{(p-1)}\, M[f(t); p]$$

Replacing p by $p + 1$, we get

(i)
$$M\left[\int_0^t f(u)du; p\right] = \frac{-1}{p}\, M[f(t); p+1]$$

$$= \frac{-1}{p}\, F(p+1)$$

(ii) By repeating the process of equation (i), we have

$$M\left[\int_0^t ds \int_0^s f(u)du; p\right] = -\frac{1}{p} M\left[\int_0^s f(u)du; p+1\right]$$

$$= -\frac{1}{p}\left[-\frac{1}{(p+1)} F(p+2)\right]$$

$$= \frac{1}{p(p+1)} F(p+2)$$

(iii) Applying principle of mathematical induction over n, we have the result

$$M[I_n f(t); p] = (-1)^n \frac{\Gamma(p)}{\Gamma(p+n)} F(p+n)$$

Example 13 Prove that

$$M\left[t^{\lambda} \int_0^{\infty} u^l\, f(tu) g(u)\, du; p\right] = F(p+\lambda) \cdot G(1-\lambda+l-p)$$

Hence, deduce that

(i) $M\left[\int_0^{\infty} f(tu) g(u)du; p\right] = F(p)\, G(1-p)$

(ii) $M\left[\int_0^\infty f(tu)\,g(u)\frac{du}{u};\,p\right] = F(p)\,G(-p)$

where, $M\{f(t);\,p\} = Fp$

Solution: Consider

$$M\left[\int_0^\infty u^l\, f(tu)\, g(u)du;\, p\right] = \int_0^\infty t^{p-1}\left[\int_0^\infty u^l\, f(tu)\, g(u)\,du\right]dt$$

$$= \int_0^\infty u^l\, g(u)du \int_0^\infty t^{p-1}\, f(tu)\, dt$$

Let $tu = s, \Rightarrow udt = ds$

$$= \int_0^\infty u^l\, g(u)du \int_0^\infty \left(\frac{s}{u}\right)^{p-1} f(s)\frac{ds}{u}$$

$$= \int_0^\infty u^{l-p}\, g(u)du \int_0^\infty s^{p-1}\, f(s)ds$$

$$\Rightarrow\ M\left[\int_0^\infty u^l\, f(tu)\, g(u)du;\, p\right] = M[g(u);\, l-p+1]\; M[f(s);\, p]$$

$$= F(p)\cdot G(l-p+1)$$

Now applying the property

$$M\left[t^\lambda \int_0^\infty u^l\, f(tu)\,g(u)\,du;\, p\right] = F(p+\lambda)\; G(1+l-p-\lambda)$$

Deduction

(i) Put $\lambda = 0 = l$, in equation (i), we have

$$M\left[\int_0^\infty f(tu)\,g(u)du;\, p\right] = F(p)\cdot G(1-p)$$

(ii) Taking $\lambda = 0$, $l = -1$ in equation (i), we have

$$M\left[\int_0^\infty f(u)\frac{gu}{u}\,du;\, p\right] = F(p)\; G(-p)$$

9.2 INVERSE MELLIN TRANSFORM

If $M[f(t); p] = F(p)$ then $f(t)$ is called the inverse Mellin transform of $F(p)$ and is defined as

$$M^{-1}[F(p); t] = f(t)$$

9.3 THE MELLIN INVERSION THEOREM

If the integral $\int_0^\infty u^{k-1} |f(t)| dt$ is bounded for some $k < 0$ and if

$$M[f(t); p] = \int_0^\infty t^{p-1} f(t) dt = F(p)$$

then the Mellin inverse transform of $F(p)$ is given by

$$f(t) = M^{-1}[F(p); t] = \frac{1}{2\pi i} \int_{C-i\infty}^{C+i\infty} t^{-p} F(p) dp$$

where $c > k$.

Proof: According to definition of complex Fourier transform $G(w)$ of the function $g(x)$, where, $-\infty < x < \infty$ is given by

$$G(w) = \int_{-\infty}^{\infty} e^{iwx} g(x) dx \qquad \text{...(i)}$$

and the inverse complex Fourier transform
$g(x)$ of $G(w)$ is given by

$$g(x) = \frac{1}{2\pi} \int_{-\infty}^{\infty} e^{iwx} G(w) dw \qquad \text{...(ii)}$$

Let $p = c + iw$ and $x = \log t$

From equations (i) and (ii), we have

$$G(ic - ip) = \int_0^\infty t^{p-c} g(\log t) \frac{dt}{t}$$

$$= \int_0^\infty t^{p-c-1} g(\log t) dt \qquad \text{...(iii)}$$

and $$g(\log t) = \frac{1}{2\pi i} \int_{C-i\infty}^{C+i\infty} t^{c-p} G(ic - ip) dp \qquad \text{...(iv)}$$

Suppose $t^{-c} g(\log t) = f(t)$ and $G(ic - ip) = F(p)$, equations (iii) and (iv) reduces to

$$F(p) = \int_0^\infty t^{p-1} f(t) dt$$

and $$f(t) = \frac{1}{2\pi i}\int_{C-i\infty}^{C+i\infty} t^{-p} F(p)dp$$

9.4 CONVOLUTION OR FALTING THEOREM FOR THE MELLIN TRANSFORM

If $F(p)$ and $G(p)$ are the Mellin transform of the functions $f(t)$ and $g(t)$ respectively. Then

$$M[f(t)\, g(t);\, p] = \frac{1}{2\pi i}\int_{C-i\infty}^{C+i\infty} F(u)\, G(p-u)du$$

Proof: By definition, we have

$$M[f(t)\, g(t);\, p] = \int_0^\infty t^{p-1} f(t)\, g(t)\, dt$$

$$= \int_0^\infty t^{p-1} g(t)\left[\frac{1}{2\pi i}\int_{C-i\infty}^{C+i\infty} u^{-p} F(u)\, du\right]dt$$

(Using Mellin Inversion Theorem)

$$= \frac{1}{2\pi i}\int_{C-i\infty}^{C+i\infty} F(u)\left[\int_0^\infty t^{p-u-1} g(t)\, dt\right]du$$

(Changing the order of integration)

or $$M[f(t)\, g(t);\, p] = \frac{1}{2\pi i}\int_{C-i\infty}^{C+i\infty} F(u)\, G(p-u)du$$

Particular Case: For $p = 1$, then

$$M[f(t)\, g(t);\, p] = \int_0^\infty t^{1-1} f(t)\, g(t)dt$$

or $$\int_0^\infty f(t)\, g(t)dt = \frac{1}{2\pi i}\int_{C-i\infty}^{C+i\infty} F(u)\, G(1-u)du$$

9.5 INVERSE MELLIN TRANSFORM OF TWO FUNCTIONS

If $F(p)$ and $G(p)$ are the Mellin transforms of the functions $f(t)$ and $g(t)$ respectively, then

$$M^{-1}\, [F(p)\, G(p);\, t] = \int_0^\infty f\left(\frac{t}{u}\right) g(u)\, \frac{du}{u}$$

Proof: By the Mellin inversion theorem, we have

$$M^{-1}[F(p)\,G(p);\,t] = \frac{1}{2\pi i}\int_{C-i\infty}^{C+i\infty} t^{-p}\,F(p)\,G(p)dp$$

$$= \frac{1}{2\pi i}\int_{C-i\infty}^{C+i\infty} t^{-p}\,F(p)\left[\int_0^\infty u^{p-1}\,g(u)\,du\right]dp$$

[By definition of $G(p)$]

$$= \int_0^\infty g(u)\left[\frac{1}{2\pi i}\int_{C-i\infty}^{C+i\infty}\left(\frac{t}{u}\right)^{-p} F(p)\,dp\right]\frac{du}{u} \qquad \text{...(i)}$$

[Changing the order of integration by replacing t by u]

Applying Mellin inversion theorem we have

$$f\left(\frac{t}{u}\right) = \frac{1}{2\pi i}\int_{C-i\infty}^{C+i\infty}\left(\frac{t}{u}\right)^{-p} F(p)\,dp \qquad \text{...(ii)}$$

Using equation (ii) in equation (i), we get

$$M^{-1}[F(p)\,G(p);\,t] = \int_0^\infty f\left(\frac{t}{u}\right) g(u)\,\frac{du}{u}$$

Example 14 If $M\{f(t);\,p\} = F(p)$, then show that

(i) $$M^{-1}\left[\frac{\cos\pi p}{2}\,\Gamma(p)\,F(1-p);\,t\right] = \sqrt{\frac{\pi}{2}}\,F_c\,\{f(x);\,t\}$$

(ii) $$M^{-1}\left[\frac{\Gamma\left(\frac{p}{2}\right)}{\Gamma\left(\frac{p+1}{2}\right)}F(p);\,t\right] = \frac{2}{\sqrt{\pi}}\int_t^\infty \frac{f(x)\,dx}{\sqrt{x^2-t^2}}$$

Solution: Using the definition of Fourier cosine transform

$$M\left[\sqrt{\frac{\pi}{2}}\,F_c\,\{f(x);\,t\};\,p\right] = M\left[\sqrt{\frac{\pi}{2}}\cdot\sqrt{\frac{2}{\pi}}\int_0^\infty f(x)\cos(xt)dx;\,p\right]$$

$$= \int_0^\infty t^{p-1}\int_0^\infty f(x)\,\cos(xt)\,dx\,dt$$

$$= \int_0^\infty f(x)\left\{\int_0^\infty t^{p-1}\cos(xt)dt\right\}dx \qquad \text{...(i)}$$

Consider $\int\limits_{0}^{\infty} t^{p-1} \cos(xt)\, dt$

Since cos(*xt*) is the real part of e^{-ixt}

$$= \text{Real part of } \int\limits_{0}^{\infty} e^{-ixt}\, t^{p-1}\, dt$$

$$= \text{Real part of } L\{t^{p-1}, ix\} = \frac{\Gamma(p)}{(ix)^p}$$

$$= \Gamma(p)\, i^{-p}\, x^{-p} \qquad \text{...(ii)}$$

By equations (i) and (ii), we get

$$M\left[\sqrt{\frac{\pi}{2}}\, F_c\,\{f(x); t\}; p\right] = \int\limits_{0}^{\infty} f(x)\, \Gamma(p)\, i^{-p}\, x^{-p}\, dx$$

$$= \text{Real part of } \Gamma(p)\left(\cos\frac{\pi}{2} + i\sin\frac{\pi}{2}\right)^{-p} \int\limits_{0}^{\infty} t^{(1-p)-1} f(x)\, dx$$

$$= \text{Real part of } \Gamma(p)\left(\cos\frac{p\pi}{2} - i\sin\frac{\pi p}{2}\right) M\{f(x);\, 1-p\}$$

$$= \Gamma(p)\cos\frac{p\pi}{2}\, F(1-p)$$

(ii) Since $M\left[\frac{2}{\sqrt{\pi}}\int\limits_{t}^{\infty}\frac{f(x)dx}{x^2 - t^2}; p\right] = \int\limits_{0}^{\infty} t^{p-1}\left\{\frac{2}{\sqrt{\pi}}\int\limits_{t}^{\infty}\frac{f(x)\,dx}{\sqrt{x^2 - t^2}}\right\}dt$

$$= \frac{2}{\sqrt{\pi}}\int\limits_{0}^{\infty} t^{p-1}\int\limits_{t}^{\infty}\frac{f(x)\,dx}{\sqrt{x^2 - t^2}}\Bigg]dt$$

$$= \frac{2}{\sqrt{\pi}}\int\limits_{0}^{\infty} f(x)\, dx \int\limits_{t=0}^{t=x}\frac{t^{p-1}}{\sqrt{x^2 - t^2}}\,dt$$

Consider $I = \int\limits_{t=0}^{t=x}\frac{t^{p-1}}{\sqrt{x^2 - t^2}}\,dt$

Let $t = x \sin t \Rightarrow dt = x\cos\theta\; d\theta$

$$= \int_0^{\pi/2} \frac{(x \sin\theta)^{p-1}}{\sqrt{x^2 - x^2 \sin^2\theta}} \, x \cos\theta \, d\theta \qquad \text{...(i)}$$

$$= \int_0^{\pi/2} x^{p-1} (\sin\theta)^{p-1} \theta \cos\theta \, d\theta$$

$$= x^{p-1} \int_0^{\pi/2} \sin\theta^{p-1} \theta \cos^0\theta \, d\theta$$

$$= \frac{x^{p-1}\,\Gamma\left(\frac{p-1+1}{2}\right)\Gamma\left(\frac{0+1}{2}\right)}{2\Gamma\left(\frac{p-1+1+0+1}{2}\right)}$$

$$= \frac{x^{p-1}\,\Gamma\left(\frac{p}{2}\right)\Gamma\left(\frac{1}{2}\right)}{2\Gamma\left(\frac{p+1}{2}\right)} = x^{p-1} \frac{\sqrt{\pi}}{2} \frac{\Gamma\left(\frac{p}{2}\right)}{\Gamma\left(\frac{p+1}{2}\right)}$$

$$= x^{p-1} \frac{\sqrt{\pi}}{2} \frac{\Gamma\left(\frac{p}{2}\right)}{\Gamma\left(\frac{p+1}{2}\right)} \qquad \text{...(ii)}$$

Now, by equations (i) and (ii), we have in equation (i)

$$= \frac{2}{\sqrt{\pi}} \int_0^{\infty} f(x) \left[x^{p-1} \frac{\sqrt{\pi}}{2} \frac{\Gamma\left(\frac{p}{2}\right)}{\Gamma\left(\frac{p+1}{2}\right)} \right] dx$$

$$= \frac{\Gamma\left(\frac{p}{2}\right)}{\Gamma\left(\frac{p+1}{2}\right)} \int_0^{\infty} x^{p-1} f(x) dx$$

$$= \frac{\Gamma\left(\frac{p}{2}\right)}{\Gamma\left(\frac{p+1}{2}\right)} M\{f(x); p\}$$

$$= \frac{\Gamma\left(\frac{p}{2}\right)}{\Gamma\left(\frac{p+1}{2}\right)} F(p)$$

EXERCISE

1. Find $M\{e^{-t}\}$. [***Ans.*** $\Gamma(p)$]
2. Find $M\{\sin t\}$. [***Ans.*** $\Gamma(p) \sin(\pi p/2)$]
3. Find $M\{\cos t\}$. [***Ans.*** $\Gamma(p) \cos(\pi p/2)$]
4. Find $M\{(1 + x^{-1})\}$. [***Ans.*** $\pi/\sin(\pi p)$]
5. Find $M\{(1 + x^{a})^{-1}\}$, $a > 0$. $\left[\textbf{\textit{Ans.}}\ \frac{\pi}{a} \text{ cosec } (\pi p/a)\right]$
6. Find $M[x^m H(t - a)]$. $\left[\textbf{\textit{Ans.}}\ \frac{-a^{p+m}}{p+m} \text{ Re } (p) < -\text{ Re } (m)\right]$

Chapter 10

The Hankel Transform

10.1 INTRODUCTION

The Hankel transform is an integral transform and was first developed by the mathematician Hermann Hankel. It is also known as Fourier–Bessel's transform as the Hankel transform over an infinite interval is corresponding to the Fourier–Bessel's series over a finite interval. We shall study about the Hankel transform in this chapter and its various properties.

10.2 THE HANKEL TRANSFORM

Definition: Let $f(t)$ be a function of the variable $0 < t < \infty$, then the Hankel transform of $f(t)$ is denoted by $H_v\{f(t); p\}$ or $F_v(p)$, can be defined as

$$H_v\{f(t); p\} = F_v(p) = \int_0^\infty tJ_v(pt)\, f(t)\, dt$$

where, $J_v(pt)$ is the Bessel function of the first kind of order v.

Here, $tJ_v(pt)$ is kernel of the Hankel transform

10.3 SOME USEFUL RESULTS OF BESSEL FUNCTIONS

1. Bessel's function of first kind

$$J_v(t) = \sum_{r=0}^{\infty} \frac{(-1)^r}{r!\,\Gamma(v+r+1)}\left(\frac{t}{2}\right)^{v+2r}$$

2. Recurrence formula for $J_v(t)$

(i) $t\dfrac{d}{dt} J_v(t) = -vJ_v(t) + tJ_{v-1}(t)$

(ii) $t\dfrac{d}{dt} J_v(t) = vJ_v(t) - tJ_{v+1}(t)$

(iii) $2J_v'(t) = J_{v-1}(t) - J_{v+1}(t)$

(iv) $2\nu J_\nu(t) = t[J_{\nu+1}(t) + J_{\nu-1}(t)]$

(v) $\frac{d}{dt}[t^\nu J_\nu(t)] = t^\nu J_{\nu-1}(t)$

(vi) $\frac{d}{dt}[t^{-\nu} J_\nu(t)] = -t^{-\nu} J_{\nu+1}(t)$

(vii) $J_0'(t) = -J_1(t)$

3. Infinite integrals for based function

(i) $\int_0^\infty e^{-at} J_0(pt)dt = (a^2 + p^2)^{-1/2}$, where $a > 0$

(ii) $\int_0^\infty e^{-at} J_1(pt)dt = \frac{1}{p} - \frac{a}{p(a^2 + p^2)^{1/2}}$

(iii) $\int_0^\infty te^{-at} J_0(pt)dt = a(a^2 + p^2)^{-3/2}$

(iv) $\int_0^\infty te^{-at} J_1(pt)dt = p(a^2 + p^2)^{-3/2}$

(v) $\int_0^\infty e^{-at} J_1(pt)\frac{dt}{t} = \frac{(a^2 + b^2)^{1/2} - a}{p}$

(vi) $\int_0^\infty \sin at\, J_0(pt)dt = \begin{cases} \frac{1}{\sqrt{a^2 - p^2}}, & 0 < p < a \\ 0 \quad , & p > a \end{cases}$

(vii) $\int_0^\infty \cos at J_0(pt)dt = \begin{cases} \frac{1}{\sqrt{p^2 - a^2}}, & p > a \\ 0 \quad , & 0 < p < a \end{cases}$

10.4 SOME ELEMENTARY PROPERTIES OF THE HANKEL TRANSFORM

1. Linear Property

If $f(t)$ and $g(t)$ are two functions and c_1, c_2 two constants, then

$$H_\nu[c_1 f(t) + c_2 g(t); p] = c_1 H_\nu[f(t); p] + c_2 H_\nu\, g(t); p]$$

Proof: We have

$$H_\nu\, [c_1 f(t) + c_2 g(t); p] = \int_0^\infty t[c_1 f(t) + c_2 g(t)] J_\nu(pt)dt$$

$$= c_1 \int_0^\infty tf(t)J_v(pt)dt + c_2 \int_0^\infty tg(t)J_v(pt)dt$$

$$= c_1\, H_v\{f(t); p\} + c_2\, H_v\{g(t): p\}$$

Note: In general, let $F_v\,(p)$ are Hankel transform for all $i = 1, 2, .., n$ of $fi(t)$ and c_i are any constants, then $H_v[C_1f_1(t) + C_2f_2(t) + ... + C_nf_n(t); p] = c_1\, F_1(p) + c_2\, F_2(p) + ... + c_n\, F_n(p)$

2. Change of Scale Property

If $\quad H_v\{ft;p\} = F_v(p)$ then

$$H_v\{f(at);p\} = \frac{1}{p^2} F_v\left(\frac{p}{a}\right),\ a \neq 0$$

Proof: By definition

$$H_v\{f(at); p\} = \int_0^\infty tJ_v(pt)\, f(at)dt$$

Let $\quad at = u \Rightarrow adt = du$

$$\therefore \quad H_v\{f(at); p\} = \int_0^\infty \left(\frac{u}{a}\right) J_v\left(p\frac{u}{a}\right) f(u)\frac{du}{a}$$

$$= \frac{1}{a^2}\int_0^\infty u\, J_v\left(p\frac{u}{a}\right) f(u)du$$

$$= \frac{1}{a^2} H_v\left\{\frac{p}{a}; u\right\} = \frac{1}{a^2}\, F_v\left(\frac{p}{a}\right) \qquad \text{where } a \neq 0$$

3. If $\quad H_v\{f(t); p\} = F_v(p)$ then

$$H_v\left\{\frac{f(t)}{t}; p\right\} = \frac{p}{2v}\int_0^\infty [F_{v-1}(p) + F_{v+1}(p)]$$

Proof:
$$H_v\left\{\frac{f(t)}{t}; p\right\} = \int_0^\infty tJ_v(pt)\, \frac{f(t)}{t}\, dt$$

$$= \int_0^\infty J_v(pt)\, f(t)dt \qquad ...(1)$$

We know that

$$2v\, J_v(t) = t[J_{v-1}(t) + J_{v+1}(t)]$$

$$\therefore \quad 2v\, J_v(pt) = pt[J_{v-1}(pt) + J_{v+1}(pt)] \qquad ...(2)$$

From equations (i) and (ii), we have

$$H_{\nu}\left[\frac{1}{t} f(t); p\right] = \frac{p}{2\nu} \int_0^{\infty} f(t)[tJ_{\nu-1}(pt) + tJ_{\nu+1}(pt)]dt$$

$$= \frac{p}{2\nu}\left[\int_0^{\infty} t\, J_{\nu-1}(pt)\, f(t)dt + \int_0^{\infty} t\, J_{\nu+1}(pt)\, f(t)dt\right]$$

$$= \frac{p}{2\nu}\, [J_{\nu-1}(p) + J_{\nu+1}(p)]$$

4. If $\quad H_{\nu}\{f(t){:}p\} = F_{\nu}(p)$, then

$$H_{\nu}\{f(t-a)\, H(t-a);\, p\} = \sum_{m=-\infty}^{\infty} \alpha_m\, F_m(p)$$

where, $\quad \alpha_m = J_{\nu-m}(p) + \dfrac{pa}{2}\left\{\dfrac{J_{\nu-m-1}\,(pa)}{m+1} + \dfrac{J_{\nu-m+1}\,(pa)}{m-1}\right]$

Proof: $H_{\nu}\,\{f(t-a)H(t-a);\, p\} = \int_0^{\infty} tf(t-a)\, H(t-a)\, J_{\nu}(pt)dt$

$$= \int_0^{\infty} tf(t-a)\cdot 0\, J_{\nu}(pt)dt + \int_0^{\infty} tf(t-a)\cdot 1\cdot J_{\nu}(pt)dt$$

(Using definition of Heaviside unit function)

$$= \int_0^{\infty} tf(t-a)\, J_{\nu}(pt)dt$$

Let $\quad t - a = u \quad \Rightarrow \quad dt = du$

$$= \int_0^{\infty} (a+u)\, f(u)\, J_{\nu}(pa + pu)du \qquad ...(3)$$

From addition theorem we know that

$$J_{\nu}(pa + pu) = \sum_{m=-\infty}^{\infty} J_m(pu)\, J_{\nu-m}(pa) \qquad ...(4)$$

Using equation (4), equation (3) reduces to

$$H_{\nu}\{f(t-a)\, H(t-a);\, p\} = \int_0^{\infty} (a+u)\left[\sum_{m=-\infty}^{\infty} J_m(pu)\, J_{\nu-m}(pa)]\, f(u)du\right]$$

$$= \sum_{m=-\infty}^{\infty} J_{v-m}(pa)\left[\int_0^\infty u J_m(pu)\, f(u)du + a\int_0^\infty J_m(pu)\, f(u)du\right]$$

$$= \sum_{m=-\infty}^{\infty} J_{v-m}(pa)\left[F_m(p) + \frac{ap}{2m}\int_0^\infty u(J_{m-1}(pu) + J_{m+1}(pu)\right] f(u)du$$

$$= \sum_{m=-\infty}^{\infty} J_{v-m}(pa)\left[F_m(p) + \frac{ap}{2m}\left\{\int_0^\infty uJ_{m-1}(pu)f(u)du\right.\right.$$

$$\left.\left. + \int_0^\infty u\, J_{m+1}(pu)\, f(u)du\right\}\right]$$

$$= \sum_{m=-\infty}^{\infty} J_{v-m}(pa)\left[F_m(p) + \frac{ap}{2m}\{F_{m-1}(p) + F_{m+1}(p)\}\right]$$

$$= \sum_{m=-\infty}^{\infty} J_{v-m}(pa)\, F_m(p) + \frac{ap}{2}\sum_{m=-\infty}^{\infty}\frac{J_{v-m}(pa)F_{m-1}(p)}{m}$$

$$+\frac{ap}{2}\sum_{m=-\infty}^{\infty}\frac{J_{v-m}(pa)F_{m+1}(p)}{m}$$

$$= \sum_{m=-\infty}^{\infty} J_{v-m}(pa)\, F_m(p) + \frac{ap}{2}\sum_{m=-\infty}^{\infty}\frac{J_{v-m-1}(pa)F_{m-1}(p)}{(m+1)}$$

$$+\frac{ap}{2}\sum_{m=-\infty}^{\infty}\frac{J_{v-m+1}(pa)F_m(p)}{(m-1)}$$

$$= \sum_{m=-\infty}^{\infty} F_m(p)\left[J_{v-m}(pa) + \frac{ap}{2}\left\{\frac{J_{v-m-1}(pa)}{(m+1)} + \frac{J_{v-m+1}(pa)}{(m-1)}\right\}\right]$$

$$= \sum_{m=-\infty}^{\infty} F_m(p)\, \alpha_m$$

Example 1 Consider $t\, J_0(pt)$ as the kernel,

Find the Hankel transform of

(i) $\dfrac{\sin at}{t}$ (ii) $\dfrac{\cos at}{t}$

Solution: (i) $H_0\left\{\dfrac{\sin at}{t}; p\right\} = \displaystyle\int_0^\infty t\, J_0(pt)\, \frac{\sin at}{t}\, dt$

$$= \int_0^\infty \sin at\, J_0(pt)dt$$

$$= -\left[\text{Imaginary part of } \int_0^\infty e^{-iat} J_0(pt)dt\right]$$

$$= -[\text{Imaginary part of } (i^2 a^2 + p^2)^{-1/2}]$$

$$= -[\text{Imaginary part of } (p^2 - a^2)^{-1/2}]$$

$$= \begin{cases} 0; & p > a \\ (a^2 - p^2)^{-1/2}; & 0 < p < a \end{cases}$$

(ii) $$H_0\left\{\frac{\cos at}{t}; p\right\} = \int_0^\infty tJ_0(pt)\, \frac{\cos at}{t}\, dt$$

$$= \int_0^\infty \cos at\, J_0(pt)dt$$

$$= \text{Real part of } (i^2 a^2 + p^2)^{-1/2}$$

$$= \text{Real part of } (p^2 - a^2)$$

$$= \begin{cases} (p^2 - a^2)^{-1/2}, & p > a \\ 0, & 0 < p < a \end{cases}$$

Example 2 Find the Hankel transform of

(i) e^{-t} (ii) $\frac{e^{-t}}{t}$ (iii) e^{-at} (iv) $\frac{e^{-at}}{t}$

by considering $t\, J_0(pt)$ as the kernel.

Solution: (i) $H_0\{e^{-t}; p\} = \int_0^\infty t\, J_0(pt)e^{-t}dt$

$$= (1 + p^2)^{-3/2}$$

(ii) $$H_0\left\{\frac{e^{-t}}{t}; p\right\} = \int_0^\infty tJ_0(pt)\frac{e^{-t}}{t}\, dt$$

$$= \int_0^\infty e^{-t} J_0(pt)\, dt$$

$$= (1 + p^2)^{-1/2}$$

(iii) $$H_0\{e^{-at}; p\} = \int_0^\infty tJ_0(pt)e^{-at}dt$$
$$= a(a^2 + p^2)^{-3/2}$$

(iv) $$H_0\left\{\frac{e^{-at}}{t}; p\right\} = \int_0^\infty t\, J_0(pt)\frac{e^{-at}}{t}\, dt$$
$$= \int_0^\infty J_0(pt)e^{-at}\, dt$$
$$= (a^2 + p^2)^{-1/2}$$

Relation between Hankel and Laplace Transform

$$H_v\{e^{-at}f(t); p\} = \int_0^\infty t\, J_v(pt)e^{-at}f(t)dt$$
$$= \int_0^\infty e^{-at}\{tj_v(pt)f(t)\}dt$$
$$= L\{tJ_v(pt)\, f(t); a\}$$

Example 3 Prove that

$$H_v\{t^{-1}e^{-at}; p\} = \frac{1}{\sqrt{a^2+p^2}}\left\{\frac{p}{a+\sqrt{a^2+p^2}}\right\}^v.$$

Solution: By the definition of the Hankel transform

$$H_v\{t^{-1}e^{-at}; p\} = \int_0^\infty tJ_v(pt)\, t^{-1}e^{-at}dt$$
$$= \int_0^\infty e^{-at}J_v(pt)dt = L\{J_v(pt); a\}$$

We know that

$$L\{J_v(pt); a\} = \frac{\{\sqrt{a^2+p^2}-a\}^v}{p^v\sqrt{a^2+p^2}}$$
$$= \frac{1}{\sqrt{a^2+p^2}}\left[\frac{\sqrt{a^2+p^2}-a}{p}\times\frac{\sqrt{a^2+p^2}+a}{\sqrt{a^2+p^2}+a}\right]^v$$

$$= \frac{1}{\sqrt{a^2+p^2}}\left[\frac{p}{a+\sqrt{a^2+p^2}}\right]^v$$

Example 4 Prove that

$$H_v\{e^{-at}; p\} = \frac{a+v\sqrt{a^2+p^2}}{(a^2+p^2)^{3/2}}\left\{\frac{p}{a+\sqrt{a^2+p^2}}\right\}^v$$

Solution: By definition of the Hankel transform

$$H_v\{e^{-at}; p\} = \int_0^\infty tJ_v(pt)e^{-at}dt$$

$$=L\{tJ_v(pt);\, a\} \qquad ...(1)$$

We know that

$$L\{J_v(pt);\, a\} = \frac{1}{\sqrt{a^2+p^2}}\left\{\frac{p}{a+\sqrt{a^2+p^2}}\right\}^v \qquad ...(2)$$

We also know that

$$L\{t^n f(t);\, p\} = (-1)^n \frac{d^n}{dp^n} F(p) \qquad ...(3)$$

By using equations (2) and (3) in equation (1), we get

$$H_v\{e^{-at}; p\} = (-1)\frac{d}{da}L\{J_v(pt);\, a\}$$

$$= (-1)\frac{d}{da}\left\{\frac{1}{\sqrt{a^2+p^2}}\left(\frac{p}{a+\sqrt{a^2+p^2}}\right)^v\right\}$$

$$= \frac{a}{(a^2+p^2)^{3/2}}\left(\frac{p}{a+\sqrt{a^2+p^2}}\right)^v + \left(\frac{p}{a+\sqrt{a^2+p^2}}\right)^v\frac{v}{(a^2+p^2)}$$

$$= \left(\frac{p}{a+\sqrt{a^2+p^2}}\right)^v\left[\frac{a}{(a^2+p^2)^{3/2}}+\frac{v}{(a^2+p^2)}\right]$$

$$= \left(\frac{p}{a+\sqrt{a^2+p^2}}\right)^v \left[\frac{a+v\sqrt{a^2+p^2}}{(a^2+p^2)^{3/2}}\right]$$

Example 5 If $v > -1$, prove that

$$H_v\{t^v e^{-at}; p\} = \frac{p^v}{(2a)^{v+1}} e^{-p^2/4a}.$$

Solution: $H_v\{t^v e^{-at^2}; p\} = \int_0^\infty tJ_v(pt)t^v e^{-at^2} dt$

$$= \int_0^\infty t^{v+1} J_v(pt) e^{-at^2} dt$$

$$= \int_0^\infty t^{v+1} \sum_{r=0}^\infty \frac{(-1)^r}{r!\,\Gamma(v+r+1)} \left(\frac{pt}{2}\right)^{v+2r} e^{-at^2} dt$$

$$= \sum_{r=0}^\infty \frac{(-1)^r}{r!\,\Gamma(v+r+1)} \left(\frac{p}{2}\right)^{v+2r} \int_0^\infty t^{2v+2r+1}\, e^{-at^2} dt$$

Let $t^2 = u \Rightarrow 2tdt = du$

$$= \sum_{r=0}^\infty \frac{(-1)^r}{r!\,\Gamma(v+r+1)} \frac{1}{2} \left(\frac{p}{2}\right)^{2v+2r} \int_0^\infty u^{v+r}\, e^{-au} du$$

$$= \sum_{r=0}^\infty \frac{(-1)^r}{r!\,\Gamma(v+r+1)} \frac{p^{v+2r}}{2^{v+2r+1}} \frac{\Gamma(v+r+1)}{a^{v+r+1}}$$

(From the definition of Gamma function)

$$= \sum_{r=0}^\infty \frac{p^v}{(2a)^{v+1}} \frac{(-1)^r}{r!} \frac{p^{2r}}{2^{2r} a^r}$$

$$= \frac{p^v}{(2a)^{v+1}} \sum_{r=0}^\infty \frac{(-1)^r (p^2)^r}{(4a)^r \cdot r!}$$

$$= \frac{p^v}{(2a)^{v+1}} \sum_{r=0}^\infty \frac{\left(-\frac{p^2}{4a}\right)^r}{r!}$$

$$= \frac{p^{\nu}}{(2a)^{\nu+1}} e^{-\frac{p^2}{4a}}.$$

Example 6 Prove that

$$H_0\left[e^{\frac{-a^2t^2}{4}} \sin(bt^2u);t\right]$$

$$= 2a^{-2}(1+\nu^2)^{-1} e^{-R^2/(1+\nu^2)}\left[\nu\cos\frac{R^2\nu}{1+\nu^2} - \sin\frac{R^2\nu}{1+\nu^2}\right]$$

where $R = \dfrac{p}{a}$, and $\nu = 4a^{-2}bu$.

Solution: Consider

$$H_0\left[e^{\frac{-a^2t^2}{4}} e^{-ibt^2u};p\right]$$

$$= \int_0^{\infty} t\, e^{-\left(\frac{a^2}{4}+ibu\right)t^2} J_0(pt)dt$$

Put $t^2 = x \Rightarrow 2t\,dt = dx$

$\therefore$

$$H_0\left[e^{\frac{-a^2t^2}{4}} e^{-ibt^2u};p\right]$$

$$= \frac{1}{2}\int_0^{\infty} e^{-\left(\frac{a^2}{4}+ibu\right)x} J_0\left(p\sqrt{x}\right)dx$$

$$= \frac{1}{2}\int_0^{\infty} e^{-\left(\frac{a^2}{4}+ibu\right)x} \sum_{r=0}^{\infty} \frac{(-1)^r\left(\frac{p\sqrt{x}}{2}\right)^{2r}}{r!\,\Gamma(r+1)} dt$$

$$= \frac{1}{2}\sum_{r=0}^{\infty} \frac{(-1)^r\left(\frac{p}{2}\right)^{2r}}{(r!)^2} \int_0^{\infty} x^r\, e^{-\left[\frac{a^2}{4}+ibu\right]x} dx$$

$$= \sum_{r=0}^{\infty} \frac{1}{2^{2r+1}} \frac{(-1)^r p^{2r}}{(r!)^2} \frac{\Gamma(r+1)}{\left[\frac{a^2}{4} + ibu\right]^{r+1}}$$

(From the definition of Gamma function)

$$= \frac{1}{2} \sum_{r=0}^{\infty} \frac{p^{2r}}{2^{2r}} \frac{(-1)^r}{r!} \frac{2^{2r+2}}{(a^2 + 4ibu)^{r+1}}$$

$$= \frac{2}{a^2} \sum_{r=0}^{\infty} \frac{p^{2r}}{a^{2r}} \frac{(-1)^r}{r!\left(1 + \frac{4ibu}{a^2}\right)^{r+1}}$$

$$= \frac{2}{a^2} \sum_{r=0}^{\infty} \frac{(-1)^r \left(\frac{p}{a}\right)^{2r} \left(1 - \frac{4ibt}{a^2}\right)^{r+1}}{r!\left[1 + \frac{16b^2t^2}{a^4}\right]^{r+1}}$$

$$= \frac{2}{a^2} \frac{\left[1 - \frac{4ibt}{a^2}\right]}{\left[1 - \frac{16b^2t^2}{a^4}\right]} \sum_{r=0}^{\infty} \frac{(-1)^r \left(\frac{p}{a}\right)^{2r} \left(1 - \frac{4ibt}{a^2}\right)^r}{r!\left[1 + \frac{16b^2t^2}{a^4}\right]^r}$$

$$= \frac{2}{a^2} \frac{(1 - iv)}{(1 + v^2)} \sum_{r=0}^{\infty} \frac{1}{r!} \left[(-1)^r \left(\frac{p^2}{a^2}\right)^r \frac{(1 - iv)^2}{(1 + v^2)^r}\right]$$

$$= \frac{2}{a^2} \frac{(1 - iv)}{(1 + v^2)} \sum_{r=0}^{\infty} \frac{1}{r!} \left[(-1)^r R^2 \left(\frac{1 - iv}{1 + v^2}\right)^r\right]$$

$$= \frac{2}{a^2} \frac{(1 - iv)}{(1 + v^2)} \exp\left[-R^2 \frac{(1 - iv)}{1 + v^2}\right]$$

$$= 2a^{-2} (1 + v^2)^{-1} (1 - iv) \exp\left(\frac{-R^2}{1 + v^2}\right) \exp\left(\frac{iR^2 v}{1 + v^2}\right)$$

$$= 2a^{-2} (1 + v^2)^{-1} (1 - iv) \exp\left(\frac{-R^2}{1 + v^2}\right) \left[\cos\frac{R^2 v}{1 + v^2} + i\sin\frac{R^2 v}{1 + v^2}\right]$$

$$\Rightarrow \qquad H_0\left[e^{-\frac{a^2t^2}{4}}(\cos bt^2u - i\sin bt^2u);p\right]$$

$$= 2a^{-2}(1+v^2)^{-1}\exp\left(\frac{-R^2}{1+v^2}\right)\left[\cos\frac{R^2v}{1+v^2}+i\sin\frac{R^2v}{1+v^2}\right.$$

$$\left.-iv\cos\frac{R^2v}{1+v^2}+v\sin\frac{R^2v}{1+v^2}\right]$$

Comparing real parts on both sides

$$H_0\left[\frac{-a^2t^2}{4}\cos bt^2u;p\right] = 2a^{-2}(1+v^2)^{-1}\exp\left(\frac{-R^2}{1+v^2}\right)\left[\cos\frac{R^2v}{1+v^2}+v\sin\frac{R^2v}{1+v^2}\right]$$

and comparing imaginary parts

$$H_0\left[e\frac{-a^2t^2}{4}\sin bt^2u;p\right] = 2a^{-2}(1+v^2)^{-1}e^{\frac{-R^2}{1+v^2}}\left[v\cos\frac{R^2v}{1+v^2}-\sin\frac{R^2v}{1+v^2}\right]$$

where, $R = \dfrac{p}{a}$ and $v = 4a^{-2}bu$

Hankel Transform of the Derivatives of a Function

Theorem: Let $F_v(p)$ be the Hankel transform of order v of the function $f(t)$ and $F_v'(p)$ be the Hankel transform of $f'(t)$. Then

$$H_v\{f'(t)\,;\,p\} = F_v'(p)$$

$$= \frac{-p}{2v}\,[(v+1)F_{v-1}(p) - (v-1)F_{v+1}(p)]$$

Proof: We know that $H_v\{f(t)\,;\,p\} = F_v(p) = \int_0^\infty tJ_v(pt)\,f(t)\,dt$

and $\qquad H_v\{f'(t)\,;\,p\} = F_v'(p) = \int_0^\infty \{tJ_v(pt)\}\dfrac{df}{dt}\,dt$

Integrating by parts assuming $\dfrac{df}{dt}$ as the second function, we have

$$F_v'(p) = \left[tJ_v(pt)f(t)\right]_0^\infty - \int_0^\infty f(t)\frac{d}{dt}\{tJ_v(pt)\}dt$$

$$= 0 - \int_0^\infty f(t)\left[J_v(pt)+t\frac{d}{dt}\{J_v(pt)\}\right]dt \qquad ...(1)$$

We know that the recurrence relation for $J_v(t)$

$$t\frac{d}{dt}J_v(t) = t\,J_{v-1}(t) - v\,J_v(t)$$

On replacing t by pt, we get

$$pt\frac{d}{pdt}\,J_v(pt) = pt\,J_{v-1}(pt) - v\,J_v(pt)$$

or
$$t\frac{d}{dt}\{J_v(pt)\} = pt\,J_{v-1}(pt) - v\,J_v(pt)$$

or
$$J_v(pt) + t\frac{d}{dt}\,\{J_v(pt)\} = J_v(pt) + pt\,J_{v-1}(pt) - v\,J_v(pt)$$

or
$$J_v(pt) + t\frac{d}{dt}\,\{J_v(pt)\} = -\,(v-1)J_v(pt) + pt\,J_{v-1}(pt) \qquad ...(2)$$

On using equation (2) in equation (1), we get

$$F_v'(p) = \int_0^\infty f(t)\{(v-1)J_v(pt) - pt\,J_{v-1}(pt)\}dt$$

$$= (v-1)\int_0^\infty f(t)J_v(pt) - p\int_0^\infty f(t)\,t\,J_{v-1}(pt)dt \qquad ...(3)$$

We know recurrence relation for $J_n(t)$,

$$2vJ_v(t) = t[J_{v-1}(t) + J_{v+1}(t)]$$

On replacing t by pt, we get

$$J_v(pt) = \frac{pt}{2v}\,[J_{v-1}(pt) + J_{v+1}(pt)] \qquad ...(4)$$

By equation (4), equation (3) reduces to

$$F_v'(p) = \frac{p(v-1)}{2v}\int_0^\infty f(t)\,t\{J_{v-1}(pt) + J_{v+1}(pt)\}dt - p\int_0^\infty f(t)\,tJ_{v-1}(pt)dt$$

$$= \frac{p(v-1)}{2v}\int_0^\infty f(t)\,tJ_{v+1}(pt)dt - \frac{p(v+1)}{2v}\int_0^\infty f(t)\,tJ_{v-1}(pt)dt$$

$$= \frac{p(v-1)}{2v}F_{v+1}(p) - \frac{p(v+1)}{2v}F_{v-1}(p)$$

or
$$F_v'(p) = -\frac{p}{2v}\,[(v+1)F_{v-1}(p) - (v-1)F_{v+1}(p)] \qquad ...(5)$$

Special Cases

(i) When $v = 1$, then equation (5) reduces to

$$F_1'(p) = -\frac{p}{2}[2F_0(p)] = -pF_0(p)$$

or $H_1\{f'(t);p\} = -p\, H_0\{f(t);\, p\}$

(ii) When $v = 2$, then equation reduces to

$$F_2'(p) = -\frac{p}{4}[3F_1(p)] - F_3(p)]$$

Theorem: Prove that

$$H_v\{f''(t)\,;\, p\} = F_v''(p)$$

$$= \frac{p^2}{4}\left[\left(\frac{v+1}{v-1}\right)F_{v-2}(p) - 2\left(\frac{v^2-3}{v^2-1}\right)F_v(p) + \left(\frac{v-1}{v+1}\right)F_{v+2}(p)\right]$$

Proof: We know that

$$H_v\{f'(t)\,;\, p\} = F_v''(p) = -p\left[\frac{(v+1)}{2v}F_{v-1}(p) - \left(\frac{(v-1)}{2v}\right)F_{v+1}(p)\right] \qquad ...(i)$$

Replacing $f(t)$ by $f'(t)$ in equation (i), we get

$$F_v''(p) = -p\left[\frac{(v+1)}{2v}F_{v-1}'(p) - \frac{(v-1)}{2v}F_{v+1}'(p)\right] \qquad ...(ii)$$

To get the value of $F'_{v-1}(p)$ and $F'_{v+1}(p)$, and on replacing v by $(v-1)$, and $v+1$ respectively in equation (i), we get

$$F'_{v-1}(p) = -p\left[\frac{v}{2(v-1)}F_{v-2}(p) - \frac{(v-2)}{2(v-1)}F_v(p)\right] \qquad ...(iii)$$

and

$$F'_{v+1}(p) = -p\left[\frac{v+2}{2(v+1)}F_v(p) - \frac{v}{2(v+1)}F_{v+2}(p)\right] \qquad ...(iv)$$

Using equations (iii) and (iv) in equation (ii), we get

$$F_v''(p) = \frac{p^2}{4}\left[\left(\frac{v+1}{v}\right)\left\{\frac{v}{v-1}F_{v-2}(p) - \left(\frac{v-2}{v-1}\right)F_v(p)\right\}\right.$$

$$-\left(\frac{\nu-1}{\nu}\right)\left\{\frac{\nu+2}{\nu+1}F_\nu(p)-\left(\frac{\nu}{\nu+1}\right)F_{\nu+2}(p)\right\}$$

$$=\frac{p^2}{4}\left[\left(\frac{\nu+1}{\nu-1}\right)F_{\nu-2}(p)-2\frac{(\nu^2-3)}{(\nu^2-1)}F_\nu(p)+\left(\frac{\nu-1}{\nu+1}\right)F_{\nu+2}(p)\right]$$

Theorem If $H_\nu\{f(t); p\} = F_\nu(p)$, then prove that

$$H_\nu\left\{\frac{d^2f}{dt^2}+\frac{1}{t}\frac{df}{dt}-\frac{\nu^2 f}{t^2};p\right\} = -p^2 F_\nu(p) \qquad ...(i)$$

Proof: $H_\nu\left\{\frac{d^2f}{dt^2}+\frac{1}{t}\frac{df}{dt}-\frac{\nu^2 f}{t^2};p\right\}$

$$=\int_0^\infty tJ_\nu(pt)\left\{\frac{d^2f}{dt^2}+\frac{1}{t}\frac{df}{dt}-\frac{\nu^2}{t^2}f\right\}dt$$

$$=\int_0^\infty J_\nu(pt)\left\{t\frac{d^2f}{dt^2}+\frac{df}{dt}\right\}\nu^2\int_0^\infty \frac{1}{t}J_\nu(pt)f(t)dt$$

$$=\int_0^\infty J_\nu(pt)\frac{d}{dt}\left\{t\frac{df}{dt}\right\}dt-\nu^2\int_0^\infty \frac{1}{t}J_\nu(pt)f(t)dt \qquad ...(ii)$$

On integration by parts of the first term of equation (ii) in R.H.S., we get

$$=\left[J_\nu(pt)t\frac{df}{dt}\right]_0^\infty-\int_0^\infty \frac{d}{dt}\{J_\nu(pt)\}t\frac{df}{dt}dt-\nu^2\int_0^\infty \frac{1}{t}J_\nu(pt)f(t)dt$$

$$=-\int_0^\infty \left\{t\frac{d}{dt}J_\nu(pt)\right\}\frac{df}{dt}dt-\nu^2\int_0^\infty \frac{1}{t}J_\nu(pt)f(t)dt$$

(Assuming that $t f'(t) \to 0$ when $t \to 0$ or $t \to \infty$)

Now, again integrating by parts the first term of above equation taking $\frac{df}{dt}$ as second function

$$=-\left[t\frac{d}{dt}\{J_\nu(pt)\}f(t)\right]_0^\infty+\int_0^\infty \frac{d}{dt}\left[t\frac{d}{dt}\{J_\nu(pt)\}\right]f(t)dt$$

$$-\nu^2\int_0^\infty \frac{1}{t}J_\nu(pt)f(t)dt$$

$$= 0+\int_0^\infty \frac{d}{dt} J_\nu(pt)f(t)dt + \int_0^\infty t\frac{d^2}{dt^2} J_\nu(pt)f(t)dt - \nu^2\int_0^\infty \frac{1}{t} J_\nu(pt)f(t)dt$$

(Assuming $t\,f(t) \to 0$ where $t \to 0$ or $t \to \infty$)

$$= \int_0^\infty \left[t\frac{d^2}{dt^2} J_\nu(pt) + \frac{d}{dt} J_\nu(pt) - \frac{\nu^2}{t} J_\nu(pt) \right] f(t)dt$$

$$= \int_0^\infty \frac{1}{t}\left[t^2\frac{d^2}{dt^2} J_\nu(pt) + t\frac{d}{dt} J_\nu(pt) - \nu^2 J_\nu(pt) \right] f(t)dt$$

Now, since $J_\nu(t)$ satisfies the Bessel's differential equation which is

$$t^2\frac{d^2y}{dt^2} + t\frac{dy}{dt} + (t^2 - \nu^2)y = 0$$

$$\Rightarrow \qquad t^2\frac{d^2}{dt^2} J_\nu(t) + t\frac{d}{dt} J_\nu(t) + (t^2 - \nu^2)(t) = 0$$

On replacing t by pt, we get

$$(p^2t^2)\frac{d^2}{p^2dt^2} J_\nu(pt) + pt\frac{d}{pdt} J_\nu(pt) + (p^2t^2 - \nu^2)J_\nu(pt) = 0$$

or $$t^2\frac{d^2}{dt^2} J_\nu(pt) + t\frac{d}{dt} J_\nu(pt) - \nu^2 J_\nu(pt) = -\,p^2t^2\, J_\nu(pt)$$

Thus, we have

$$H_\nu\left\{\frac{d^2f}{dt^2} + \frac{1}{t}\frac{d}{dt}\frac{-\nu^2}{t^2} f; p\right\}$$

$$= \int_0^\infty \frac{1}{t}\,[-\,p^2t^2\, J_\nu(pt)]\, f(t)dt$$

$$= -\,p^2 \int_0^\infty t J_\nu(pt)\, f(t)dt$$

$$= -\,p^2\, H_\nu\,\{f(t);\, p\} = -\,p^2\, F_\nu(p)$$

Special cases:

(i) For $\nu = 0$, we have

$$\int_0^\infty \left(\frac{d^2f}{dt^2} + \frac{1}{t}\frac{df}{dt}\right) t\, J_0(pt)dt = -\,p^2\, F_0(p)$$

where $F_0(p) = H_0\{f(t); p\}$ = Hankel transform of function $f(t)$ of order zero.

(ii) For $v = 1$, we have

$$\int_0^\infty \left[\frac{d^2 f}{dt^2} + \frac{1}{t}\frac{df}{dt} - \frac{1}{t^2} f\right] t\, J_1(pt)dt = -\, p^2 F_1(p)$$

where, $F_1(p) = H_1\{f(t); p\}$ = Hankel transform of function $f(t)$ of order one.

10.5 INVERSION FORMULA FOR THE HANKEL TRANSFORM

If $F_v(p)$ is the Hankel transform of the function $f(t)$ then,

$$F_v(p) = \int_0^\infty t\, J_v(pt)\, f(t)dt,$$

then,

$$f(t) = \int_0^\infty p\, J_v(pt)\, F_v(p)dp$$

is called the inversion formula for the Hankel transform of $F_v(p)$ and can be written as

$$f(t) = H_v^{-1}\,\{F_v(p);t\}$$

Proof: If $F(p)$ denotes the complex Fourier transform of $f(t)$, i.e.,

$$F(p) = \int_{-\infty}^{\infty} f(t)e^{ipt}\, dt$$

then $f(t)$ can be written as

$$f(t) = \frac{1}{2\pi}\int_{-\infty}^{\infty} F(p)e^{-ipt}\, dt$$

Extending the above two results to cover functions of two variables, we have

$$F(u, \xi) = \int_{-\infty}^{\infty} f(t, u)\, e^{i(vt+wu)}dt\, du \qquad \text{...(i)}$$

Then,

$$f(t, u) = \frac{1}{4\pi^2}\int_{-\infty}^{\infty} F(v, w)\, e^{-i(vt+wu)}dv\, dw \qquad \text{...(ii)}$$

Substituting $t = r\cos\theta$, $u = r\sin\theta$, $v = p\cos\alpha$ and $w = p\sin\alpha$ in equations (i) and (ii), we get

$$F(p, \alpha) = \int_0^\infty r\,dr \int_0^{2\pi} f(r, \theta)\, e^{ipr\cos(\theta-\alpha)}\, dv \qquad \text{....(iii)}$$

and

$$f(r, \theta) = \frac{1}{4\pi^2}\int_0^\infty p\,dp \int_0^{2\pi} F(p, \alpha)\, e^{-ipr\cos(\theta-\alpha)}\, d\alpha \qquad \text{...(iv)}$$

Let us suppose that $f(r, \theta) = f(r)\, e^{-is\theta}$ ...(v)

Then equation (iii) reduces to

$$F(p, \alpha) = \int_0^\infty r f(r) dr \int_0^{2\pi} e^{i[-s\theta + pr\cos[\theta - \alpha]} d\theta \qquad \text{...(vi)}$$

Now, let $\phi = \alpha - \theta - \dfrac{\pi}{2}$ then

$$\int_0^{2\pi} e^{i[-s\theta + pr\cos(\theta-\alpha)]} d\theta = \int_0^{2\pi} e^{i\left[s\left(\phi - \alpha + \frac{\pi}{2}\right) + pr\cos\left(\phi - \frac{\pi}{2}\right)\right]} d\phi$$

$$= \int_0^{2\pi} e^{is\left(\frac{\pi}{2} - \alpha\right)} e^{i\left[s\phi + pr\cos\left(\phi + \frac{\pi}{2}\right)\right]} d\phi$$

$$= e^{is\left(\frac{\pi}{2} - \alpha\right)} \int_0^{2\pi} e^{i[s\phi - pr\sin\phi]} d\phi$$

$$= e^{is\left(\frac{\pi}{2} - \alpha\right)} 2\pi J_v(pr) \qquad \text{...(vii)}$$

$$\left[\text{We know that } J_v(pr) = \frac{1}{2\pi} \int_0^{2\pi} e^{i(s\phi - pr\sin\phi)} d\phi\right]$$

Using equation (vii), equation (vi) reduces to

$$F(p, \alpha) = \int_0^\infty f(r) \cdot r e^{is\left(\frac{\pi}{2} - \alpha\right)} 2\pi J_v(pr)\, dr$$

$$= 2\pi e^{is\left(\frac{\pi}{2} - \alpha\right)} \int_0^\infty f(r) \cdot r J_v(pr)\, dr$$

$$= 2\pi e^{is\left(\frac{\pi}{2} - \alpha\right)} F_v(p) \qquad \text{...(viii)}$$

where, $F_v(p)$ is the Hankel transform of the function $f(r)$.

Now using equations (v) and (viii) in equation (iv), we get

$$f(r) e^{-is\theta} = \frac{1}{4\pi^2} \int_0^\infty p\, dp \int_0^{2\pi} 2\pi e^{is\left(\frac{\pi}{2} - \alpha\right)} F_v(p)\, e^{-ipr\cos(\theta - \alpha)}\, dx$$

$$= \frac{1}{2\pi}\int_0^{\infty} pF_v(p)dp \int_0^{2\pi} e^{is\left[s\left(\frac{\pi}{2}-\alpha\right)-pr\cos(\theta-\alpha)\right]} d\alpha \qquad ...(ix)$$

Now, let $\psi = \theta-\alpha+\frac{\pi}{2}$, then

$$\int_0^{2\pi} e^{i\left[s\left(\frac{\pi}{2}-\alpha\right)-pr\cos(\theta-\alpha)\right]} d\alpha$$

$$= \int_0^{2\pi} e^{is(\psi-\theta)-ipr\cos\left(\frac{\pi}{2}-\alpha\right)} d\psi$$

$$= e^{-is\theta}\int_0^{2\pi} e^{i[s\psi-pr\sin\psi)} d\psi$$

$$= e^{-is\theta}\, 2\pi\, J_v(pr) \qquad ...(x)$$

Using the value of the integral from equation (x) in equation (ix), we have

$$f(r)e^{-is\theta} = \frac{1}{2\pi}\int_0^{\infty} pF_v(p)\, e^{-is\theta}\, 2\pi\, J_v(pr)dp$$

$$= e^{-is\theta}\int_0^{\infty} pF_v(p)\, J_v(pr)dp$$

or
$$f(r) = \int_0^{\infty} pF_v(p)\, J_v(pr)dp \qquad ...(xi)$$

Equation (xi) is the required inversion formula for the Hankel transform.

10.6 PARSEVAL'S THEOREM FOR HANKEL TRANSFORMS

Theorem: If $F_v(p)$ and $G_v(p)$ ard the Hankel transforms of the functions $f(t)$ and $g(t)$ respectively, then

$$\int_0^{\infty} t\, f(t)\, g(t)dt = \int_0^{\infty} pF_v(p)\, G_v(p)dp \qquad ...(1)$$

Proof: We know that

$$F_v(p) = \int_0^{\infty} tJ_v(pt)f(t)dt \qquad ...(2)$$

and
$$G_v(p) = \int_0^{\infty} tJ_v(pt)\, g(t)dt \qquad ...(3)$$

Using equation (3) in R.H.S. of equation (4)

$$\int_0^\infty p F_v(p) F_v(p)dp = \int_0^\infty p F_v(p)\left[\int_0^\infty tJ_v(pt)g(t)dt\right]dp$$

Changing the order of integration

$$= \int_0^\infty t g(t)\left[\int_0^\infty pF_v(p)J_v(pt)dp\right]dt$$

$$= \int_0^\infty t g(t) f(t)dt \text{ [By inversion formula]}$$

$$= \int_0^\infty p F_v(p) G_v(p)dp = \int_0^\infty tf(t)\, g(t)dt$$

Example 7 Find $H_0^{-1}\left\{\frac{e^{-ap}}{p};t\right\}$.

Solution: By the definition of inverse Hankel transform

$$H_0^{-1}\left\{\frac{e^{-ap}}{p};t\right\} = \int_0^\infty pJ_0(pt)\frac{e^{-ap}}{p}dp$$

$$= \int_0^\infty e^{-ap} J_0(pt)dp$$

$$= (a^2 + t^2)^{-1/2}$$

Example 8 Find the Hankel transform of order zero of $\frac{1}{t}$ and then apply the inversion formula to set the original function.

Solution: Since $H_0\left\{\frac{1}{t};p\right\} = \int_0^\infty tJ_0(pt)\frac{1}{t}\,dt$

$$= \int_0^\infty J_0(pt)dp$$

We know that $\int_0^\infty e^{-ap} J_0(pt)dt = (a^2 + t^2)^{-1/2}$, put $a = 0$

$$\int_0^\infty J_0(pt)dp = \frac{1}{p}$$

Now, we apply the inversion formula

$$f(t) = \int_0^\infty p J_0(pt)\, H_0(p)dp$$

$$= \int_0^\infty p J_0(pt)\, \frac{1}{p}\, dp$$

$$= \int_0^\infty J_0(pt)dp = (0^2 + t^2)^{-1/2} = \frac{1}{t}$$

Example 9 If $H_\nu^{-1}\{F_\nu(p); t\} = f(t)$, then show that

$$H_\nu^{-1}\{F_\nu(ap); t\} = \frac{1}{a^2} f\left(\frac{t}{a}\right), \quad a > 0.$$

Solution: By inversion formula we know that

$$H_\nu^{-1}\{F_\nu(p); t\} = \int_0^\infty p J_\nu(pt)\, F_\nu(p)dp$$

$$\therefore \qquad H_\nu^{-1}\{F_\nu(ap); t\} = \int_0^\infty p J_\nu(pt)\, F_\nu(p)dp$$

Let $ap = u$ so that $a\, dp = du$

$$\int_0^\infty \left(\frac{u}{a}\right) J_\nu\left(t\frac{u}{a}\right) F_\nu(u)\frac{du}{a}$$

$$= \frac{1}{a^2}\int_0^\infty uJ_\nu\left(\frac{u}{a}t\right) F_\nu(u)du = \frac{1}{a^2} f\left(\frac{t}{a}\right)$$

Example 10 Find the Hankel transform of $\frac{\partial^2 f}{dx^2}$ of order ν *wr* to variable t, where, f is a function of x and t.

Solution: $H_\nu\left\{\frac{\partial^2 f}{\partial x^2}; p\right\} = \int_0^\infty tJ_r(pt)\frac{\partial^2 f}{\partial x^2}dt$

$$= \frac{\partial^2}{\partial x^2}\int_0^\infty f(x, t)t\, J_\nu|pt)\, dt$$

$$= \frac{\partial^2}{\partial x^2} f(x, t)t\, J_\nu|pt)\, dt$$

Example 11 Find the Hankel transform of $f(t) = e^{-at}$, $a > 0$ taking $tJ_0/pt)$ as the kernel and show that

$$H_o\left\{\frac{1}{(a^2 + t^2)^{3/2}}; p\right\} = \frac{e^{-ap}}{a}, \; a > 0$$

Solution: $H_o\left\{e^{-at};p\right\}=\int_0^\infty t\,J_0(pt)e^{-at}\,dt$

$$= a(a^2+p^2)^{-3/2}$$

$$\int_0^\infty a(a^2+p^2)^{-3/2}\,pJ_0(pt)\,dp = e^{-at}$$

Interchanging t and p, it follows that

$$\int_0^\infty tJ_0(pt)\,(a^2+t^2)^{-3/2}\,dt = \frac{e^{-ap}}{a}$$

$$\therefore \qquad H_0\left\{\frac{1}{(a^2+t^2)^{3/2}};p\right\} = \frac{e^{-ap}}{a},\ a>0$$

Example 12 Prove that

$$H_v\left\{e^{v-1}\frac{\partial}{\partial e}e^{1-v}f(e);\xi\right\} = \xi H_{v-1}\{f(e);\xi\}$$

Provided that

$$\lim_{e\to 0} e^{v+1}f(e) = 0 \text{ and } \lim_{e\to\infty} e^{1/2}f(e) = 0$$

Solution: By definition, we have

$$H_v\left[e^{v-1}\frac{\partial}{\partial e}\{e^{1-v}f(e)\};\zeta\right]$$

$$= \int_0^\infty e\,e^{v-1}J_v(\xi e)\frac{\partial}{\partial e}\{e^{1-v}f(e)\}de$$

$$= \int_0^\infty \{e^vJ_v(\xi e)\}\frac{\partial}{\partial e}\{e^{1-v}f(e)\}de$$

Integrating by parts by taking $e^vJ_v(e\xi)$ as first integral

$$= \left[e^vJ_v(e\xi)e^{1-v}f(e)\right]_0^\infty - \int_0^\infty e^v\xi\,J_{v-1}(e\xi)e^{1-v}f(e)de$$

$$\left[\because \frac{d}{dt}[t^2J_v(pt)] = t^2pJ_{v-1}(pt)\right]$$

$$= \left[eJ_v(e\xi)f(e)\right]_0^\infty - \xi\int_0^\infty e\,J_{v-1}(e\xi)f(e)de$$

The first term vanishes provided that $f(e)$ is such that

$$\lim e^{v+1} f(e) = 0 \text{ and } \lim_{e\to\infty} e^{1/2} f(e) = 0$$

Hence, $\left[e^{v-1}\dfrac{\partial}{\partial e}\{e^{1-v} f(e);\xi\}\right] = -\xi\, H_{v-1}\{f(e);\xi\}$

EXERCISE

1. Find the Hankel transform of $\dfrac{\sin at}{t}$, considering $tJ_1(pt)$ as the kernel.

$$\left[\textbf{Ans. } \begin{cases} a/[p(p^2-a^2)^{1/2}] & , \quad p>a \\ 0 & , \quad 0<p<a \end{cases}\right]$$

2. Evaluate Hankel transform of zero order for $t^2H(a-t)$.

$$\left[\textbf{Ans. } \frac{a^2}{p^2}\left[2J_0(pa)+\left(ap-\frac{4}{ap}\right)J_1(ap)\right]\right]$$

3. Evaluate Hankel transform of v order for t^{s-1}.

$$\left[\textbf{Ans. } \frac{2^s\Gamma\left(\frac{1}{2}s+\frac{1}{2}v+\frac{1}{2}\right)}{p^{s+1}\Gamma\left(\frac{1}{2}v-\frac{1}{s}+\frac{1}{2}\right)}, -1-v<s<1+v\right]$$

4. Find Hankel transform of

(i) $\dfrac{\sin at}{t}$.

$$\left[\textbf{Ans. } \begin{cases} 0 & , \quad p>a \\ (a^2-b^2)^{-1/2} & , \quad 0<p<a \end{cases}\right]$$

(ii) e^{-at}/t^2.

$$\left[\textbf{Ans. } \frac{(a^2+p^2)^{1/2}-a}{p}\right]$$

5. Find, $H_v^{-1}\left\{\dfrac{e^{-ap}}{p}\right\}$. [**Ans.** $(a^2+t^2)^{-1/2}$]

6. Find the Hankel transform of order zero of $1/t$ and then apply the inversion formula to set the original function.

[**Ans.** $1/p$]

Multiple-Choice Questions

1. The value of $L\{\sin t/2\}$ is
   (a) $\tan^{-1} 1/p$ (b) $\tan^{-1} 2/p$
   (c) $\tan^{-1} p$ (d) None of the above
2. The Laplace transform of cosh at is
   (a) $p/(p^2 - a^2)$ (b) $a/(p^2 + a^2)$
   (c) $1/(p^2 - a^2)$ (d) None of the above
3. The Laplace transform of e^{2t} is
   (a) $1/(p - 2)$ (b) $L/(p - 2)^2$
   (c) $1/p^2$ (d) None of the above
4. The value of $L\{t \sin at\}$ is
   (a) $a/(p^2 + a^2)$ (b) $2ap/(p^2 + a^2)$
   (c) $2\,ap/(p^2 + a^2)^2$ (d) None of the above
5. The value of $\int_0^\infty J_0(t)dt$ is
   (a) 1 (b) 0
   (c) –1 (d) None of the above
6. The value of $\int_0^\infty \frac{\sin t}{t} dt$ is
   (a) 0 (b) $\pi/2$
   (c) 1 (d) None of the above

7. If $L\left\{\frac{\sin t}{t}\right\} = \tan^{-1} 1/p$, then $L(\sin 3t/t)$ is equal to

(a) $\tan^{-1}(3/p)$ (b) $\frac{1}{3}\tan^{-1}\left(\frac{3}{p}\right)$

(c) $\tan^{-1}(3/p)$ (d) None of the above

8. If $L\{F(t)\} = f(p)$, then $L\{F(at)\}$ is equal to

(a) $f(p/a)$ (b) $\frac{1}{a}f(p/a)$

(c) $\frac{1}{a}f(p)$ (d) None of the above

9. The Laplace transform of $F(t) = 1$ is equal to

(a) p (b) 1

(c) $1/p$, $p > 0$ (d) None of the above

10. If $L\{F(t)\} - f(p)$, then $L\{tF(t)\}$ is equal to

(a) $-f'(p)$ (b) $f'(p)$

(c) $f''(p)$ (d) None of the above

11. The value of $L^{-1}(1/p^4)$ is equal to

(a) $t^4/4!$ (b) $t^3/3!$

(c) $t^5/5!$ (d) None of the above

12. The value of $L^{-1}(1/p^{5/2})$ is equal to

(a) $\frac{4}{3}t^{3/2}$ (b) $\frac{4}{3}t\sqrt{\frac{t}{\pi}}$

(c) $t^3/3!$ (d) None of the above

13. The value of $L^{-1}(1/\sqrt{p})$ is equal to

(a) $\frac{1}{\sqrt{\pi t}}$ (b) $\frac{1}{\sqrt{\pi}}t$

(c) $\frac{2}{\sqrt{\pi}}t$ (d) None of the above

14. The value of $\int_0^\infty e^{-t^2} dt$ is equal to

(a) $2\sqrt{\pi}$

(b) $\frac{1}{2}\sqrt{\frac{\pi}{2}}$

(c) π

(d) None of the above

15. Inverse Laplace transform of $\frac{1}{p^2 + a^2}$ is equal to

(a) $\sin at$

(b) $\cos at$

(c) $\frac{1}{a} \sin at$

(d) None of the above

16. $L^{-1} \left\{ \frac{1}{(p-1)^3} \right\}$ is equal to

(a) $\frac{1}{2} e^t t^2$

(b) $\frac{1}{2} e^t t^3$

(c) $\frac{1}{4} e^{2t} t^2$

(d) None of the above

17. If $L^{-1}\{F(t)\} = f(p)$, then $L^{-1}\{f(p-q)\}$ is equal to

(a) $e^{at} F(t)$

(b) $\frac{1}{a} F(t/a)$

(c) $(-1)^n t^n F(t)$

(d) None of the above

18. If $L^{-1}\{f(p)\} = F(t)$, then $L^{-1} \left\{ \frac{f(p)}{p} \right\}$ is equal to

(a) $(-1)^n t^n F(t)$

(b) $F'(t)$

(c) $\int_0^t F(u)\, du$

(d) None of the above

19. If $L^{-1}\{f(p)\} = F(t)$, then $L^{-1}\{f(ap)\}$ is equal to

(a) $a\, F(at)$

(b) $F(at)$

(c) $F(t/a)$

(d) None of the above

20. If $L^{-1}\{f(p)\}$ is equal to $F(t)$, then $L^{-1}\{p(b)\}$ is equal to when $F(0) = 0$.

(a) $(-n)^n\, t^n\, F(t)$

(b) $F'(t)$

(c) $F(t)/t$

(d) None of the above

21. If $y(x, t)$ is a function of x and t, then $L\left\{\frac{\partial y}{\partial x}\right\}$ is equal to

(a) $\frac{\partial y}{\partial x}$ (b) $\frac{\partial^2 y}{\partial x^2}$

(c) $\frac{\partial^2 y}{\partial y^2}$ (d) None of the above

22. The value of $F(t)$ is equal to ..... . If its Fourier since transform is e^{-ap}/p.

(a) $\sqrt{\frac{2}{\pi}}\tan^{-1}\frac{t}{a}$ (b) $\sqrt{\frac{2}{\pi}}\tan^{-1}\frac{a}{t}$

(c) $\sqrt{\frac{2}{\pi}}\tan^{-1}\frac{2t}{a}$ (d) None of the above

23. If $\bar{f}(p)$ is the complex Fourier transform of $F(x)$, thus complex Fourier transform of $F(at)$ is equal to

(a) $\frac{1}{a}\bar{f}(p/a)$ (b) $\bar{f}(p/a)$

(c) $\frac{1}{a}\bar{f}(p)$ (d) None of the above

24. The infinite Fourier transform of e^{-t} is equal to

(a) $\sqrt{\frac{2}{\pi}}.\frac{1}{1+p^2}$ (b) $\sqrt{\frac{2}{\pi}}.\frac{p}{1+p^2}$

(c) $\sqrt{\frac{2}{\pi}}.\frac{1}{p}$ (d) None of the above

25. The finite Fourier transform of t/π is equal to

(a) $\frac{(-1)^p}{p^2}$ (b) $\frac{(-1)^{p+1}}{p^2}$

(c) $\frac{(-1)^{p-1}}{p}$ (d) None of the above

26. The finite Fourier transform of t is equal to

(a) $\frac{\pi}{p}(-1)^{p+1}$ (b) $\frac{\pi}{p}(-1)^p$

(c) $\frac{\pi}{p^2}(-1)^{p+1}$ (d) None of the above

27. $L^{-1}\left\{\dfrac{p}{2p^2-8}\right\}$ is equal to

(a) sinh $2t$ (b) cosh $2t$

(c) $\dfrac{1}{2}$ sinh $2t$ (d) None of the above

28. Inverse Laplace transform of $1(p-a)$, $p > a$, is equal to

(a) t^2 (b) t^3

(c) e^{2t} (d) None of the above

29. Consider the function $f(x) = x$ on the interval $[-\pi, \pi]$, which of the following is the Fourier series of $f(x)$?

(a) $\sum\limits_{n=1}^{\infty} \dfrac{2}{n} \sin nx$ (b) $\sum\limits_{n=1}^{\infty} \dfrac{2}{n} \cos nx$

(c) $\sum\limits_{n=1}^{\infty} \dfrac{1}{n} \cos nx$ (d) None of the above

30. The sum of the series $1 + \dfrac{k}{3} + \dfrac{k \cdot (k+1)}{3 \cdot 6} + \dfrac{k(k+1)(k+2)}{3 \cdot 6 \cdot 9} + \ldots$ is

(a) $(1/3)^k$ (b) $(3/2)^k$

(c) $(2/3)^k$ (d) None of the above

31. The Fourier series for $f(x) = \sin^2 x$ defined over the interval $[-\pi, \pi]$ is

(a) $1 + \sin 2x$ (b) $\dfrac{1}{2} - \dfrac{\cos 2x}{2}$

(c) $\dfrac{1}{3} - \dfrac{\sin 2x}{2}$ (d) None of the above

32. The sum of the series $\dfrac{1}{1^2} + \dfrac{1}{2^2} + \dfrac{1}{3^2} + \ldots$ is equal to

(a) $\pi^2/12$ (b) $\pi^2/6$

(c) $\pi^2/8$ (d) None of the above

33. The finite Fourier sine transform of $\dfrac{2}{\pi}$ is

(a) $\dfrac{(-1)^{p+1}}{p}$ (b) $\dfrac{1}{p}$

(c) $\dfrac{(-1)^{p+1}}{p^2}$ (d) None of the above

34. The finite Fourier transform of $f(x) = x$ is

(a) $\frac{\pi}{p}(-1)^{p+1}$ (b) $\frac{\pi}{p^2}(-1)^{p+1}$

(c) $\frac{\pi}{p}(-1)^{p}$ (d) None of the above

35. What is $f(x)$ if its infinite Fourier sine transform is $\frac{e^{-ap}}{p}$

(a) $\sqrt{\frac{2}{\pi}}\tan x/a$ (b) $\sqrt{\frac{2}{\pi}}\sin^{-1}\frac{x}{a}$

(c) $\sqrt{\frac{2}{\pi}}\tan^{-1}\frac{x}{a}$ (d) None of the above

36. The infinite Fourier sine transform of e^{-x} is

(a) $\frac{p}{1-p^2}\sqrt{\frac{\pi}{2}}$ (b) $\frac{p}{1+p^2}\sqrt{\frac{2}{\pi}}$

(c) $\frac{p+2}{1+p^3}\sqrt{\frac{2}{\pi}}$ (d) None of the above

37. The series $\sum_{n=0}^{\infty}\frac{1}{4^n}(x-1)^{2n}$ converges for

(a) (–1, 3) (b) (–2, 2)

(c) (1, 2) (d) None of the above

38. If the function $f(x)$ is even, then, which of the Fourier coefficient is zero

(a) a_0 (b) a_n

(c) b_n (d) None of the above

39. If the function $f(x)$ is odd, which of the only coefficient is non-zero:

(a) a_0 (b) a_n

(c) b_n (d) None of the above

40. If $y(x, t)$ is a function of x and t, then $L\left[\frac{\partial y}{\partial x}\right]$ is

(a) $\partial^2\bar{y}/\partial x^2$ (b) $\partial\bar{y}/\partial x$

(c) $\partial\bar{y}/\partial x^2$ (d) None of the above

41. The solution of $(D + 1)\, y = 0,\ t > 0$, given that $y = y_0$ when $t = 0$ is

(a) $y_0 e^{-t}$ (b) $2y_0 e^{2t}$

(c) $3y_0 e^{-2t}$ (d) None of the above

42. An integral equation of the form $F(t) = y(t) + \int_0^t K(t-u)\, F(u)\, du$ is known as

(a) Cauchy's integral equation

(b) Integral equation of convolution type

(c) Integro-differential equation

(d) None of the above

43. $L^{-1}\,(1/p^4)$ is equal to

(a) $t^4/4!$ (b) $t^5/5!$

(c) $t^3/3!$ (d) None of the above

44. The value of $L^{-1}\{1/p^{5/2}\}$ is

(a) $\frac{4}{3} t\sqrt{t/\pi}$ (b) $t^{5/2}$

(c) $t^{3/2}$ (d) None of the above

45. Inverse Laplace transform of $1/p^2 + a^2$ is

(a) $\frac{1}{a}\sin at$ (b) $\frac{1}{a}\cos at$

(c) $\frac{1}{a}\tan at$ (d) None of the above

46. If $L^{-1}\,\{f(p)\} = f(t)$, then $L^{-1}\left\{\frac{f(p)}{p}\right\}$ is equal to

(a) $f'(t)$ (b) $\int_0^t F(u)\,du$

(c) $F''(t)$ (d) None of the above

47. If $L^{-1}\,\{f(p)\} = F(t)$, then $L^{-1}\{f(p-a)\}$ is equal to

(a) $e^{-at}\,F(t)$ (b) $F(t)$

(c) $e^{at}\,F(t)$ (d) None of the above

48. If $L^{-1}\,\{f(p)\} = F(t)$, and $F(0) = 0$, then $L^{-1}\{pf(p)\}$ is equal to

(a) $F'(t)$ (b) $F''(t)$

(c) $F'''(t)$ (d) None of the above

Answers to Multiple-Choice Questions

1. (d)	2. (a)	3. (a)	4. (c)	5. (a)	6. (b)	7. (a)	8. (b)
9. (c)	10. (a)	11. (b)	12. (b)	13. (a)	14. (b)	15. (c)	16. (a)
17. (a)	18. (c)	19. (d)	20. (d)	21. (d)	22. (a)	23. (a)	24. (b)
25. (d)	26. (a)	27. (c)	28. (d)	29. (a)	30. (b)	31. (b)	32. (c)
33. (a)	34. (a)	35. (c)	36. (b)	37. (a)	38. (c)	39. (c)	40. (b)
41. (a)	42. (b)	43. (c)	44. (a)	45. (a)	46. (b)	47. (c)	48. (a)

Index